T0271028

Heavy Tails
and Copulas
Topics in Dependence Modelling in Economics and Finance

Heavy Tails
and Copulas
Topics in Dependence Modelling in Economics and Finance

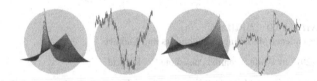

Rustam Ibragimov

Imperial College London, UK

Artem Prokhorov

University of Sydney, Australia

World Scientific

NEW JERSEY · LONDON · SINGAPORE · BEIJING · SHANGHAI · HONG KONG · TAIPEI · CHENNAI · TOKYO

Published by

World Scientific Publishing Co. Pte. Ltd.

5 Toh Tuck Link, Singapore 596224

USA office: 27 Warren Street, Suite 401-402, Hackensack, NJ 07601

UK office: 57 Shelton Street, Covent Garden, London WC2H 9HE

British Library Cataloguing-in-Publication Data
A catalogue record for this book is available from the British Library.

HEAVY TAILS AND COPULAS
Topics in Dependence Modelling in Economics and Finance

ISBN 978-981-4689-79-3

Desk Editor: Jiang Yulin

Typeset by Stallion Press
Email: enquiries@stallionpress.com

Printed in Singapore

To the memory of Kamil and grandmother Masguda

R.I.

To $A^2 + I$

A.P.

Preface

The idea of putting together a book on copulas and heavy tails has been brewing in our conversations for several years. Both of us have been working on various problems in this field and we felt a monograph covering some of these results could have value. There is a number of excellent and comprehensive treatments of copulas or heavy tails, with a statistical, mathematical, and risk management perspective. This book is different in that it provides a unified approach to handling *both* copulas and heavy tails, and it takes an economics and finance perspective.

We are thinking of a diverse readership for this book. First, it is academic and business readers, practitioners and theoreticians, who work with copula models and heavy tailed data. The benefit here is to have various results under one title as opposed to scattered across academic journals and to outline leads for promising research directions and useful applications. Second, it is graduate and advanced undergraduate students especially in econometrics and finance, but also in statistics, risk management and actuarial sciences, who look for a deeper understanding of dependence and heavy tails for their theses and degrees. The level of mathematical rigor is that of a research paper but we tried to make the book readable for a PhD or Master's student, with some parts suitable for senior undergraduate and honors students.

This book is based on recent and on-going research by the authors and their coauthors. Specifically, Chapter 2 draws on Ibragimov

(2009b). Some of the results reviewed therein are also presented in Section 2.1.1 of the recent book by Ibragimov *et al.* (2015) that deals with the analysis of models in economics and finance to heavy-tailedness. Chapter 3 draws on de la Peña *et al.* (2006); de la Peña *et al.* (2004); Medovikov and Prokhorov (2016). Chapter 4 is based on Ibragimov and Walden (2011); and Ibragimov and Prokhorov (2016). Chapter 5 is based on Prokhorov and Schmidt (2009); Burda and Prokhorov (2014) and Hill and Prokhorov (2016). Chapter 6 draws on Prokhorov and Schmidt (2009); Huang and Prokhorov (2014); and Prokhorov *et al.* (2015).

These are fairly recent papers and the topics can be viewed as part of the state-of-the-art in the area. More importantly, this book is not equal to the sum of the papers. The reasons are that, first, we do not use entire papers as chapters — many proofs are omitted and some technical details are dropped, targeting a wider audience and assuming that interested readers will look them up in the original. Second, we provide a leitmotif for each chapter that shows how we think the chapters are linked into a logical and readable sequence. The ultimate goal is to provide a framework for thinking about fat tails and copulas in economics and finance, rather than to review the content of the papers.

R. M. Ibragimov, London, 2016
A. B. Prokhorov, Sydney, 2016

Foreword

The copula is a generally applicable and flexible tool for handling multivariate non-Gaussian dependence. Sklar's theorem implies that a multivariate distribution function can be written as the composition of the copula function with univariate cumulative distribution functions as arguments. Hence for multivariate modelling, one can separate the modelling of univariate margins from the dependence structure as summarized by the copula. This is especially useful if univariate margins have heavy tails and/or joint tail probabilities have more dependence than Gaussian dependence.

In this book, probabilistic properties are studied on the effect of heavy-tailedness and joint tail dependence on risk measures such as Value-at-Risk, and these properties have relevance to portfolio diversification. Theory and tools are presented so that under some dependence assumptions, bounds on such quantities as option prices can be obtained, and the effect of the strength of dependence can be studied.

In practical data analysis using copulas, any parametric model being used is misspecified to some extent. Without a physical or stochastic basis, "true" multivariate distributions cannot be expected to have simple forms, but flexible parametric constructions, such as vine and factor models for dependence, might provide good approximations. Generally, a copula model might be satisfactory if it has relevant dependence and univariate/joint tail properties, or matches some "generalized" moments. The latter chapters of this book have results on estimation procedures that might be robust to a small degree of model misspecification.

This book differentiates itself from other books with "copula" in the title with its viewpoint via economic theory. I commend the authors for writing this book and bringing together useful research in heavy tails and copula dependence, with orientation to economics and finance. It should help to stimulate further research on the theme, and I look forward to seeing future developments.

Harry Joe
University of British Columbia
Vancouver, Canada

Acknowledgments

We thank our esteemed coauthors for collaboration and for their permission to include joint work in this book and we thank our colleagues in our field and at our institutions for interest, comments and support. The names we would like to list in the acknowledgements are, alphabetically, Axel Bücher, Martin Burda, Victor de la Peña, Prosper Dovonon, Gordon Fisher, Jonathan Hill, Wanling Huang, Marat Ibragimov, Dwight Jaffee, Di Liu, Ivan Medovikov, Ulrich Müller, Adrian Pagan, Valentyn Panchenko, Tommaso Proietti, Shaturgun Sharakhmetov, Ulf Schepsmeier, Peter Schmidt, Johan Walden, Halbert White and Yajing Zhu. The list is inevitably incomplete and we apologize for any omissions.

Parts of this book were written while Artem Prokhorov was on sabbatical at St. Petersburg State University and University of New South Wales and while Rustam Ibragimov was visiting Innopolis University (Kazan, Russia) and Kazan (Volga Region) Federal University. We wish to thank the hosting faculties at these institutions for hospitality and support. We thank James Diaz and William Liu for excellent research assistance and Irina Agafonova and Ilyas Ibragimov for proofreading earlier drafts.

The results that form the core of the book were presented at numerous conferences and department seminars — too numerous to list — and we thank the seminar and conference participants for their input.

Financial support through grants from FQRSC (Le Fonds québécois de la recherche sur la société et la culture), SSHRC (the Social Sciences and Humanities Research Council of Canada), NSF (the National Science Foundation) and RSF (the Russian Science Foundation, Project No. 16-18-10432) for various and non-overlapping parts of this research is gratefully acknowledged.

Contents

Chapter 1

Introduction and Overview

In this chapter, we set the stage by defining the subject matter of the book and describing the main tools used to study it. We also outline the structure of the book.

1.1. Crises, contagion and other features of modern economic and financial data

Modern economics, finance, risk management and insurance deal with data that is correlated, heterogeneous and/or heavy-tailed in some usually unknown fashion. When we say that data is heavy-tailed, or fat-tailed, we mean that it has a large proportion of relatively big fluctuations, where 'large' and 'relatively big' refer to proportions and fluctuations that would characterize a normally distributed random variable. These large fluctuations tend to happen simultaneously across various markets, even though individual markets usually behave differently, i.e., are heterogeneous.

Consider, for example, stock market returns during October 2008. In only a few days between October 6 and 10, the S&P500 — a US stock price index of the 500 largest companies — lost about 15%. If S&P500 was normally distributed, this event would happen no more often than once in a million years. Now look at the other world markets. During the same week, the FTSE100 — a key European stock index — lost about 14%, while the Nikkei 225 — a key Asian stock index — lost about 21%. Similar, and even larger, drops happened earlier, for example on October 19, 1987, the so-called

Black Monday, but within a single day (see, e.g., Stock and Watson
(2006), Section 2.4). Using estimates of the mean and standard deviation of the indices, it is possible to show that if the returns were
normally distributed, the probability of such drops would be of order
10^{-107}, i.e., no more than the inverse of a googol (10^{100}). One can
also add that 10^{-107} is much smaller than the probability of choosing
a particular atom from all atoms in the observable universe as their
number is estimated to be 10^{80}!

Similar to stock market returns, crucial deviations from normality are observed for many other key financial and economic indicators
and variables, including income and wealth, losses from natural disasters, firm and city sizes, operational risks and many others (see, e.g.,
the reviews by, Embrechts *et al.* (1997); McNeil *et al.* (2005); Gabaix
(2009); Ibragimov *et al.* (2015)). When the number of extreme events
is abnormally high, we refer to such distributions as heavy-tailed and
when such events coincide across seemingly independent markets and
produce market crashes, we call this asymmetric tail dependence.
Distributions of financial returns are typically asymmetric because
the number of extreme negative events — abnormal drops — tends
to be higher than the number of positive events — abnormal jumps.

Evidence of heavy-tailedness and asymmetric tail dependence
have been amply documented in equity markets (see, e.g., Ang and
Chen (2002); Longin and Solnik (2001)), in foreign exchange markets (see, e.g., Patton (2006); Ibragimov *et al.* (2013)), especially
surrounding various crises such as the Latin American debt crisis of
1982, the Asian currency crisis of 1997, the North American subprime
lending crisis of 2008, etc. (see, e.g., Rodriguez (2007); Horta *et al.*
(2010)).

Tail dependence in financial markets often takes the form of financial contagion, which is usually described as periods when declining
prices and increased volatility spread among economic and financial markets causing markets that usually have little or no correlation to behave very similarly, often contrary to the fundamentals
(see, e.g., Hamao *et al.* (1990); Lin *et al.* (1994); Longin and Solnik
(2001); Mierau and Mink (2013)). Incidents of unfounded contagion
are puzzling because they imply some sort of irrationality on the

part of market participants — they cannot be explained using standard risk management strategies and optimal portfolio choices. For this reason, traditional explanations were based on various types of market imperfections, such as liquidity and coordination problems, information asymmetry, information cost and performance compensation factors (see, e.g., Dungey *et al.* (2005); Dornbusch *et al.* (2000); Kyle and Xiong (2001), for surveys).

This book provides an econometric treatment of such events. That is, it seeks to build a general framework for analyzing such events using statistical methods and models of relevance to economics and finance. There will be no economic models of crises or contagion; instead, we will look at the distributional and dependence characteristics of financial and economic data that may give rise to the described behavior and at modern methods of statistical and econometric analysis suitable for such data. The aim is to provide a framework for thinking about contagion statistically and econometrically and to survey the state-of-the-art econometric tools used in the setting of tail-dependent, heterogenous, and heavy-tailed data.

The two key distributional features here are heavy-tails — to accommodate excessive volatility or excess kurtosis — and copulas — to model tail dependence and contagion. In other words, we will examine models and methods used for the analysis of multivariate economic and financial data, whose copulas accommodate non-zero tail dependence and whose univariate distributions are diverse, heavy tailed and have relatively small and possibly unequal values of the tail index.

1.2. Econometric tools for modern financial and economic data

1.2.1. *Multivariate distributions and copulas*

From an econometric point of view, the complicated nature of financial time series originates from the statistical properties of disturbances affecting financial markets. These properties are captured by their cumulative distribution functions, or cdf's. Individual behavior of a single financial indicator is represented by a univariate cdf, while joint behavior of multiple indices — a particular focus of our analysis — is characterized by a multivariate cdf.

Let $F_k : \mathbf{R} \to [0,1]$, $k = 1, \ldots, d$, be one-dimensional cdf's, also known as marginal cdf's or simply marginals, and let $\xi_1, \ldots, \xi_d$ be independent r.v.'s on some probability space $(\Omega, \Im, P)$ with $\mathbb{P}(\xi_k \le x_k) = F_k(x_k)$, $x_k \in \mathbf{R}$, $k = 1, \ldots, d$. A multivariate cdf $F(x_1, \ldots, x_d)$, $x_i \in \mathbf{R}$, $i = 1, \ldots, d$, with given marginals F_k, is a function satisfying the following conditions:

(a) $F(x_1, \ldots, x_d) = \mathbb{P}(X_1 \le x_1, \ldots, X_d \le x_d)$ for some r.v.'s $X_1, \ldots, X_d$ on a probability space $(\Omega, \Im, P)$;

(b) the one-dimensional marginal cdf's of F are $F_1, \ldots, F_d$;

(c) F is absolutely continuous with respect to $dF(x_1) \ldots dF_d(x_d)$ in the sense that there exists a Borel function $G : \mathbf{R}^d \to [0, \infty)$ such that

$$F(x_1, \ldots, x_d) = \int_{-\infty}^{x_1} \cdots \int_{-\infty}^{x_d} G(t_1, \ldots, t_d) dF_1(t_1) \ldots dF_d(t_1).$$

In addition to joint distributions of d random variables (r.v.'s), we are often interested in the distribution functions of subsets of these variables. Let $F(x_{j_1}, \ldots, x_{j_k})$, $1 \le j_1 < \cdots < j_k \le d$, $k = 2, \ldots, d$, stand for a k-dimensional marginal cdf of $F(x_1, \ldots, x_d)$. It represents the joint behavior of k out of d r.v.'s if $k > 1$ and it represents the individual univariate marginal cdf's $F(x_j)$ if $k = 1$.

Copulas are functions that allow us, by a celebrated theorem due to Sklar (1959), to represent a joint distribution of random variables (r.v.'s) as a function of marginal distributions.[1] Copulas, therefore, capture dependence properties of the data generating process (more precisely, they reflect all the dependence properties that are invariant to increasing transformations of data).

We start with a formal definition of copulas and the formulation of Sklar's (1959) theorem (see e.g., Embrechts *et al.* (2002); Nelsen (2006); McNeil *et al.* (2005)).

---

[1]The concept of copulas is closely related to the probability integral transformation (see Rosenblatt (1952); Gouriéroux and Monfort (1979) and Section 4 in Breymann *et al.* (2003)) and to Fréchet classes of joint distributions (see Chapter 3 in Joe (1997) and Chapter 2 in Joe (2014)).

Definition 1.1. A function $C : [0,1]^d \rightarrow [0,1]$ is called a d-dimensional **copula** if it satisfies the following conditions:

1. $C(u_1, \ldots, u_d)$ is increasing in each component u_i.
2. $C(u_1, \ldots, u_{k-1}, 0, u_{k+1}, \ldots, u_d) = 0$ for all $u_i \in [0,1]$, $i \neq k$, $k = 1, \ldots, d$.
3. $C(1, \ldots, 1, u_i, 1, \ldots, 1) = u_i$ for all $u_i \in [0,1]$, $i = 1, \ldots, d$.
4. For all $(a_1, \ldots, a_d)$, $(b_1, \ldots, b_d) \in [0,1]^d$ with $a_i \leq b_i$,

$$\sum_{i_1=1}^{2} \cdots \sum_{i_d=1}^{2} (-1)^{i_1 + \cdots + i_d} C(x_{1i_1}, \ldots, x_{di_d}) \geq 0,$$

where $x_{j1} = a_j$ and $x_{j2} = b_j$ for all $j \in \{1, \ldots, d\}$. Equivalently, C is a d-dimensional copula if it is a joint cdf of d r.v.'s each of which is uniformly distributed on $[0,1]$.

Copulas and related concepts have been applied to a wide range of problems in economics, finance and risk management (see, among others; Cherubini *et al.* (2004, 2012) and references therein; Patton (2006); McNeil *et al.* (2005); Hu (2006); the review by de la Peña *et al.* (2006); Granger *et al.* (2006); Patton (2012)).

We will use the word *copula* to denote the function (cdf) C as described above. When that cdf has a density, we will call it a *copula density*. We now give its formal definition.

Definition 1.2. A copula $C : [0,1]^d \rightarrow [0,1]$ is called **absolutely continuous** if, when considered as a joint cdf, it has a joint density given by $c(u_1, \ldots, u_d) := \partial C^d(u_1, \ldots, u_d)/\partial u_1 \ldots \partial u_d$.

Proposition 1.1. (Sklar, 1959, pp. 229–230). *If $X_1, \ldots, X_d$ are r.v.'s defined on a common probability space, with the one-dimensional cdf's $F_{X_k}(x_k) = \mathbb{P}(X_k \leq x_k)$ and the joint cdf $F_{X_1,\ldots,X_d}(x_1, \ldots, x_d) = \mathbb{P}(X_1 \leq x_1, \ldots, X_d \leq x_d)$, then there exists a d-dimensional copula $C_{X_1,\ldots,X_d}(u_1, \ldots, u_d)$ such that*

$$F_{X_1,\ldots,X_d}(x_1, \ldots, x_d) = C_{X_1,\ldots,X_d}(F_{X_1}(x_1), \ldots, F_{X_d}(x_d))$$

for all $x_k \in \mathbf{R}$, $k = 1, \ldots, d$. If the univariate marginal cdf's $F_{X_1}, \ldots, F_{X_d}$ are all continuous, then the copula is unique and can *be obtained via inversion:*

$$C_{X_1,\ldots,X_d}(u_1,\ldots,u_d) = F_{X_1,\ldots,X_d}(F_{X_1}^{-1}(u_1),\ldots,F_{X_d}^{-1}(u_d)), \quad (1.1)$$

where $F_{X_k}^{-1}(u_k) = \inf\{x : F_{X_k}(x) \geq u_k\}$. *Otherwise, the copula is uniquely determined at points* $(u_1,\ldots,u_d)$, *where* u_k *is in the range of* F_k, $k = 1,\ldots,d$.

R.v.'s $X_1,\ldots,X_d$ with copula $C(u_1,\ldots,u_d)$ are jointly independent if and only if C is the product copula:

$$C(u_1,\ldots,u_d) = u_1 \ldots u_d. \quad (1.2)$$

Well-studied examples of copulas are given by, for example, Clayton, Gumbel and Frank copulas (see, e.g., Joe (1997, 2014); Nelsen (2006)). Taking in (1.1) F to be a d-dimensional normal cdf with linear correlation matrix R:

$$F(x) = \Phi_R^d(x) = \int_{-\infty}^{x_1} \cdots \int_{-\infty}^{x_d} \phi_{d,R}(x)dx, \quad (1.3)$$

with $\phi_{d,R}(x) = 1/((2\pi)^{d/2}|R|^{1/2})\exp(-\frac{1}{2}xR^{-1}x)$, one obtains the well-known normal, or Gaussian, copula $C_R^d(u_1,\ldots,u_d)$:

$$C_R^d(u_1,\ldots,u_d) = \Phi_R^d(\Phi^{-1}(u_1),\ldots,\Phi^{-1}(u_d)), \quad (1.4)$$

where $\Phi(x)$ denotes the standard normal univariate cdf. In the bivariate case, the normal copula can be written as:

$$C_\rho(u_1,u_2) = \int_{-\infty}^{\Phi^{-1}(u_1)} \int_{-\infty}^{\Phi^{-1}(u_2)} \frac{1}{2\pi\sqrt{1-\rho^2}} e^{-\frac{u_1^2 - 2\rho u_1 u_2 + u_2^2}{2(1-\rho^2)}} du_1 du_2,$$

$$(1.5)$$

where ρ is the linear correlation coefficient of the corresponding bivariate normal distribution.

Let $\nu > 0$ and let F be a d-dimensional Student-t cdf $t_{\nu,R}^d$ with ν degrees of freedom, a linear correlation matrix R and location parameter fixed at 0. That is, $F = t_{\nu,R}^d$ is the joint cdf of the random vector $\sqrt{\nu}Y/\sqrt{S}$, where $Y \sim \mathcal{N}_d(0,R)$ has a d-dimensional normal distribution with correlation matrix R and $S \sim \chi^2(\nu)$ is a chi-square r.v. with

ν degrees of freedom that is independent of Y. Formula (1.1) then gives a d-dimensional t-copula with correlation matrix R:

$$C^t_{\nu,R}(u_1, \ldots, u_d) = t^d_{\nu,R}(t^{-1}_\nu(u_1), \ldots, t^{-1}_\nu(u_d)), \qquad (1.6)$$

where $t_\nu(x)$ denotes the cdf of a univariate Student-t distribution with ν degrees of freedom.

In the bivariate case, a t-copula takes the form

$$C^t_{\nu,\rho}(u_1, u_2) = \int_{-\infty}^{t^{-1}_\nu(u_1)} \int_{-\infty}^{t^{-1}_\nu(u_2)} \frac{1}{2\pi\sqrt{1-\rho^2}}$$

$$\times \left(1 + \frac{u_1^2 - 2\rho u_1 u_2 + u_2^2}{\nu(1-\rho^2)}\right)^{-\frac{\nu+2}{2}} du_1 du_2, \qquad (1.7)$$

where $\rho \in (-1, 1)$.

Most applications of copulas in economics have used the "converse" part of Sklar's theorem. That is, you have a set of marginal cdf's $F_{X_1}, \ldots, F_{X_d}$ implied by some model, but you want a joint cdf $F_{X_1, \ldots, X_d}$. So you pick a copula and it generates a joint cdf consistent with the marginals. Lee (1983) appears to be the earliest application of this approach in econometrics.

Copulas seem to have received more attention in the finance literature than in economics. They are used to model dependence in financial time series (e.g., Patton (2006); Breymann *et al.* (2003)) and in risk management applications (Embrechts *et al.* (2002, 2003); McNeil *et al.* (2005)). Cherubini *et al.* (2004, 2012) and Bouyé *et al.* (2000) cover a wide range of copula applications in finance.

However, use of copulas in other subfields of econometrics has been growing. Smith (2003) incorporates a copula in selectivity models and provides applications to labor supply and duration of hospitalization; Cameron *et al.* (2004) use a copula to develop a bivariate count data model with an application to the number of doctor visits. Zimmer and Trivedi (2006) use copulas in a selection model with count data. Trivedi and Zimmer (2007) consider benefits of copula-based estimation relative to simulation-based approaches. Choroś *et al.* (2010) review estimation methods for copula models. Fan and Patton (2014) provide a review of copula uses in economics.

Copulas are attractive because of an invariance property. They are invariant under strictly increasing transformations of r.v.'s with continuous univariate cdf's.

Proposition 1.2. *Let X_k, $1 \leq k \leq d$, be r.v.'s with continuous univariate marginal cdf's F_{X_k} and a copula C. If $f_k : \mathbf{R} \to \mathbf{R}$, $1 \leq k \leq d$, are strictly increasing functions, then the r.v.'s $Y_k = f_k(X_k)$ have the same copula C.*

From Propositions 1.1 and 1.2, it follows that copulas can be obtained from any joint distribution as a result of transforming the initial r.v.'s into their marginal cdfs. Essentially, they are joint distributions with uniform marginals, useful because given the marginals, they represent the dependence in the joint distribution.

In the case of r.v.'s X_k, $1 \leq k \leq d$, with continuous cdf's F_k the probability integral transforms $U_k = F_k(X_k), 1 \leq k \leq d$, are the uniform r.v.'s that form the margins of C. So, equivalently, C can be defined as a joint cdf of d r.v.'s, each of which is uniform on $[0, 1]$. The fact that we can model F_k separately from modelling the dependence between F_k's is what makes copulas natural in the analysis of dependent heavy tailed marginals.

A well known property of the copula function is that it is bounded by the Frechet-Hoeffding bounds, which correspond to extreme positive and extreme negative dependence. For a bivariate copula, let X_1 be a fixed increasing function of X_2, then the copula of (X_1, X_2) can be written as $\min(u_1, u_2)$ and this is the upper bound for bivariate copulas. Now let X_1 be a fixed decreasing function of X_2; then the copula of (X_1, X_2) can be written as $\max(u_1 + u_2 - 1, 0)$. So the two extreme cases of comonotonicity and countermonotonicity are nested within the copula framework, at least for the bivariate case. Joe (1997, 2014) and Nelsen (2006) provide excellent introductions to copulas.

Conversely, given marginals and a copula, one can construct a joint distribution, which will have the given marginals. This property of copulas makes them a natural tool in the analysis of heavy-tailed distributions, where the marginals will have a power-law form, while dependence between them will be captured by a copula.

Intuitively, we can think of a copula as a function that operates on fractions. Suppose we have a sample of observations on two r.v.'s $(x_{1i}, x_{2i}), i = 1, \ldots, N$. Let (n_{1i}, n_{2i}) denote the ranks of each $x_{ki}, k = 1, 2$, among the available observations of that variable. For example, if $x_1 = (0.1, 0.24, -0.5)$ then $n_1 = (2, 1, 3)$. Now fractions $\frac{n_{ki}}{N}$ can be viewed as values of F_{X_k} evaluated at x_{ki}. And a copula is a distribution over such fractions. Obviously, nothing will change if we change (x_{1i}, x_{2i}) as long as the change does not affect (n_{1i}, n_{2i}). This is why copulas represent dependence which is invariant to rank-preserving transformations.

1.2.2. *Heavy tailed stable and power law distributions*

A natural next question is how to model the marginal distributions $F_k, k = 1, \ldots, d$, so that they exhibit fat tails. A number of frameworks have been proposed to model heavy-tailedness, including stable distributions and their truncated versions, Pareto distributions, multivariate t-distributions, mixtures of normals, power exponential distributions, ARCH processes, mixed diffusion jump processes, variance gamma and normal inverse Gamma distributions (see, e.g., Cover and Thomas (2012), and references therein).

Arguably the most common framework is to model heavy tailed distributions as a power law family. The literature on such distributions goes back at least to Mandelbrot (1960, 1963) and Fama (1965b). It has by now become common in financial econometrics to use the tail index of a power law to measure thickness of its tails (see, e.g., Embrechts *et al.* (1997); Mandelbrot (1997); McNeil *et al.* (2005); Peters and Shevchenko (2015); Ibragimov *et al.* (2015)).

Definition 1.3. A r.v. X follows a **power law** distribution with tail index α if

$$\mathbb{P}(|X| > x) \asymp x^{-\alpha}, \tag{1.8}$$

where $\alpha > 0$ and $f(x) \asymp g(x)$ means that $0 < c \leq f(x)/g(x) \leq C < \infty$, for large x, for constants c and C.

The tail index characterizes the heaviness, or the rate of decay, of the relevant marginal distribution, assuming it obeys a power law in the tails. Because the tail probability of the r.v. X in Definition 1.3 is a power function, this permits modelling distributions with rates of tail decay that are much slower than exponential rate of the Normal distribution.

The tail index governs the likelihood of observing large deviations and large downfalls in the r.v. X: a smaller tail index means slower rate of decay, which means that this likelihood is higher. When the tail index is less than two, the likelihood is so big that the second moment of the r.v. X becomes infinite, implying that its variance is either infinite or undefined; when the tail index is less than one, the absolute first moment of X is infinite (and the mean of the r.v. is infinite or undefined). More generally, if X follows power law then absolute moments of X are finite if and only if their order is less than the tail index α. That is,

$$\mathbb{E}|X|^p < \infty \quad \text{if} \quad p < \alpha; \quad \mathbb{E}|X|^p = \infty \quad \text{if} \quad p \geq \alpha.$$

Many distributions can be viewed as special cases of power law, at least for asymptotically large X's. This includes Student-t, Cauchy, Levy and Pareto and other stable distributions with parameter $\alpha < 2$. We will say that a risk has *extremely* heavy or fat tails if $\alpha < 1$, and *moderately* heavy or fat tails if $\alpha > 1$.

The power law is asymptotic with respect to X, i.e., it is defined for large values of X. A wide class of power law distributions is given by the stable family.

Definition 1.4. For $0 < \alpha \leq 2$, $\sigma > 0$, $\beta \in [-1, 1]$ and $\mu \in \mathbf{R}$, a r.v. X follows **a stable distribution** denoted by $S_\alpha(\sigma, \beta, \mu)$ with the characteristic exponent (index of stability) α, the scale parameter σ, the symmetry index (skewness parameter) β and the location parameter μ if its characteristic function can be written as follows:

$$\mathbb{E}(e^{ixX}) = \begin{cases} \exp\left\{i\mu x - \sigma^\alpha |x|^\alpha (1 - i\beta \text{sign}(x) \tan(\pi\alpha/2))\right\}, & \alpha \neq 1, \\ \exp\left\{i\mu x - \sigma |x|(1 + (2/\pi)i\beta \text{sign}(x) \ln |x|)\right\}, & \alpha = 1, \end{cases}$$

$$(1.9)$$

where $x \in \mathbf{R}$, $i^2 = -1$ and $sign(x)$ is the sign of x defined by $sign(x) = 1$ if $x > 0$, $sign(0) = 0$ and $sign(x) = -1$ otherwise.

In what follows, we write $X \sim S_\alpha(\sigma, \beta, \mu)$, if the r.v. X has the stable distribution $S_\alpha(\sigma, \beta, \mu)$.

The index of stability α characterizes the heaviness (the rate of decay) of the tails of stable distributions $S_\alpha(\sigma, \beta, \mu)$. In particular, if $X \sim S_\alpha(\sigma, \beta, \mu)$ with $\alpha \in (0, 2)$ then its distribution satisfies power law (1.8), so in this case, stable distributions can be viewed as a special case of power law. This implies that the p-th absolute moments $\mathbb{E}|X|^p$ of a r.v. $X \sim S_\alpha(\sigma, \beta, \mu)$, $\alpha \in (0, 2)$ are finite if $p < \alpha$ and infinite otherwise.

The symmetry index β characterizes the skewness of the distribution. Stable distributions with $\beta = 0$ are symmetric about the location parameter μ. The stable distributions with $\beta = \pm 1$ and $\alpha \in (0, 1)$ (and only they) are one-sided, the support of these distributions is the semi-axis $[\mu, \infty)$ for $\beta = 1$ and is $(-\infty, \mu]$ for $\beta = -1$ (in particular, the Lévy distribution with $\mu = 0$ is concentrated on the positive semi-axis for $\beta = 1$ and on the negative semi-axis for $\beta = -1$). In the case $\alpha > 1$, the location parameter μ is the mean of the distribution $S_\alpha(\sigma, \beta, \mu)$.

The scale parameter σ is a generalization of the concept of standard deviation; it coincides with the standard deviation in the special case of Gaussian distributions ($\alpha = 2$).

Definition 1.5. Distributions $S_\alpha(\sigma, \beta, \mu)$ are called **strictly stable** if $\mu = 0$ for $\alpha \neq 1$ and $\beta = 0$ for $\alpha = 1$.

Theorem 1.1. *If* $X_i \sim S_\alpha(\sigma, \beta, \mu)$, $\alpha \in (0, 2]$, *are i.i.d. strictly stable then, for all* $c_i \geq 0$, $i = 1, \ldots, n$, *such that* $\sum_{i=1}^n c_i \neq 0$,

$$\sum_{i=1}^n c_i X_i / \left(\sum_{i=1}^n c_i^\alpha \right)^{1/\alpha} \overset{d}{=} X_1 \qquad (1.10)$$

Equation (1.10) is known as the convolution property of stable distributions and is implied by Definition 1.4 and product decomposition of characteristic functions of linear combinations of stable r.v.'s under independence. From the property, it follows that stable distributions are closed under portfolio formation. For a detailed review of the properties of stable and power-law distributions, the reader is referred to Zolotarev (1986), Uchaikin and Zolotarev (1999), Bouchaud and Potters (2004) and Borak *et al.* (2005).

Although there are a number of approaches to heavy-tailedness modelling available in the literature, stable heavy-tailed distributions exhibit several properties that make them appealing in applications. Most importantly, stable distributions provide natural extensions of the Gaussian law since they are the only possible limits for appropriately normalized and centered sums of i.i.d. r.v.'s. This property is useful in representing heavy-tailed financial data as cumulative outcomes of market agents' decisions in response to information they amass. In addition, stable distributions are flexible to accommodate both heavy-tailedness and skewness. Furthermore, their multivariate extensions allow us to model certain kinds of dependence among the risks or returns in consideration (see Chapter 4).

Empirical estimates document values of α ranging from below one to above four for many key economic and financial variables (see, e.g., Loretan and Phillips (1994); Rachev and Mittnik (2000); Gabaix *et al.* (2003, 2006); Rachev *et al.* (2005); Jansen and Vries (1991); McCulloch (1997); Chavez-Demoulin *et al.* (2006); Silverberg and Verspagen (2007); Ibragimov *et al.* (2013, 2015), and references therein). Mandelbrot (1963) presented evidence that historical daily changes of cotton prices have the tail index $\alpha \approx 1.7$, and thus have infinite variances. Using different models and statistical techniques, subsequent research reported the following estimates of the tail parameters α for returns on various stocks and stock indices: $3 < \alpha < 5$ (Jansen and Vries (1991)); $2 < \alpha < 4$ (Loretan and Phillips (1994)); $1.5 < \alpha < 2$ (McCulloch (1996, 1997)); $0.9 < \alpha < 2$ (Rachev and Mittnik (2000)).

Gabaix *et al.* (2003, 2006) find that the returns on many stocks and stock indices have the tail exponent $\alpha \approx 3$, while the distributions of trading volume and the number of trades on financial markets obey the power laws (1.8) with $\alpha \approx 1.5$ and $\alpha \approx 3.4$, respectively. Moreover, they find that tail exponents for financial and economic time series are similar in different countries (see also Lux (1996); Guillaume *et al.* (1997)). Gabaix *et al.* (2003, 2006) propose a model in which the latter power laws are implied by trading of large market participants, namely, the largest mutual funds whose sizes have tail exponent $\alpha \approx 1$.

Power laws (1.8) with $\alpha \approx 1$ (also known as the Zipf law) have been found for firm sizes (Axtell, 2001) and city sizes (Gabaix, 1999a,b).

According to Ibragimov *et al.* (2013) (see also the discussion in Ibragimov *et al.* (2015)), in contrast to developed markets, the tail indices of several emerging country exchange rates may be smaller than two, implying infinite variances.

Scherer *et al.* (2000) and Silverberg and Verspagen (2007) report the tail indices α to be considerably less than one for financial returns from technological innovations. As discussed by Nešlehova *et al.* (2006) and Peters and Shevchenko (2015), tail indices less than one are observed for empirical loss distributions of a number of operational risks.

Ibragimov *et al.* (2009) show that standard seismic theory implies that the distributions of economic losses from earthquakes have heavy tails with tail indices $\alpha \in [0.6, 1.5]$ that can thus be significantly less than one. These estimates follow from power laws for magnitudes of earthquakes. Similar analysis also holds for economic losses from other natural disasters with heavy-tailed physical characteristics surveyed by Ibragimov *et al.* (2009).

Rachev *et al.* (2005, Chapter 11) discuss and review the vast literature that supports heavy-tailedness and the stable Paretian hypothesis (with $1 < \alpha < 2$) for equity and bond return distributions.

Thus, power-law and stable distributions with a low tail index are very common and provide a natural building block for modelling economic and financial markets affected by crises and economic and financial variables exhibiting large fluctuations or outliers.

One should note here that commonly used approaches to inference on the tail indices, such as OLS log-log rank-size regressions and Hill's estimator, are strongly biased in small samples and are very sensitive to deviations from power laws (1.8) in the form of regularly varying tails (see, among others Embrechts *et al.* (1997); Gabaix and Ibragimov (2011)). In particular, these procedures tend to overestimate the tail index of heavy-tailed stable distributions when $\alpha < 2$ and the sample size is typical for applications (see, e.g., McCulloch (1997)). Therefore, point estimates of the tail index greater than one do not necessarily exclude heavy-tailedness with infinite means and true values $\alpha < 1$ in the same way as point estimates of the tail exponent greater than two do not necessarily exclude stable regimes with infinite variances.

1.3. Robustness to heavy tails and to copula misspecification

1.3.1. *Robustness of models to heavy tails*

Recent studies have shown that heavy-tailedness is of key importance for the reliability of conclusions arising from many models in economics, finance, risk management and insurance (see, e.g., Ibragimov (2009b); Ibragimov and Walden (2007); Gabaix (2009); Ibragimov and Prokhorov (2016), and references therein). The property of a model's prediction to remain valid even when risks are allowed to have heavy tails is known as *robustness* of the model to heavy tails. The state-of-the-art in this work is that many mainstream economic and financial models are *not* robust to heavy tails — their implications are reversed when the tail index is extremely low.

An important example of model (non) robustness is the analysis of diversification and optimal portfolio choice in Value-at-Risk (VaR) models. The key finding here is that while diversification is preferable for moderately heavy-tailed independent risks with tail index greater than one, diversification increases risk in the case of extremely heavy-tailed risks with the tail index smaller than one (Ibragimov *et al.*, 2015). Similar results are available for bounded risks concentrated on a sufficiently large interval: for such cases, diversification may increase risk up to a certain portfolio size and then reduce risk. Ibragimov *et al.* (2009) demonstrate how this analysis can be used to explain abnormally low levels of reinsurance among insurance providers in markets for catastrophic insurance. Ibragimov *et al.* (2011) show how to analyze the recent financial crisis as a case of excessive risk sharing between banks when risks are extremely heavy-tailed. These key results help explain a variety of observations, which may seem counterintuitive or irrational when viewed from the conventional, thin-tailed risk management perspective but unfortunately, most of these results are limited to the case of independent data.

1.3.2. *Robustness of methods to heavy tails and to copula misspecification*

Parallel to the study of *model* robustness to heavy tailedness, there have been many new results on robust *estimation* (see, e.g., Aguilar and Hill (2015); Hill and Prokhorov (2016); Hill (2015a,b); Prokhorov and Schmidt (2009)). A method is robust (to misspecification, to extremely heavy tails, etc.) when it does not lose some desirable properties when the assumptions that are used to motivate it (correct specification, moderately heavy tails, etc.) are violated. Similarly to the *model* robustness, the recurring theme here is that most popular estimation methods such as the traditional MLE, GMM, GEL are not robust to copula misspecification or to extremely heavy tails. They are inconsistent and inefficient under heavy tails and copula misspecification.

At the same time, recent scholarly articles and popular press have been concerned with the use of misspecified copulas in pricing financial assets and in representing tail-dependence between them (see, e.g., Zimmer (2012); Rodriguez (2007)). This interest has been stimulated by the link discovered between the large-scale mispricing of collaterized debt obligations (CDOs) and the financial crisis of 2008. Portfolio mispricing has been tied with the use of a misspecified copula. For example, Zimmer (2012) echoes earlier newspaper articles (see, e.g., "The formula that felled Wall St," by S. Jones, *Financial Times*, April 24, 2009) and argues that the massive CDO mispricing is related to the use of a specific parametric copula widely adopted by the Wall Street. The Gaussian copula, considered, e.g., by Li (2000), is known to be incapable of modelling tail dependence. The CDO prices obtained using it did not reflect the risk of joint defaults to the extent they took place in 2008.

Suitable *copula correctness*, or goodness-of-fit, tests would permit early detection of an inappropriate dependence structure used in pricing and investment decisions. Similarly, powerful tests of *copula robustness* would allow us to preserve the desirable statistical properties of the commonly used estimators — as long as a misspecified copula is robust, estimators based on it are consistent. It is of course desirable that such tests themselves be *robust* to extremely heavy tails.

1.4. Plan for the book

We start with *Chapter 2. Portfolio Diversification under Independent Fat Tailed Risks*. It lays the foundation for studying robustness of financial models to heavy tails by formulating results on limits of diversification for independent risks with power law distributions.

Chapter 3. From Independence to Dependence via Copulas and U-statistics offers a discussion of how to come up with arbitrary multivariate distributions, copulas, and dependence structures. It builds on Chapter 2 using independence as a starting point. It also discusses some more or less commonly used dependence measures, which include independence as a special case.

Chapter 4. Limits of Diversification under Fat Tails and Dependence discusses how heavy tails and various kinds of dependence combine to produce flexible distributions capable of modelling tail-dependence, asymmetry and heavy tails. It turns out that the same limits of diversifications apply for dependent heavy tailed risks as in Chapter 2.

Chapter 5. Robustness of Econometric Methods to Copula Misspecification and Heavy Tails introduces likelihood-based estimation and discusses what happens if we use a misspecified copula. In some cases, we can rescue consistency and say important things about efficiency — if we use a robust parametric copula. Alternatively we can use nonparametric copula estimators, which are inherently robust because they use no assumption of a correctly specified parametric copula. Or we can robustify conventional methods by, say, trimming the extreme observations.

Chapter 6. *Copula Tests Using Information Matrix* provides an application of the arguments in Chapter 5. It covers several recent tests of validity and robustness of parametric copulas, that involve information contained in a copula.

Chapter 7. *Summary and Conclusion* discusses several directions of ongoing research that naturally follow from the topics covered in this book.

Chapters 2, 3 and 4 deal with robustness of models, e.g., of VaR models. Chapters 5 and 6 deal with robustness of methods, e.g., MLE. This forms an implicit divide of the book into two parts, modelling and estimation. The types of robustness we consider are against extremely heavy tails and against copula misspecification. The first part focuses on the former type of robustness, the second on the latter.

Chapter 2

Portfolio Diversification under Independent Fat Tailed Risks

This chapter looks at Value-at-Risk (VaR) of a diversified portfolio when its components are independent and have a heavy tailed distribution. We use a notion of diversification based on majorization theory and show that the conventional wisdom that portfolio diversification reduces risk does not hold for extremely heavy-tailed returns.[1]

2.1. Introduction

VaR models are frequently used in economics, finance and risk management because they provide useful alternatives to the traditional expected utility framework (see, e.g., McNeil *et al.* (2005); Bouchaud and Potters (2004); Szegö (2004) and Rachev *et al.* (2005) for a review of the VaR and related risk measures). Expected utility comparisons are not readily available under heavy-tailedness since moments of the risks or returns in consideration become infinite. The VaR analysis is thus, in many regards, a unique approach to portfolio choice and riskiness comparisons that does not impose restrictions on heavy-tailedness of the risks.

VaR minimization is an example of many models in economics, finance and risk management that are based on majorization results for linear combinations of random variables. The majorization

---

[1]Some of the results reviewed in the chapter were presented in Ibragimov *et al.* (2015).

relation is a formalization of the concept of diversity in the components of vectors. Over the past decades, majorization theory has found applications in disciplines ranging from statistics, probability theory and economics to mathematical genetics, linear algebra and geometry (see the seminal book on majorization theory and its application by Marshall and Olkin and its second edition Marshall *et al.* (2011) and references therin). A number of papers in economics use majorization and related concepts in the analysis of income inequality and its effects on the properties of economic models (see, among others, the reviews in Marshall and Olkin (1979); Marshall *et al.* (2011) and Ibragimov and Ibragimov (2007) and references therein). Lapan and Hennessy (2002) and Hennessy and Lapan (2003) applied majorization theory to analyze the portfolio allocation problem. Bouchaud and Potters (2004, Chapter 12) present a detailed discussion of portfolio choice under various distributional and dependence assumptions and a discussion of diversification measures, including the asymptotic results in the VaR framework for heavy-tailed power law distributions.

In this chapter, we discuss majorization results for linear combinations of heavy-tailed r.v.'s and we use them to study portfolio diversification and VaR. We discuss a precise formalization of the concept of portfolio diversification on the basis of majorization pre-ordering — see Section 2.4 (see also Ibragimov *et al.* (2015)). We further discuss how the stylized fact that portfolio diversification decreases risk is reversed for a wide class of distributions — Theorem 2.2. The class of distributions for which this is the case is the class of extremely heavy-tailed distributions. In terms of power law distributions introduced in Section 1.2.2, this happens when the tail index $\alpha < 1$. For these distributions, diversification actually leads to an increase, rather than a decrease, in portfolio riskiness.

On the other hand, the conventional wisdom that diversification reduces risk continues to hold as long as distributions are moderately heavy-tailed — Theorem 2.1. In the power law family, this is the case when $\alpha > 1$. We also obtain sharp bounds on the portfolio VaR under heavy-tailedness — Thereoms 2.5–2.6 — and discuss implications of the results for the analysis of coherency of VaR.

We model heavy-tailedness using the framework of independent stable distributions and their convolutions. More precisely, the class of moderately heavy-tailed distributions is modelled using convolutions of stable distributions with (different) indices of stability greater than one. Similarly, the results for extremely heavy-tailed cases are presented and proven using the framework of convolutions of stable distributions with characteristic exponents less than one.

The proof of the results in the benchmark case of convolutions of independent stable distributions exploits several symmetries in the problem. First, the property that i.i.d. stable distributions are closed under convolutions — relation (1.10), together with positive homogeneity of VaR (relation a3 in Section 2.3), allows us to reduce the portfolio VaR analysis for i.i.d. stable risks to comparisons of functions of portfolio weights that are Schur-convex or Schur-concave and are thus symmetric in their arguments (see Section 2.4.1 for definitions of Schur-convex and Schur-concave functions and their symmetry property in (2.3)).

As discussed in Section 2.4.3, the results in Section 2.4.2 also hold for heterogenous risks and skewed heavy-tailed risks. Furthermore, Ibragimov and Walden (2007) demonstrate that they hold for a wide class of bounded r.v.'s concentrated on a sufficiently large interval with distributions given by truncations of stable and α-symmetric ones.

Besides the analysis of portfolio diversification under heavy-tailedness, the results on portfolio VaR comparisons and analogous results on majorization properties of linear combinations of r.v.'s have a number of other applications. These applications include the study of robustness and efficiency of linear estimators, the study of firm growth when firms can invest in information about their markets, the study of optimal multi-product strategies of a monopolist, as well as the study of inheritance in mathematical evolutionary theory. In all these studies, models are robust to heavy-tailedness assumptions as long as the distributions entering these assumptions are moderately heavy-tailed. But the implications of these models are reversed for distributions with extremely heavy tails (see Ibragimov *et al.* (2015) and references therein).

The chapter is organized as follows. Section 2.2 contains notation and definitions of the classes of heavy-tailed distributions used in this and some of the following chapters. Section 2.3 discusses the definition of VaR and coherent risk measures and summarizes some relevant properties of VaR needed for the analysis. Sections 2.4.1–2.4.3 discuss the definition of majorization pre-ordering used in formalization of the concept of portfolio diversification and present the main results of this chapter, with extensions. Section 2.5 makes some concluding remarks.

Appendix A1 summarizes auxiliary results on VaR comparisons and unimodality properties for log-concave and stable distributions used in the derivation of the main results. We put the main proofs in Appendix A2. More detailed proofs can be found in Ibragimov (2009b).

2.2. Notation and classes of distributions

In this chapter, a univariate density $f(x)$, $x \in \mathbf{R}$, will be referred to as symmetric (about zero) if $f(x) = f(-x)$ for all $x > 0$. In addition, as usual, an absolutely continuous distribution of a r.v. X with the density $f(x)$ will be called symmetric if $f(x)$ is symmetric (about zero).[2] For two r.v.'s X and Y, we write $X \stackrel{d}{=} Y$ if X and Y have the same distribution.

A r.v. X with density $f(x)$, $x \in \mathbf{R}$, and the convex distribution support $\Omega = \{x \in \mathbf{R} : f(x) > 0\}$ is log-concavely distributed if $\log f(x)$ is concave in $x \in \Omega$, that is, if for all $x_1, x_2 \in \Omega$, and any $\lambda \in [0,1]$, $f(\lambda x_1 + (1 - \lambda)x_2) \geq (f(x_1))^\lambda (f(x_2))^{1-\lambda}$ (see An, 1998; Bagnoli and Bergstrom, 2005). Examples of log-concave distributions include normal, uniform, exponential, Gamma distribution $\Gamma(\alpha, \beta)$ with shape parameter $\alpha \geq 1$, Beta distribution $\mathcal{B}(a, b)$ with $a \geq 1$ and $b \geq 1$, and Weibull distribution $\mathcal{W}(\gamma, \alpha)$ with shape parameter $\alpha \geq 1$.

The class of log-concave distributions is closed under convolution and has many other appealing properties that have been utilized in a number of works in economics and finance (see surveys by Karlin (1968); Marshall and Olkin (1979); An (1998); Bagnoli and

---

[2]This concept of (univariate) symmetry is not to be confused with joint α-symmetric, spherical distributions, or radially symmetric copulas discussed in Sections 4.4.1 and 5.4.4, which capture dependence among *components* of random vectors.

Bergstrom (2005)). However, such distributions cannot be used in the study of heavy-tailedness phenomena since any log-concave density is extremely light-tailed: in particular, if a r.v. X is log-concavely distributed, then its density has at most an exponential tail, that is, $f(x) = O(\exp(-\lambda x))$ for some $\lambda > 0$, as $x \to \infty$ and all the power moments $\mathbb{E}|X|^\gamma$, $\gamma > 0$, of the r.v. are finite (see An (1998), Corollary 1). Throughout the chapter, $\mathcal{LC}$ denotes the class of symmetric log-concave distributions ($\mathcal{LC}$ stands for "log-concave").

As before, for $0 < \alpha \le 2$, $\sigma > 0$, $\beta \in [-1, 1]$ and $\mu \in \mathbf{R}$, we denote by $S_\alpha(\sigma, \beta, \mu)$ stable distributions with characteristic exponent (index of stability) α, scale parameter σ, symmetry index (skewness parameter) β and location parameter μ (see Section 1.2.2 for a discussion of these parameters). Its characteristic function is given in (1.4) and a closed form expression for the density $f(x)$ of $S_\alpha(\sigma, \beta, \mu)$ is available in the following cases (and only in those cases):

(1) $\alpha = 2$ (Gaussian distributions);
(2) $\alpha = 1$ and $\beta = 0$ (Cauchy distributions with densities $f(x) = \sigma/(\pi(\sigma^2 + (x - \mu)^2))$);
(3) $\alpha = 1/2$ and $\beta = \pm 1$ (Lévy distributions that have densities $f(x) = (\sigma/(2\pi))^{1/2} \exp(-\sigma/(2x))x^{-3/2}$, $x \ge 0$; $f(x) = 0$, $x < 0$, where $\sigma > 0$, and their shifted versions).

Degenerate distributions correspond to the limiting case $\alpha = 0$. The p-th absolute moments $\mathbb{E}|X|^p$ of a r.v. $X \sim S_\alpha(\sigma, \beta, \mu)$, $\alpha \in (0, 2)$ are finite if $p < \alpha$ and are infinite otherwise.

For $0 < r < 2$, we denote by $\overline{\mathcal{CS}}(r)$ the class of distributions which are convolutions of symmetric stable distributions $S_\alpha(\sigma, 0, 0)$ with characteristic exponents $\alpha \in (r, 2]$ and $\sigma > 0$ (here and below, $\mathcal{CS}$ stands for "convolutions of stable"; the overline indicates that convolutions of stable distributions with indices of stability *greater* than the threshold value r are taken). That is, $\overline{\mathcal{CS}}(r)$ consists of distributions of r.v.'s X such that, for some $k \ge 1$, $X = Y_1 + \ldots + Y_k$, where Y_i, $i = 1, \ldots, k$, are independent r.v.'s such that $Y_i \sim S_{\alpha_i}(\sigma_i, 0, 0)$, $\alpha_i \in (r, 2]$, $\sigma_i > 0$.

Further, for $0 < r \le 2$, $\underline{\mathcal{CS}}(r)$ stands for the class of distributions which are convolutions of symmetric stable distributions $S_\alpha(\sigma, 0, 0)$ with indices of stability $\alpha \in (0, r)$ and $\sigma > 0$ (the underline indicates considering stable distributions with indices of stability *less* than the threshold value r). That is, $\underline{\mathcal{CS}}(r)$ consists of distributions of r.v.'s X such that, for some $k \geq 1$, $X = Y_1 + \ldots + Y_k$, where Y_i, $i = 1, \ldots, k$, are independent r.v.'s such that $Y_i \sim S_{\alpha_i}(\sigma_i, 0, 0)$, $\alpha_i \in (0, r)$, $\sigma_i > 0$, $i = 1, \ldots, k$.

Finally, we denote by $\overline{\mathcal{CSLC}}$ the class of convolutions of distributions from the classes $\mathcal{LC}$ and $\overline{\mathcal{CS}}(1)$. That is, $\overline{\mathcal{CSLC}}$ is the class of convolutions of symmetric distributions which are either log-concave or stable with characteristic exponents greater than one ($\mathcal{CSLC}$ is the abbreviation of "convolutions of stable and log-concave"). In other words, $\overline{\mathcal{CSLC}}$ consists of distributions of r.v.'s X such that $X = Y_1 + Y_2$, where Y_1 and Y_2 are independent r.v.'s with distributions belonging to $\mathcal{LC}$ or $\overline{\mathcal{CS}}(1)$.

All the classes $\mathcal{LC}$, $\overline{\mathcal{CSLC}}$, $\overline{\mathcal{CS}}(r)$ and $\underline{\mathcal{CS}}(r)$ are closed under convolutions. In particular, the class $\overline{\mathcal{CSLC}}$ coincides with the class of distributions of r.v.'s X such that, for some $k \geq 1$, $X = Y_1 + \ldots + Y_k$, where Y_i, $i = 1, \ldots, k$, are independent r.v.'s with distributions belonging to $\mathcal{LC}$ or $\overline{\mathcal{CS}}(1)$.

A linear combination of independent stable r.v.'s with the *same* characteristic exponent α also has a stable distribution with the same α. However, in general, this does not hold in the case of convolutions of stable distributions with *different* indices of stability. Therefore, the class $\overline{\mathcal{CS}}(r)$ of *convolutions* of symmetric stable distributions with *different* indices of stability $\alpha \in (r, 2]$ is wider than the class of *all* symmetric stable distributions $S_\alpha(\sigma, 0, 0)$ with $\alpha \in (r, 2]$ and $\sigma > 0$. Similarly, the class $\underline{\mathcal{CS}}(r)$ is wider than the class of *all* symmetric stable distributions $S_\alpha(\sigma, 0, 0)$ with $\alpha \in (0, r)$ and $\sigma > 0$.

Clearly, $\overline{\mathcal{CS}}(1) \subset \overline{\mathcal{CSLC}}$ and $\mathcal{LC} \subset \overline{\mathcal{CSLC}}$. It should also be noted that the class $\overline{\mathcal{CSLC}}$ is wider than the class of (two-fold) convolutions of log-concave distributions with stable distributions $S_\alpha(\sigma, 0, 0)$ with $\alpha \in (1, 2]$ and $\sigma > 0$.

By definition, for $0 < r_1 < r_2 \leq 2$, the following inclusions hold: $\overline{\mathcal{CS}}(r_2) \subset \overline{\mathcal{CS}}(r_1)$ and $\underline{\mathcal{CS}}(r_1) \subset \underline{\mathcal{CS}}(r_2)$. Cauchy distributions $S_1(\sigma, 0, 0)$ are at the dividing boundary between the classes $\underline{\mathcal{CS}}(1)$ and $\overline{\mathcal{CS}}(1)$ (and between the classes $\underline{\mathcal{CS}}(1)$ and $\overline{\mathcal{CSLC}}$). Similarly, for $r \in (0, 2)$, stable distributions $S_r(\sigma, 0, 0)$ with the characteristic exponent $\alpha = r$ are at the dividing boundary between the classes $\underline{CS}(r)$ and $\overline{CS}(r)$. Further, normal distributions $S_2(\sigma, 0, 0)$ are at the dividing boundary between the class $\mathcal{LC}$ of log-concave distributions and the class $\underline{CS}(2)$ of convolutions of symmetric stable distributions with indices of stability $\alpha < 2$. More precisely, the Cauchy distributions $S_1(\sigma, 0, 0)$ are the only ones that belong to all the classes $\underline{CS}(r)$ with $r > 1$ and all the classes $\overline{CS}(r)$ with $r < 1$. Stable distributions $S_r(\sigma, 0, 0)$ are the only ones that belong to all the classes $\underline{CS}(r')$ with $r' > r$ and all the classes $\overline{CS}(r')$ with $r' < r$. Normal distributions are the only distributions belonging to the class $\mathcal{LC}$ and all the classes $\overline{CS}(r)$ with $r \in (0, 2)$.

We write $X \sim \mathcal{LC}$ (respectively, $X \sim \overline{CS\mathcal{LC}}$, $X \sim \overline{CS}(r)$ or $X \sim \underline{CS}(r)$) if the distribution of the r.v. X belongs to the class $\mathcal{LC}$ (respectively, $\overline{CS\mathcal{LC}}$, $\overline{CS}(r)$ or $\underline{CS}(r)$).

The properties of stable distributions discussed earlier imply that the p-th absolute moments $\mathbb{E}|X|^p$ of a r.v. $X \sim \overline{CS}(r)$, $r \in (0, 2)$, are finite if $p \leq r$. On the other hand, the r.v.'s $X \sim \underline{CS}(r)$, $r \in (0, 2]$ have infinite moments of order r: $\mathbb{E}|X|^r = \infty$. In particular, the distributions of r.v.'s X from the class $\underline{CS}(1)$ are extremely heavy-tailed in the sense that their first moments are infinite: $\mathbb{E}|X| = \infty$. In contrast, the distributions of r.v.'s X in $\overline{CS\mathcal{LC}}$ are moderately heavy-tailed in the sense that they have finite means: $\mathbb{E}|X| < \infty$.

2.3. Value-at-Risk (VaR): Definition and main properties

Given a loss probability $q \in (0, 1)$ and a r.v. (risk) X we denote by $\mathrm{VaR}_q(X)$ the VaR of X at level q, that is, its $(1 - q)$-quantile

$$\mathrm{VaR}_q(X) = \inf\{z \in \mathbf{R} : \mathbb{P}(X > z) \leq q\}.$$

We interpret the positive values of risks X as a risk holder's losses.

Let $\mathcal{X}$ be a certain linear space of r.v.'s X defined on a probability space $(\Omega, \Im, P)$. We assume that $\mathcal{X}$ contains all degenerate r.v.'s $X \equiv a \in \mathbf{R}$. According to the definition of Artzner *et al.* (1999) (also see Nešlehová *et al.* (2006); Frittelli and Gianin (2002)), a functional

$\mathcal{R} : \mathcal{X} \to R$ is said to be a *coherent* measure of risk if it satisfies the following axioms:

a1 (Monotonicity) $\mathcal{R}(X) \geq \mathcal{R}(Y)$ for all $X, Y \in \mathcal{X}$ such that $Y \leq X$ (a.s.), that is, $\mathbb{P}(X \leq Y) = 1$.

a2 (Translation invariance) $\mathcal{R}(X + a) = \mathcal{R}(X) + a$ for all $X \in \mathcal{X}$ and any $a \in \mathbf{R}$.

a3 (Positive homogeneity) $\mathcal{R}(\lambda X) = \lambda \mathcal{R}(X)$ for all $X \in \mathcal{X}$ and any $\lambda \geq 0$.

a4 (Subadditivity) $\mathcal{R}(X + Y) \leq \mathcal{R}(X) + \mathcal{R}(Y)$ for all $X, Y \in \mathcal{X}$. In some papers (see Fölmer and Schied, 2002; Frittelli and Gianin, 2002), the axioms of positive homogeneity and subadditivity are replaced by the following weaker axiom of convexity:

a5 (Convexity) $\mathcal{R}(\lambda X + (1 - \lambda)Y) \leq \lambda \mathcal{R}(X) + (1 - \lambda)\mathcal{R}(Y)$ for all $X, Y \in \mathcal{X}$ and any $\lambda \in [0, 1]$ (clearly, a5 follows from a3 and a4).

It is easy to verify that $\mathrm{VaR}_q(X)$ satisfies the axioms of monotonicity, translation invariance and positive homogeneity a1, a2 and a3. However, as follows from the counterexamples constructed by Artzner *et al.* (1999) and Nešlehova *et al.* (2006), in general, it fails to satisfy the subadditivity and convexity properties a4 and a5.

2.4. Majorization, diversification and (non-)coherency of VaR

2.4.1. *Majorization of random vectors and diversification of portfolio riskiness*

We start by introducing the notion of majorization of vectors. Let $\mathbf{R}_+$ stand for $\mathbf{R}_+ = [0, \infty)$. A vector $a \in \mathbf{R}_+^n$ is said to be majorized by a vector $b \in \mathbf{R}^n$, written $a \prec b$, if $\sum_{i=1}^k a_{[i]} \leq \sum_{i=1}^k b_{[i]}$, $k = 1, \ldots, n-1$, and $\sum_{i=1}^n a_{[i]} = \sum_{i=1}^n b_{[i]}$, where $a_{[1]} \geq \ldots \geq a_{[n]}$ and $b_{[1]} \geq \ldots \geq b_{[n]}$ denote the components of a and b in decreasing order.

In economics of income inequality, the majorization ordering defined in previous paragraph is also known as Lorenz dominance (see Marshall and Olkin (1979); Arnold (1987); Saposnik (1993);

Tong (1994); Marshall *et al.* (2011)). The relation $a \prec b$ implies that the components of the vector a are less diverse than those of b.

In this context, it is easy to see that the following relations hold (see Marshall and Olkin (1979), p.7; Marshall *et al.* (2011)):

$$\left(\sum_{i=1}^{n} a_i/n, \ldots, \sum_{i=1}^{n} a_i/n\right) \prec (a_1, \ldots, a_n)$$

$$\prec \left(\sum_{i=1}^{n} a_i, 0, \ldots, 0\right), \quad a \in \mathbf{R}_+^n, \quad (2.1)$$

for all $a \in \mathbf{R}_+^n$. In particular,

$$(1/(n+1), \ldots, 1/(n+1), 1/(n+1)) \prec (1/n, \ldots, 1/n, 0), \quad n \geq 1. \quad (2.2)$$

It is also immediate that if $a = (a_1, \ldots, a_n) \prec b = (b_1, \ldots, b_n)$, then $(a_{\pi(1)}, \ldots, a_{\pi(n)}) \prec (b_{\pi(1)}, \ldots, b_{\pi(n)})$ for all permutations π of the set $\{1, \ldots, n\}$.

A function $\phi : \mathbf{R}_+^n \to \mathbf{R}$ is called *Schur-convex* (respectively, *Schur-concave*) if $(a \prec b) \implies (\phi(a) \leq \phi(b))$ (respectively, $(a \prec b) \implies (\phi(a) \geq \phi(b))$ for all $a, b \in \mathbf{R}_+^n$. If, in addition, $\phi(a) < \phi(b)$ (respectively, $\phi(a) > \phi(b)$) whenever $a \prec b$ and a is not a permutation of b, then ϕ is said to be *strictly* Schur-convex (respectively, *strictly* Schur-concave) on A. Evidently, if $\phi : \mathbf{R}_+^n \to \mathbf{R}$ is Schur-convex or Schur-concave, then

$$\phi(a_1, \ldots, a_n) = \phi(a_{\pi(1)}, \ldots, a_{\pi(n)}) \quad (2.3)$$

for all $(a_1, \ldots, a_n) \in \mathbf{R}_+^n$ and all permutations π of the set $\{1, \ldots, n\}$.

For $w = (w_1, \ldots, w_n) \in \mathbf{R}_+^n$, denote by Z_w the return on the portfolio of risks $X_1, \ldots, X_n$ with weights w. Several results in the chapter require the assumption that $\sum_{i=1}^{n} w_i = 1$ for the portfolio weights w_i, $i = 1, \ldots, n$. If this is the case, we write that w belongs to the simplex $\mathcal{I}_n = \{w = (w_1, \ldots, w_n) : w_i \geq 0, i = 1, \ldots, n, \sum_{i=1}^{n} w_i = 1\} : w \in \mathcal{I}_n$.

Denote $\underline{w} = (1/n, 1/n, \ldots, 1/n) \in \mathcal{I}_n$ and $\overline{w} = (1, 0, \ldots, 0) \in \mathcal{I}_n$. The expressions $\text{VaR}_q(Z_{\underline{w}})$ and $\text{VaR}_q(Z_{\overline{w}})$ are, thus, the VaRs of the portfolio with equal weights and of the portfolio consisting of only one return (risk), respectively.

Suppose that $v = (v_1, \ldots, v_n) \in \mathbf{R}_+^n$ and $w = (w_1, \ldots, w_n) \in \mathbf{R}_+^n$, $\sum_{i=1}^n v_i = \sum_{i=1}^n w_i$, are the weights of two portfolios. If $v \prec w$, it is natural to think about the portfolio with weights v as being more diversified than that with weights w so that, for example, the portfolio with equal weights $\underline{w}$ is most diversified and the portfolio with weights $\overline{w}$ consisting of one risk is least diversified among all the portfolios with weights $w \in \mathcal{I}_n$ (in this regard, the notion of one portfolio being more or less diversified than another is, in some sense, the opposite of that for vectors of weights for the portfolio).

Heterogeneity in distributions of risks may require altering of formalizations of portfolio diversification using majorization. Section 2.4.3 contains some suggestions on these extensions motivated by VaR analysis for skewed and non-identically distributed risks.

Observe that the above formalization of diversification is applicable in the case where some components of v or w are zero. Thus, it allows us to compare portfolios with possibly unequal number of risks. For instance, portfolio diversification orderings implied by majorization relations (2.2) are intuitive. It is natural to think about the portfolio of X_i, $i = 1, \ldots, n+1$, with weights $(1/(n+1), \ldots, 1/(n+1), 1/(n+1)) \in \mathcal{I}_{n+1}$ as being more diversified than the portfolio of X_i, $i = 1, \ldots, n$, with weights $(1/n, \ldots, 1/n) \in \mathcal{I}_n$ since the former portfolio contains an additional non-redundant risk X_{n+1}.

Examples of strictly Schur-convex functions $\phi : \mathbf{R}_+^n \to \mathbf{R}$ are given by $\phi(w_1, \ldots, w_n) = \sum_{i=1}^n w_i^2$ and, more generally, by $\phi_\alpha(w_1, \ldots, w_n) = \sum_{i=1}^n w_i^\alpha$ for $\alpha > 1$. The functions $\phi_\alpha(w_1, \ldots, w_n)$ are strictly Schur-concave for $\alpha < 1$ (see Marshall and Olkin (1979); Marshall *et al.* (2011), Propositions 3.C.1.a). The functions ϕ_α have the same form as measures of diversification considered by Bouchaud and Potters (2004, Chapter 12, p. 205).

2.4.2. *Subadditivity of VaR*

A simple example where diversification is preferable is provided by the standard case with normal risks. Let $n \geq 2$, $q \in (0, 1/2)$, and let

$X_1, \ldots, X_n \sim S_2(\sigma, 0, 0)$ be i.i.d. symmetric normal r.v.'s. Then, for the portfolio of X_i's with equal weights $\underline{w} = (1/n, 1/n, \ldots, 1/n)$ we have $Z_{\underline{w}} = (1/n) \sum_{i=1}^{n} X_i \overset{d}{=} (1/\sqrt{n}) X_1$. Consequently, by positive homogeneity of the VaR given by property a3 in Section 2.3,

$$\mathrm{VaR}_q(Z_{\underline{w}}) = (1/\sqrt{n})\mathrm{VaR}_q(X_1) = (1/\sqrt{n})\mathrm{VaR}_q(Z_{\overline{w}}) < \mathrm{VaR}_q(Z_{\overline{w}}).$$

That is, the value at risk of the most diversified portfolio with equal weights $\underline{w}$ is less than that of the least diversified portfolio with weights $\overline{w}$ consisting of only one risk Z_1.

Theorem 2.1 shows that similar results hold for all moderately heavy-tailed risks X_i with arbitrary weights $w = (w_1, \ldots, w_n) \in \mathbf{R}_+^n$. In these settings, diversification of a portfolio of X_i's leads to a decrease in VaR of the portfolio return $Z_w = \sum_{i=1}^{n} w_i X_i$.

Theorem 2.1. *Let $q \in (0, 1/2)$ and let X_i, $i = 1, \ldots, n$, be i.i.d. risks such that $X_i \sim \overline{\mathcal{CSLC}}$, $i = 1, \ldots, n$. Then*

(i) $VaR_q(Z_v) < VaR_q(Z_w)$ *if $v \prec w$ and v is not a permutation of w (in other words, the function $\psi(w, q) = VaR_q(Z_w)$ is strictly Schur-convex in $w \in \mathbf{R}_+^n$).*

(ii) *In particular, $VaR_q(Z_{\underline{w}}) < VaR_q(Z_w) < VaR_q(Z_{\overline{w}})$ for all $q \in (0, 1/2)$ and all weights $w \in \mathcal{I}_n$ such that $w \neq \underline{w}$ and w is not a permutation of $\overline{w}$.*

These results, along with their analogues under certain kinds of dependence provided in Theorems 4.4, substantially generalize the riskiness analysis for uniform (equal weights) portfolios of independent stable risks considered, among others, by Fama (1965a); Ross (1976) and Samuelson (1967). The formalizations of portfolio diversification using majorization pre-ordering allow us to obtain comparisons of riskiness for portfolios of heavy-tailed and possibly dependent risks with arbitrary, rather than equal, weights.

Let us now illustrate the settings where diversification is suboptimal in the VaR framework. Let $q \in (0, 1)$ and let $X_1, \ldots, X_n$ be i.i.d. risks with a Lévy distribution $S_{1/2}(\sigma, 1, 0)$ with $\alpha = 1/2$, $\beta = 1$ and the density $f(x) = (\sigma/(2\pi))^{1/2} \exp(-\sigma/(2x)) x^{-3/2}$. Using (1.10) with $\alpha = 1/2$ for the portfolio of X_i's with equal weights $w_i = 1/n$, we get $Z_{\underline{w}} = (1/n) \sum_{i=1}^n X_i \stackrel{d}{=} nX_1$. Consequently, by positive homogeneity of the VaR (see property a3 in Section 2.3),

$$\text{VaR}_q(Z_{\underline{w}}) = n\text{VaR}_q(X_1) = n\text{VaR}_q(Z_{\overline{w}}) > \text{VaR}_q(Z_{\overline{w}}).$$

Thus, the VaR of the least diversified portfolios with weights $\overline{w}$ that consists of only one risk is less than the value at risk of the most diversified portfolio with equal weights $\underline{w}$.

Theorem 2.2 shows that similar conclusions hold for portfolio VaR comparisons with arbitrary weights $w = (w_1, \ldots, w_n) \in \mathbf{R}_+^n$ under the general assumption that the distributions of risks $X_1, \ldots, X_n$ are extremely heavy-tailed. In such settings, the results in Theorem 2.1 are reversed and diversification of a portfolio increases its VaR.

Theorem 2.2. *Let $q \in (0, 1/2)$ and let X_i, $i = 1, \ldots, n$, be i.i.d. risks such that $X_i \sim \underline{CS}(1)$, $i = 1, \ldots, n$. Then*

(i) *$VaR_q(Z_v) > VaR_q(Z_w)$ if $v \prec w$ and v is not a permutation of w (in other words, the function $\psi(w, q) = VaR_q(Z_w)$, is strictly Schur-concave in $w \in \mathbf{R}_+^n$).*

(ii) *In particular, $VaR_q(Z_{\overline{w}}) < VaR_q(Z_w) < VaR_q(Z_{\underline{w}})$ for all $q \in (0, 1/2)$ and all weights $w \in \mathcal{I}_n$ such that $w \neq \underline{w}$ and w is not a permutation of $\overline{w}$.*

Let us now consider a portfolio in the borderline case $\alpha = 1$ which corresponds to i.i.d. risks $X_1, \ldots, X_n$ with a symmetric Cauchy distribution $S_1(\sigma, 0, 0)$. As discussed in Section 2.2, these distributions are exactly at the dividing boundary between the class $\overline{CSLC}$ in Theorem 2.1 and the class $\underline{CS}(1)$ in Theorem 2.2. Using (1.10) with $\alpha = 1$ we have, for all $w = (w_1, \ldots, w_n) \in \mathcal{I}_n$, $Z_w = \sum_{i=1}^n w_i X_i \stackrel{d}{=} X_1$. Consequently, for all $q \in (0, 1)$,

$$\text{VaR}_q(Z_w) = \text{VaR}_q(X_1)$$

and it is independent of w and is the same for all portfolios of risks X_i with weights $w \in \mathcal{I}_n$, $i = 1, \ldots, n$. Thus, diversification of such a portfolio has no effect on its riskiness. Similarly, for general weights

$w = (w_1, \ldots, w_n) \in \mathbf{R}^n_+$, property (1.10) with $\alpha = 1$ implies $Z_w = \sum_{i=1}^n w_i X_i \overset{d}{=} (\sum_{i=1}^n w_i) X_1$. Thus,

$$\mathrm{VaR}_q(Z_w) = \left(\sum_{i=1}^n w_i \right) \mathrm{VaR}_q(X_1)$$

is independent of w so long as $\sum_{i=1}^n w_i$ is fixed.

Consequently, $\mathrm{VaR}_q(Z_w)$ is both Schur-convex (as in Theorem 2.1) and Schur-concave (as in Theorem 2.2) in $w \in \mathbf{R}^n_+$ for i.i.d. risks $X_i \sim S_\alpha(\sigma, 0, 0)$ that have symmetric Cauchy distributions with $\alpha = 1$ (see Proschan (1965); Marshall and Olkin (1979), p. 374; Marshall *et al.* (2011), p. 492, for similar properties of tail probabilities of Cauchy distributions). We note that from the proofs of Theorems 2.1–2.2 and this property of Cauchy distributions it follows that the theorems continue to hold for convolutions of distributions from the classes $\overline{\mathcal{CSLC}}$ and $\underline{\mathcal{CS}}(1)$ with Cauchy distributions $S_1(\sigma, 0, 0)$.

2.4.3. *Extensions to heterogeneity and skewness*

We can extend the results in Theorems 2.1–2.2 to the case of skewed and heterogeneous risks.

Let $\sigma_1, \ldots, \sigma_n \geq 0$ be some scale parameters and let $X_i \sim S_\alpha(\sigma_i, \beta, 0)$, $\alpha \in (0, 2]$, be independent not necessarily identically distributed stable risks. Further, let $Z_{\tilde{w}} = \sum_{i=1}^n w_{[i]} X_i$ be the return on the portfolio with weights $\tilde{w} = (w_{[1]}, \ldots, w_{[n]})$, where, as in Section 2.4.1, $w_{[1]} \geq \ldots \geq w_{[n]}$ denote the components of the vector $w = (w_1, \ldots, w_n) \in \mathbf{R}^n_+$ in decreasing order (a certain ordering in the components of the vector of weights w is necessary for the extensions of the majorization results to the case of non-identically distributed r.v.'s since property (2.3) holds for all functions $\phi : \mathbf{R}^n_+ \to \mathbf{R}$ that are Schur-convex or Schur-concave).

Observe that, in the case of identically distributed risks X_i, $i = 1, \ldots, n$, with $\sigma_1 = \ldots = \sigma_n$, the risk $Z_{\tilde{w}}$ has the same distribution as $Z_w = \sum_{i=1}^n w_i X_i$ and, thus, $\mathrm{VaR}_q(Z_{\tilde{w}}) = \mathrm{VaR}_q(Z_w)$. That is, the VaR comparisons for $Z_{\tilde{w}}$ with the ordered weights $\tilde{w}$ under the above distributional assumptions cover the case of portfolio returns Z_w with identically distributed but possibly skewed risks.

In proofs of the following theorems, we use the Schur-convexity and Schur-concavity properties of functions of the following form:

$$\chi(c_1, \ldots, c_n) = \sum_{i=1}^{n} \sigma_i^\alpha c_{[i]}^\alpha, \tag{2.4}$$

$\alpha > 0$, where σ_i, $i = 1, \ldots, n$, are the scale parameters of the risks in consideration. These properties are similar to the definitions of p-majorization for the vectors $(c_1^\alpha, \ldots, c_n^\alpha)$ with $p = (\sigma_1^\alpha, \ldots, \sigma_n^\alpha)$ (see Marshall and Olkin (1979); Marshall *et al.* (2011), Chapter 14). This suggests that extensions of the majorization pre-ordering such as p-majorization may be useful in formalizations of the concept of diversification for portfolios of heterogenous risks.

Let $Q \sim S_\alpha(1, \beta, 0)$.

Theorem 2.3. *Let* $0 < q < \mathbb{P}(Q > 0)$ *and let* $X_1, \ldots, X_n$ *be independent risks such that* $X_i \sim S_\alpha(\sigma_i, \beta, 0)$, *where* $\alpha \in (1, 2]$, $\sigma_1 \geq \ldots \geq \sigma_n > 0$ *and* $\beta \in [-1, 1]$. *Then*

(i) $VaR_q(Z_{\tilde{v}}) < VaR_q(Z_{\tilde{w}})$ *if* $v \prec w$ *and* v *is not a permutation of* w *(in other words, the function* $\psi(w, q) = VaR_q(Z_{\tilde{w}})$ *is strictly Schur-convex in* $w \in \mathbf{R}_+^n$*).*

(ii) *In particular,* $VaR_q(Z_{\underline{w}}) < VaR_q(Z_{\tilde{w}}) < VaR_q(Z_{\overline{w}})$ *for all* $q \in (0, 1/2)$ *and all weights* $w \in \mathcal{I}_n$ *such that* $w \neq \underline{w}$ *and* w *is not a permutation of* $\overline{w}$.

Theorem 2.4. *Let* $0 < q < \mathbb{P}(Q > 0)$ *and let* $X_1, \ldots, X_n$ *be independent risks such that* $X_i \sim S_\alpha(\sigma_i, \beta, 0)$, *where* $\alpha \in (0, 1)$, $\sigma_n \geq \ldots \geq \sigma_1 > 0$ *and* $\beta \in [-1, 1]$. *Then*

(i) $VaR_q(Z_{\tilde{v}}) > VaR_q(Z_{\tilde{w}})$ *if* $v \prec w$ *and* v *is not a permutation of* w *(in other words, the function* $\psi(w, q) = VaR_q(Z_w)$, *is strictly Schur-concave in* $w \in \mathbf{R}_+^n$*).*

(ii) *In particular,* $VaR_q(Z_{\overline{w}}) < VaR_q(Z_{\tilde{w}}) < VaR_q(Z_{\underline{w}})$ *for all* $q \in (0, 1/2)$ *and all weights* $w \in \mathcal{I}_n$ *such that* $w \neq \underline{w}$ *and* w *is not a permutation of* $\overline{w}$.

Theorems 2.3 and 2.4 are extensions of Theorems 2.1 and 2.2 to the case of skewed and heterogenous risks. We also present, in Theorems 2.5–2.6, sharp VaR bounds for portfolios of such risks. Theorems

2.5–2.6 further provide VaR comparisons implied by majorization between vectors of powers of portfolio weights.

Theorem 2.5. *Let* $r \in (0, 2]$, $0 < q < \mathbb{P}(Q > 0)$ *and let* $X_1, \ldots, X_n$ *be independent risks such that* $X_i \sim S_\alpha(\sigma_i, \beta, 0)$, *where* $\alpha \in (r, 2]$, $\sigma_1 \geq \ldots \geq \sigma_n > 0$, $\beta \in [-1, 1]$, $\beta = 0$ *for* $\alpha = 1$. *Then*

(i) $VaR_q(Z_{\tilde{v}}) < VaR_q(Z_{\tilde{w}})$ *if* $(v_1^r, \ldots, v_n^r) \prec (w_1^r, \ldots, w_n^r)$ *and* $(v_1^r, \ldots, v_n^r)$ *is not a permutation of* $(w_1^r, \ldots, w_n^r)$ *(that is, the function* $\psi(w, q) = VaR_q(Z_{\tilde{w}})$, $w \in \mathbf{R}_+^n$, *is strictly Schur-convex in* $(w_1^r, \ldots, w_n^r)$*).*

(ii) *The following sharp bounds hold:*

$$n^{1-1/r} \left(\sum_{i=1}^n w_i^r \right)^{1/r} VaR_q(Z_{\underline{w}}) < VaR_q(Z_{\tilde{w}})$$

$$< \left(\sum_{i=1}^n w_i^r \right)^{1/r} VaR_q(Z_{\overline{w}})$$

for all $q \in (0, 1/2)$ *and all weights* $w \in \mathcal{I}_n$ *such that* $w \neq \underline{w}$ *and* w *is not a permutation of* $\overline{w}$.

Theorem 2.6. *Let* $r \in (0, 2]$, $0 < q < \mathbb{P}(Q > 0)$ *and let* $X_1, \ldots, X_n$ *be independent risks such that* $X_i \sim S_\alpha(\sigma_i, \beta, 0)$, *where* $\alpha \in (0, r)$, $\sigma_n \geq \ldots \geq \sigma_1 > 0$, $\beta \in [-1, 1]$, $\beta = 0$ *for* $\alpha = 1$. *Then*

(i) $VaR_q(Z_{\tilde{v}}) > VaR_q(Z_{\tilde{w}})$ *if* $(v_1^r, \ldots, v_n^r) \prec (w_1^r, \ldots, w_n^r)$ *and* $(v_1^r, \ldots, v_n^r)$ *is not a permutation of* $(w_1^r, \ldots, w_n^r)$ *(that is, the function* $\psi(w, q) = VaR_q(Z_{\underline{w}})$, $w \in \mathbf{R}_+^n$ *is strictly Schur-concave in* $(w_1^r, \ldots, w_n^r)$*).*

(ii) *The following sharp bounds hold:*

$$\left(\sum_{i=1}^n w_i^r \right)^{1/r} VaR_q(Z_{\overline{w}}) < VaR_q(Z_{\tilde{w}})$$

$$< n^{1-1/r} \left(\sum_{i=1}^n w_i^r \right)^{1/r} VaR_q(Z_{\underline{w}})$$

for all $q \in (0, 1/2)$ *and all weights* $w \in \mathcal{I}_n$ *such that* $w \neq \underline{w}$ *and* w *is not a permutation of* $\overline{w}$.

Finally, Theorem 2.7 provides an extension of Theorems 2.1 and 2.2 to classes $\overline{CS}(r)$ and $\underline{CS}(r)$.

Theorem 2.7. *Parts* (i) *and* (ii) *of Theorem* 2.5 *hold for all* $q \in (0, 1/2)$ *and i.i.d. risks* $X_i \sim \overline{CS}(r)$, $i = 1, \ldots, n$. *Parts* (i) *and* (ii) *of Theorem* 2.6 *hold for all* $q \in (0, 1/2)$ *and i.i.d. risks* $X_i \sim \underline{CS}(r)$, $i = 1, \ldots, n$.

Moreover, using conditioning arguments, it is possible to show that the extensions provided by Theorems 2.3–2.6 also hold in the case of random scale parameters σ_i. Similarly, extensions of Theorems 4.4–4.5 can be stated for dependent heterogenous skewed risks, including convolutions (4.19) of common shock models (4.50) with skewed non-identically distributed risks Y_{ij}.

It is worth making a few remarks. First, the proofs of the above results rest on Proposition 2.5 in Appendix A1, which implies that VaR comparisons for risks with symmetric unimodal densities are closed under convolutions. In particular, if X_1, X_2 and Y_1, Y_2 are independent risks with symmetric unimodal densities such that, for $i = 1, 2$, and all $q \in (0, 1/2)$, $\text{VaR}_q(X_i) < \text{VaR}_q(Y_i)$, then $\text{VaR}_q(X_1 + X_2) < \text{VaR}_q(Y_1 + Y_2)$ for all $q \in (0, 1/2)$. The densities of the returns Z_w on the portfolios of risks in the classes $\overline{CSLC}$, $\overline{CS}(r)$, $r \in (0, 2]$, and $\underline{CS}(r)$, $r \in (0, 2]$, are symmetric and unimodal as shown in Appendix A1. The above VaR and unimodality properties provide one of the key arguments in the proof of Theorems 2.1 and 2.2.

Second, from Theorem 2.1 it follows that, if X_1 and X_2 are i.i.d. risks such that $X_i \sim \overline{CSLC}$, $i = 1, 2$, then $\text{VaR}_q(X_1 + X_2) < \text{VaR}_q(X_1) + \text{VaR}_q(X_2)$ and $\text{VaR}_q(\lambda X_1 + (1 - \lambda)X_2) < \lambda \text{VaR}_q(X_1) + (1 - \lambda)\text{VaR}_q(X_2)$ for all $q \in (0, 1/2)$ and any $\lambda \in (0, 1)$. That is, VaR exhibits subadditivity and convexity and is thus a coherent measure of risk for the class $\overline{CSLC}$ (see Section 2.3 for the definition of coherent risk measures and coherency axioms in the case of VaR).

On the other hand, Theorem 2.2 implies that $\text{VaR}_q(X_1) + \text{VaR}_q(X_2) < \text{VaR}_q(X_1 + X_2)$ and $\lambda \text{VaR}_q(X_1) + (1 - \lambda)\text{VaR}_q(X_2) < \text{VaR}_q(\lambda X_1 + (1 - \lambda)X_2)$ for all $q \in (0, 1/2)$, $\lambda \in (0, 1)$ and i.i.d. risks $X_1, X_2 \sim \underline{CS}(1)$. Consequently, subadditivity and convexity are always violated for risks with extremely heavy-tailed distributions. In this case, VaR is not a coherent risk measure even under independence which is often viewed as "the worst case scenario" for diversification failure.

Third, from the counterexamples constructed by Artzner *et al.* (1999) and Nešlehova *et al.* (2006), it follows that VaR in general fails to satisfy the subadditivity and convexity properties. From the analysis similar to Examples 6 and 7 of Nešlehova *et al.* (2006) and Chapter 12 of Bouchaud and Potters (2004), it follows that subadditivity of VaR holds for distributions with power-law tails (1.8) and sufficiently small values of the loss probability q if $\alpha > 1$. Subadditivity is violated for power-law distributions (1.8) and sufficiently small values of the loss probability q if $\alpha < 1$.

More generally, following Example 7 of Nešlehova *et al.* (2006) and Section 12.1.2 of Bouchaud and Potters (2004), let X_1 and X_2 be two i.i.d. risks with regularly varying heavy tails: $\mathbb{P}(X_1 > x) = L(x)/x^\alpha$, where $\alpha > 0$ and $L(x)$ is a slowly varying at infinity function, that is $L(\lambda x)/L(x) \to 1$, as $x \to +\infty$, for all $\lambda > 0$ (see Embrechts *et al.* (1997); Zolotarev (1986), p. 8). Using the property that $\lim_{x \to +\infty} \mathbb{P}(X_1 + X_2 > x)/\mathbb{P}(X_1 > x/2^{1/\alpha}) = 1$ (see Lemma 1.3.1 of Embrechts *et al.* (1997) and Section 12.1.2 of Bouchaud and Potters (2004)), one gets that $\lim_{q \to 0} \mathrm{VaR}_q(X_1 + X_2)/(\mathrm{VaR}_q(X_1) + \mathrm{VaR}_q(X_2)) = 2^{1/\alpha - 1}$. Consequently, the subadditivity property holds for the VaR *asymptotically* as $q \to 0$ if $\alpha > 1$ and is violated as $q \to 0$ if $\alpha < 1$. The implications of Theorems 2.1 and 2.2 for the VaR coherency are qualitatively different from the counterexamples available in the literature and the above asymptotic considerations. This is because the VaR comparisons hold *regardless* of the value of q and are valid for the *whole* wide classes of heavy-tailed risks.

Fourth, it is well-known that if r.v.'s X and Y are such that $\mathbb{P}(X > x) \leq \mathbb{P}(Y > x)$ for all $x \in \mathbf{R}$, then $\mathbb{E}U(X) \leq \mathbb{E}U(Y)$ for all increasing functions $U : \mathbf{R} \to \mathbf{R}$ for which the expectations exist (see Shaked and Shanthikumar (2007), pp. 3–4). In addition, for risks Z_v and Z_w with symmetric distributions, $\mathrm{VaR}_q(Z_v) \leq \mathrm{VaR}_q(Z_w)$ for all $q \in (0, 1/2)$ if and only if $\mathbb{P}(|Z_v| > x) \leq \mathbb{P}(|Z_w| > x)$ for all $x > 0$.[3] These properties, together with Theorems 2.1

---

[3]Taking the absolute values here and inside the function U in the comparisons that follow is needed because, for Z_v and Z_w with symmetric distributions, the

and 2.2, imply majorization comparisons for expectations of (risk measure) functions of linear combinations of heavy-tailed r.v.'s. For instance, we get that if $U\colon \mathbf{R}_+ \to \mathbf{R}$ is an increasing function, then, assuming existence of the expectations, the function $\varphi(w) = \mathbb{E}U(|\sum_{i=1}^n w_i X_i|)$, $w \in \mathbf{R}_+^n$, is Schur-convex in $(w_1^r, \ldots, w_n^r)$ for i.i.d. risks $X_i \sim \overline{\mathcal{CS}}(r)$ and is Schur-concave in $(w_1^r, \ldots, w_n^r)$ for i.i.d. risks $X_i \sim \underline{\mathcal{CS}}(r)$. In particular, $\mathbb{E}U(|n^{1-1/r}(\sum_{i=1}^n w_i^r)^{1/r} Z_{\underline{w}}|) \leq \mathbb{E}U(|Z_w|) \leq \mathbb{E}U(|(\sum_{i=1}^n w_i^r)^{1/r} Z_{\overline{w}}|)$ for i.i.d. risks $X_i \sim \overline{\mathcal{CS}}(r)$ and $\mathbb{E}U(|(\sum_{i=1}^n w_i^r)^{1/r} Z_{\overline{w}}|) \leq \mathbb{E}U(|Z_w|) \leq \mathbb{E}U(|n^{1-1/r}(\sum_{i=1}^n w_i^r)^{1/r} Z_{\underline{w}}|)$ for i.i.d. risks $X_i \sim \underline{\mathcal{CS}}(r)$. By results in Section 2.4.3 we also get that the function $\varphi(w)$, $w \in \mathbf{R}_+^n$ is Schur-concave in $(w_1^2, \ldots, w_n^2)$ if $X_1, \ldots, X_n$ are i.i.d. risks such that $X_i \sim \underline{\mathcal{CS}}(2)$ or $X_i \sim S_\alpha(\sigma, \beta, 0)$ for some $\sigma > 0$, $\beta \in [-1, 1]$ and $\alpha \in (0, 2)$.

The above results extend and complement those obtained by Efron (1969) and Eaton (1970) (see also Marshall and Olkin (1979), pp. 361–365; Marshall *et al.* (2011), pp. 476–477) who studied the classes of functions $U : \mathbf{R} \to \mathbf{R}$ and r.v.'s $X_1, \ldots, X_n$ for which Schur-concavity of $\varphi(w)$, $w \in \mathbf{R}_+^n$, in $(w_1^2, \ldots, w_n^2)$ holds. Further, we obtain that $\varphi(w)$ is Schur-convex in $w \in \mathbf{R}_+^n$ under the assumptions of Theorem 2.1 and is Schur-concave in $w \in \mathbf{R}_+^n$ under the assumptions of Theorem 2.2. It is important to note here that in the case of increasing *convex* functions $U : \mathbf{R}_+ \to \mathbf{R}$ and r.v.'s $X_1, \ldots, X_n$ satisfying the assumptions of Theorem 2.2, the expectations $\mathbb{E}U(|\sum_{i=1}^n w_i X_i|)$ are infinite for all $w \in \mathbf{R}_+^n$ (since the function $(f(x) - f(0))/x$ is increasing in $x > 0$ by Marshall and Olkin (1979), p. 453; Marshall *et al.* (2011), p. 643). Therefore, the last result does not contradict the well-known fact (see Marshall and Olkin (1979), p. 361; Marshall *et al.* (2011), p. 476) that the function $\mathbb{E}f(\sum_{i=1}^n w_i X_i)$ is Schur-convex in $(w_1, \ldots, w_n) \in \mathbf{R}$ for all i.i.d. r.v.'s $X_1, \ldots, X_n$ and convex functions $f : \mathbf{R} \to \mathbf{R}$ as it might seem on the first sight.

Fifth, Theorems 2.1 and 2.2 imply corresponding results on majorization properties of the tail probabilities $\xi(w, x) = \mathbb{P}(\sum_{i=1}^n$

---

value at risk comparisons $\mathrm{VaR}_q(Z_v) \leq \mathrm{VaR}_q(Z_w)$, $q \in (0, 1/2)$, imply the opposite inequalities for the tail probabilities with negative x : $\mathbb{P}(Z_v > x) \geq \mathbb{P}(Z_w > x)$ for $x < 0$.

$w_i X_i > x$), $x > 0$, of linear combinations of heavy-tailed r.v.'s $X_1, \ldots, X_n$. These implications generalize the results in the seminal work by Proschan (1965) who showed that the tail probabilities $\xi(w, x)$ are Schur-convex in $w = (w_1, \ldots, w_n) \in \mathbf{R}_+^n$ for all $x > 0$ for i.i.d. r.v.'s $X_i \sim \mathcal{LC}$, $i = 1, \ldots, n$.[4] Schur-convexity of $\xi(w, x)$ for $X_i \sim \mathcal{LC}$ implies that the value at risk comparisons in Theorem 2.1 hold for i.i.d. log-concavely distributed risks, as stated in Proposition 2.6 in Appendix A1. The results in Proschan (1965) have been applied to the analysis of many problems in statistics, econometrics, economic theory, mathematical evolutionary theory and other fields. One should note here that applicability of these majorization results and their analogs for other classes of distributions to portfolio value at risk theory has not been recognized in the above literature even in the case of i.i.d. log-concavely distributed risks.

A number of papers in probability and statistics have focused on extension of Proschan's results (see, among others, Chan *et al.* (1989); Jensen (1997); Ma (1998) and the review in Tong (1994)). However, in all the studies that dealt with generalizations of the results, the majorization properties of the tail probabilities were of the same type as in Proschan (1965). Namely, the results gave extensions of Proschan's results concerning *Schur-convexity* of the tail probabilities $\xi(a, x)$, $x > 0$, to classes of r.v.'s more general than those considered in Proschan (1965). Analogues of Theorems 2.1 and 2.2 for the tail probabilities $\xi(a, x)$, on the other hand, provide the first general results concerning *Schur-concavity* of $\xi(a, x)$, $x > 0$, for certain wide classes of r.v.'s. According to these results, the class of distributions for which Schur-convexity of the tail probabilities $\xi(a, x)$ is replaced by their Schur-concavity is precisely the class of distributions with extremely heavy-tailed densities.

---

[4]The main results in Proschan (1965) are reviewed in Section 12.J of Marshall and Olkin (1979). The work by Proschan (1965) is presented, in a rearranged form, in Section 11 of Chapter 7 of Karlin (1968). Peakedness results in Karlin (1968) and Proschan (1965) are formulated for "PF2 densities," which is the same as "log-concave densities." The results by Proschan (1965) and the probabilistic results in this chapter are also reviewed in Section 12.J of Marshall *et al.* (2011) (in the statement (ii) of the results for convolutions of symmetric stable distributions on page 492 of the book, the condition $\alpha \leq 1$ needs to be replaced by $\alpha \geq 1$.)

2.5. Concluding remarks

This chapter has shown limits of diversification. Majorization theory is useful here for the analysis of when diversification actually reduces riskiness. In the framework of portfolio VaR, diversification is inferior under extreme heavy-tailedness in portfolio components.

We presented the results that demonstrate that portfolio diversification is optimal if the risks in consideration are moderately heavy-tailed and that these results continue to hold under heterogeneity and skewness, that is for risks that are not necessarily identically distributed. At the same time, heterogeneity may naturally require alternative formalizations of diversification. For instance, the results for non-identically distributed risks presented in Theorems 2.3–2.7 suggest that analogs of Cheng's $p-$majorization (see Marshall and Olkin (1979); Marshall *et al.* (2011), Chapter 14) may be useful in formalization of portfolio diversification for heterogenous risks.

There is a growing range of empirical applications of these seemingly counterintuitive results. Ibragimov *et al.* (2009) demonstrate how this analysis can be used to explain abnormally low levels of reinsurance among insurance providers in markets for catastrophic insurance. Ibragimov *et al.* (2011) show how to analyze the recent financial crisis as a case of excessive risk sharing between banks when risks are extremely heavy-tailed. Gabaix (2009) provides a survey of empirical applications in economics and finance.

By definition, classes $\overline{\mathcal{CS}}(r)$ in Theorem 2.7 include i.i.d. normal distributions which are log-concave, as indicated in Section 2.2. It is not known, in general, whether the results in that theorem with $r \neq 1$ also hold for a sufficiently wide subclass of i.i.d. log-concave distributions. If this is the case, then the analysis in this chapter would allow one to extend the results in Theorem 2.7 to convolutions of distributions in $\overline{\mathcal{CS}}(r)$ also in this subclass. Some results related to this problem with $r = 2$ are provided by the properties of Schur-concavity of the expectations $\varphi(w) = \mathbb{E}U(|\sum_{i=1}^{n} w_i X_i|)$ in $(w_1^2, \ldots, w_n^2)$ for certain classes of functions U and risks $X_1, \ldots, X_n$ considered above but a more comprehensive analysis of these issues is left for further research.

Additionally, the value-at-risk and expected utility comparisons are closely linked: in particular, non-diversification results for truncations of extremely heavy-tailed distributions continue to hold in the expected utility framework if investor's utility function becomes convex at any point in the domain of large losses (convexity of utility functions in the loss domain is one of the key foundations of Prospect theory, see Kahneman and Tversky (1979); it also effectively arises if there is limited liability).

There have been papers (see among others, Acerbi and Tasche (2002); Tasche (2002)) that recommend to use the Expected Shortfall (ES) as a coherent alternative to VaR. However, ES, which is defined as the average of the worst losses of a portfolio, requires existence of first moments of risks to be finite. It is not difficult to see that existence of means of the risks in considerations is also required for finiteness of coherent spectral measures of risk (see, e.g., Acerbi (2002); Cotter and Dowd (2006)) that generalize ES. So the VaR comparisons presented in this chapter remain very relevant.

In Chapter 4, we will return to VaR comparisons of this chapter and will show that similar conclusions hold under certain types of dependence.

2.6. Appendix A1: VaR and unimodality properties of log-concave and stable distributions

This appendix summarizes auxiliary VaR and unimodality results for log-concave and stable distributions needed for the analysis in this chapter.

Definition 2.1. A density function f is said to be unimodal if there exists $c \in \mathbf{R}$ such that f is nondecreasing on $(-\infty, c)$ and is nonincreasing on (c, ∞). An absolutely continuous r.v. or distribution with density f is said to be unimodal if f is unimodal.

Proposition 2.1. (Theorem 2.7.6 in Zolotarev (1986), p. 134) *Each stable r.v.* $X \sim S_\alpha(\sigma, \beta, \mu)$ *is unimodal.*

Proposition 2.2 is a consequence of Theorem 1.10 of Dharmadhikari and Joag-Dev (1988, pp. 18–20), see also An (1998).

Proposition 2.2. (Dharmadhikari and Joag-Dev (1988); An (1998)) *Any log-concave density is unimodal.*

Proposition 2.3 below easily follows from a result due to R. Askey, see Theorem 4.1 of Gneiting (1998) and the proof of the proposition in Appendix A2.

Proposition 2.3. *If* $(X_1, \ldots, X_n)$ *has an absolutely continuous* α-*symmetric distribution with the c.f. generator* $\phi \in \Phi$, *then the density of the r.v.* $Z_w = \sum_{i=1}^{n} w_i X_i$ *is symmetric and unimodal for all* $w_i \in \mathbf{R}$.

Proposition 2.4. (Theorem 1.6 in Dharmadhikari and Joag-Dev (1988), p. 13) *The convolution of two symmetric unimodal densities is unimodal.*

Proposition 2.5 is an analogue of Lemma and Theorem 1 in Birnbaum (1948) and Theorem 3.D.4 on p. 173 of Shaked and Shanthikumar (2007) in terms of strict VaR comparisons. Its proof in Appendix A2 follows the same lines as in Birnbaum (1948).

Proposition 2.5. *Let* $X_1, \ldots, X_n$ *and* $Y_1, \ldots, Y_n$ *be two sets of independent r.v.'s, all having symmetric unimodal densities. Suppose that* $VaR_q(X_i) < VaR_q(Y_i)$, $i = 1, \ldots, n$, *for all* $q \in (0, 1/2)$. *Then* $VaR_q(\sum_{i=1}^{n} X_i) < VaR_q(\sum_{i=1}^{n} Y_i)$ *for all* $q \in (0, 1/2)$.

Proposition 2.6 follows from majorization results for tail probabilities of linear combinations of log-concavely distributed r.v.'s derived by Proschan (1965), see the arguments in Appendix A2.

Proposition 2.6. *Theorem* 2.1 *holds for i.i.d. risks* $X_1, \ldots, X_n$ *such that* $X_i \sim \mathcal{LC}$, $i = 1, \ldots, n$.

2.7. Appendix A2: Proofs of theorems and propositions

Proof of Theorems 2.1 and 2.2. Let $\alpha \in (0, 2]$, $\sigma > 0$, and let $v = (v_1, \ldots, v_n) \in \mathbf{R}_+^n$ and $w = (w_1, \ldots, w_n) \in \mathbf{R}_+^n$ be two vectors of portfolio weights such that $(v_1, \ldots, v_n) \prec (w_1, \ldots, w_n)$ and $(v_1, \ldots, v_n)$ is not a permutation of $(w_1, \ldots, w_n)$ (clearly, $\sum_{i=1}^{n} v_i \neq 0$ and $\sum_{i=1}^{n} w_i \neq 0$). Let $X_1, \ldots, X_n$ be i.i.d. risks such that

$X_i \sim S_\alpha(\sigma, 0, 0)$, $i = 1, \ldots, n$. From (1.10) it follows that if $c = (c_1, \ldots, c_n) \in \mathbf{R}_+^n$, $\sum_{i=1}^n c_i \neq 0$, then $Z_c = \sum_{i=1}^n c_i X_i =^d$ $(\sum_{i=1}^n c_i^\alpha)^{1/\alpha} X_1$. Using positive homogeneity of the value at risk (property a3 in Section 2.3), we thus obtain that, for all $q \in (0, 1/2)$,

$$\mathrm{VaR}_q(Z_c) = \mathrm{VaR}_q(X_1) \left(\sum_{i=1}^n c_i^\alpha \right)^{1/\alpha}. \tag{2.5}$$

Proposition 3.C.1.a in Marshall and Olkin (1979) and Marshall *et al.* (2011) implies that the function $h(c_1, \ldots, c_n) = \sum_{i=1}^n c_i^\alpha$ is strictly Schur-convex in $(c_1, \ldots, c_n) \in \mathbf{R}_+^n$ if $\alpha > 1$ and is strictly Schur-concave in $(c_1, \ldots, c_n) \in \mathbf{R}_+^n$ if $\alpha < 1$. Therefore, we have $\sum_{i=1}^n v_i^\alpha < \sum_{i=1}^n w_i^\alpha$, if $\alpha > 1$ and $\sum_{i=1}^n w_i^\alpha < \sum_{i=1}^n v_i^\alpha$, if $\alpha < 1$. This, together with (2.5), implies that, for all $q \in (0, 1/2)$,

$$\mathrm{VaR}_q(Z_v) < \mathrm{VaR}_q(Z_w) \tag{2.6}$$

if $\alpha > 1$, and

$$\mathrm{VaR}_q(Z_v) > \mathrm{VaR}_q(Z_w) \tag{2.7}$$

if $\alpha < 1$. This completes the proof of parts (i) of Theorems 2.1 and 2.2 in the case of i.i.d. stable risks $X_i \sim S_\alpha(\sigma, 0, 0)$, $i = 1, \ldots, n$.

Let now $X_1, \ldots, X_n$ be i.i.d. risks such that $X_i \sim \overline{\mathcal{CSLC}}$, $i = 1, \ldots, n$. By definition, $X_i = \gamma Y_{i0} + \sum_{j=1}^k Y_{ij}$, $i = 1, \ldots, n$, where $\gamma \in \{0, 1\}$, $k \geq 0$, $Y_{i0} \sim \mathcal{LC}$, $i = 1, \ldots, n$, and $(Y_{1j}, \ldots, Y_{nj})$, $j = 0, 1, \ldots, k$, are independent vectors with i.i.d. components such that $Y_{ij} \sim S_{\alpha_j}(\sigma_j, 0, 0)$, $\alpha_j \in (1, 2]$, $\sigma_j > 0$, $i = 1, \ldots, n$, $j = 1, \ldots, k$. From (2.6) and the results in Proschan (1965) for tail probabilities of log-concavely distributed r.v.'s (see Proposition 2.6), it follows that, for all $q \in (0, 1/2)$ and all $j = 0, 1, \ldots, k$, $\mathrm{VaR}_q\left(\sum_{i=1}^n v_i Y_{ij} \right) < \mathrm{VaR}_q\left(\sum_{i=1}^n w_i Y_{ij} \right)$. The densities of the r.v.'s Y_{i0}, $i = 1, \ldots, n$, are symmetric and unimodal by Proposition 2.2. In addition, the densities of the r.v.'s Y_{ij}, $i = 1, \ldots, n$, $j = 1, \ldots, k$, are symmetric and unimodal by Proposition 2.1. Using Proposition 2.4, we conclude that the densities of the r.v.'s $\sum_{i=1}^n v_i Y_{ij}$ and $\sum_{i=1}^n w_i Y_{ij}$, $j = 0, 1, \ldots, k$, are symmetric and unimodal as well. By Proposition 2.5 we thus obtain

$$\mathrm{VaR}_q(Z_v) = \mathrm{VaR}_q\left(\sum_{i=1}^n v_i X_i\right) = \mathrm{VaR}_q\left(\gamma\sum_{i=1}^n v_i Y_{i0} + \sum_{j=1}^k\sum_{i=1}^n v_i Y_{ij}\right)$$

$$< \mathrm{VaR}_q\left(\gamma\sum_{i=1}^n w_i Y_{i0} + \sum_{j=0}^k\sum_{i=1}^n w_i Y_{ij}\right)$$

$$= \mathrm{VaR}_q\left(\sum_{i=1}^n w_i X_i\right) = \mathrm{VaR}_q(Z_w).$$

This completes the proof of part (i) of Theorem 2.1.

Part (i) of Theorem 2.2 may be proven in a completely similar way, with the use of relations (2.7) instead of Proposition 2.6 and inequalities (2.6). The bounds in parts (ii) of Theorems 2.1 and 2.2 follow from their parts (i) and majorization comparisons (2.1). Sharpness of the bounds in parts (ii) of the theorems follows from the property that, as follows from the discussion following Theorem 2.2, the bounds become equalities in the limit as $\alpha \to 1$ for i.i.d. risks $X_i \sim S_1(\sigma, 0, 0)$, $i = 1, \ldots, n$, with symmetric Cauchy distributions.

Proof of Proposition 2.3. By definition in (4.16), the c.f. of the r.v. $Z_w = \sum_{i=1}^n w_i X_i$ satisfies $\mathbb{E}\exp(itZ_w) = f(|t|)$, $t \in \mathbf{R}$, where $f(s) = \phi(h(w)s)$, $s \in \mathbf{R}_+$, and $h(w) = \sum_{i=1}^n |w_i|^\alpha$. Since the c.f. is real, the density of Z_w is symmetric. From Theorem 4.1 in Gneiting (1998), $f(|t|)$, $t \in \mathbf{R}$, is a c.f. of a unimodal distribution if the function $f : \mathbf{R}_+ \to \mathbf{R}$ is in $\Phi : f \in \Phi$. Evidently, this holds since $\phi \in \Phi$.

Proof of Proposition 2.5. It suffices to provide the arguments in the case $n = 2$; the case of general n then follows by induction and Proposition 2.4. Let X_1, X_2 and Y_1, Y_2 be two sets of independent r.v.'s with symmetric unimodal densities f_1, f_2 and g_1, g_2. Suppose that $\mathrm{VaR}_q(X_i) < \mathrm{VaR}_q(Y_i)$, $i = 1, 2$, for all $q \in (0, 1/2)$ or, equivalently,

$$\mathbb{P}(X_i > s) < \mathbb{P}(Y_i > s), \quad i = 1, 2, \quad \text{for all } s > 0. \qquad (2.8)$$

As in the proof of Lemma in Birnbaum (1948), for all $x > 0$, we have

$$\mathbb{P}(Y_1 + Y_2 > x) - \mathbb{P}(X_1 + X_2 > x) = I_1(x) + I_2(x), \qquad (2.9)$$

where

$$I_1(x) = \int_0^{+\infty} [\mathbb{P}(X_1 > s) - \mathbb{P}(Y_1 > s)][f_2(x + s) - f_2(x - s)]ds,$$

$$I_2(x) = \int_0^{+\infty} [\mathbb{P}(X_2 > s) - \mathbb{P}(Y_2 > s)][g_1(x + s) - g_1(x - s)]ds.$$

As shown in the proof of Lemma by Birnbaum (1948), from symmetry and unimodality of f_2 and g_1 it follows that $f_2(x + s) - f_2(x - s) \le 0$ and $g_1(x + s) - g_1(x - s) \le 0$ for all $x \ge 0$ and $s \ge 0$. This, together with (2.8) and (2.9), implies that $\mathbb{P}(Y_1 + Y_2 > x) - \mathbb{P}(X_1 + X_2 > x) \ge 0$ for all $x > 0$. In addition, $\mathbb{P}(Y_1 + Y_2 > x) = \mathbb{P}(X_1 + X_2 > x)$ for some $x > 0$ if and only if $f_2(x + s) = f_2(x - s)$ and $g_1(x + s) = g_1(x - s)$ for all $s \ge 0$, which is impossible. Thus, $\mathbb{P}(Y_1 + Y_2 > x) - \mathbb{P}(X_1 + X_2 > x) > 0$ for all $x > 0$, or, equivalently, $\mathrm{VaR}_q(X_1 + X_2) < \mathrm{VaR}_q(Y_1 + Y_2)$ for all $q \in (0, 1/2)$.

Proof of Proposition 2.6. Let $X_1, \ldots, X_n$ be i.i.d. risks such that $X_i \sim \mathcal{LC}$, $i = 1, \ldots, n$. From Theorem 2.3 in Proschan (1965), it follows that, for any $x > 0$, the function $\xi(w, x) = \mathbb{P}(\sum_{i=1}^n w_i X_i > x)$ is strictly Schur-convex in $w = (w_1, \ldots, w_n) \in \mathbf{R}_+^n$. This implies that if $v \prec w$ and v is not a permutation of w, then, for all $q \in (0, 1/2)$, $q = \mathbb{P}(Z_v > \mathrm{VaR}_q(Z_v)) < \mathbb{P}(Z_w > \mathrm{VaR}_q(Z_v))$. Consequently, $\mathrm{VaR}_q(Z_v) < \mathrm{VaR}_q(Z_w)$ for all $q \in (0, 1/2)$, and thus part (i) of Theorem 2.1 holds for i.i.d. $X_i \sim \mathcal{LC}$. Part (ii) of the theorem for i.i.d. $X_i \sim \mathcal{LC}$ follows from part (i) for i.i.d. $X_i \sim \mathcal{LC}$ and majorization comparisons (2.1).

Proof of Theorems 2.5–2.6. Let $r, \alpha \in (0, 2]$, $\sigma_1, \ldots, \sigma_n > 0$, and let $v = (v_1, \ldots, v_n) \in \mathbf{R}_+^n$ and $w = (w_1, \ldots, w_n) \in \mathbf{R}_+^n$ be two vectors of portfolio weights such that $(v_1^r, \ldots, v_n^r) \prec (w_1^r, \ldots, w_n^r)$ and $(v_1^r, \ldots, v_n^r)$ is not a permutation of $(w_1^r, \ldots, w_n^r)$ (similar to the proof of Theorems 2.1 and 2.2, these assumptions evidently imply $\sum_{i=1}^n v_i \ne 0$ and $\sum_{i=1}^n w_i \ne 0$). Let $X_1, \ldots, X_n$ be independent risks such that $X_i \sim S_\alpha(\sigma_i, 0, 0)$, $i = 1, \ldots, n$. Similar to the proof of Theorems 2.1 and 2.2, we note that from (1.10) it follows that if $c = (c_1, \ldots, c_n) \in \mathbf{R}_+^n$, $\sum_{i=1}^n c_i \ne 0$, then $\sum_{i=1}^n c_{[i]} X_i / (\sum_{i=1}^n c_{[i]}^\alpha \sigma_i)^{1/\alpha} \sim S_\alpha(1, \beta, 0)$. Using positive homogeneity of VaR (see property a3 in

Section 2.3), we thus obtain that, for all $0 < q < \mathbb{P}(Q > 0)$,

$$\text{VaR}_q \left(\sum_{i=1}^n c_{[i]} X_i \right) = \text{VaR}_q(Q) \left(\sum_{i=1}^n c_{[i]}^\alpha \sigma_i \right)^{1/\alpha}. \tag{2.10}$$

By Theorem 3.A.4 in Marshall and Olkin (1979) and Marshall *et al.* (2011) the function $\chi(c_1, \ldots, c_n)$ defined in (2.4) is strictly Schur-convex in $(c_1, \ldots, c_n) \in \mathbf{R}_+^n$ if $\alpha > 1$ and $\sigma_1 \geq \ldots \geq \sigma_n \geq 0$ and is strictly Schur-concave in $(c_1, \ldots, c_n) \in \mathbf{R}_+^n$ if $\alpha < 1$ and $\sigma_n \geq \ldots \geq \sigma_1 \geq 0$ (see also Propositions 3.H.2.b and 4.B.7 in Marshall and Olkin (1979) and Marshall *et al.* (2011)).

Therefore, we have $\sum_{i=1}^n v_{[i]}^\alpha \sigma_i^\alpha = \sum_{i=1}^n (v_{[i]}^r)^{\alpha/r} \sigma_i^\alpha < \sum_{i=1}^n (w_{[i]}^r)^{\alpha/r} \sigma_i^\alpha = \sum_{i=1}^n w_{[i]}^\alpha \sigma_i^\alpha$, if $\alpha/r > 1$ and $\sigma_1 \geq \ldots \geq \sigma_n > 0$. Similarly, $\sum_{i=1}^n w_{[i]}^\alpha \sigma_i^\alpha = \sum_{i=1}^n (w_{[i]}^r)^{\alpha/r} \sigma_i^\alpha < \sum_{i=1}^n (v_{[i]}^r)^{\alpha/r} \sigma_i^\alpha = \sum_{i=1}^n v_{[i]}^\alpha \sigma_i^\alpha$, if $\alpha/r < 1$ and $\sigma_n \geq \ldots \geq \sigma_1 > 0$. This, together with (2.10), implies that, for all $q \in (0, 1/2)$, $\text{VaR}_q(Z_{\tilde{v}}) < \text{VaR}_q(Z_{\tilde{w}})$ if $\alpha > r$ and $\sigma_1 \geq \ldots \geq \sigma_n \geq 0$ and $\text{VaR}_q(Z_{\tilde{v}}) > \text{VaR}_q(Z_{\tilde{w}})$ if $\alpha < r$ and $\sigma_n \geq \ldots \geq \sigma_1 \geq 0$. This completes the proof of parts (i) of Theorems 2.5 and 2.6.

The bounds in parts (ii) of Theorems 2.5 and 2.6 follow from their parts (i) and the property that, by majorization comparisons (2.1), $\left(\sum_{i=1}^n w_i^r/n, \ldots, \sum_{i=1}^n w_i^r/n \right) \prec (w_1^r, \ldots, w_n^r) \prec \left(\sum_{i=1}^n w_i^r, 0, \ldots, 0 \right)$ for all portfolio weights $w \in \mathbf{R}_+^n$ and all $r \in (0, 2]$. Sharpness of the bounds in the theorems follows from the property that the bounds become equalities in the limit as $\alpha \to r$ for i.i.d. stable risks $X_i \sim S_r(\sigma, 0, 0)$.

Theorems 2.3 and 2.4 are consequences of Theorems 2.5 and 2.6 with $r = 1$.

Proof of Theorem 2.7. Suppose that $X_1, \ldots, X_n$ are i.i.d. risks such that $X_i \sim \overline{\mathcal{CS}}(r)$, $i = 1, \ldots, n$. By definition of the class $\overline{\mathcal{CS}}(r)$, there exist independent r.v.'s Y_{ij}, $i = 1, \ldots, n$, $j = 1, \ldots, k$, such that $Y_{ij} \sim S_{\alpha_j}(\sigma_j, 0, 0)$, $\alpha_j \in (0, r)$, $\sigma_j > 0$, $i = 1, \ldots, n$, $j = 1, \ldots, k$, and $X_i = \sum_{j=1}^k Y_{ij}$, $i = 1, \ldots, n$. Using Theorem 2.3 for the r.v.'s $Y_{ij} \sim S_{\alpha_j}(\sigma_j, 0, 0)$, $\alpha_j \in (r, 2]$, $\sigma_j > 0$, $i = 1, \ldots, n$, with $\beta = 0$ and the implied equality $\mathbb{P}(Q > 0) = 1/2$, we conclude that, for all

$q \in (0, 1/2)$ and all $j = 1, \ldots, k$,

$$\mathrm{VaR}_q \left(\sum_{i=1}^{n} v_i Y_{ij} \right) < \mathrm{VaR}_q \left(\sum_{i=1}^{n} w_i Y_{ij} \right). \tag{2.11}$$

The densities of the r.v.'s Y_{ij}, $i = 1, \ldots, n$, $j = 1, \ldots, k$, are symmetric and unimodal by Proposition 2.1. Therefore, Proposition 2.4 implies that the densities of the r.v.'s $\sum_{i=1}^{n} v_i Y_{ij}$, $j = 1, \ldots, k$, and $\sum_{i=1}^{n} w_i Y_{ij}$, $j = 1, \ldots, k$, are symmetric and unimodal as well. By Proposition 2.5, this, together with relations (2.11), implies that $\mathrm{VaR}_q(Z_v) = \mathrm{VaR}_q(\sum_{j=1}^{k} \sum_{i=1}^{n} v_i Y_{ij}) < \mathrm{VaR}_q(\sum_{j=1}^{k} \sum_{i=1}^{n} w_i Y_{ij}) = \mathrm{VaR}_q(Z_w)$ for all $q \in (0, 1/2)$. Therefore, part (i) of Theorem 2.3 continues to hold for all $q \in (0, 1/2)$ and i.i.d. risks $X_i \sim \overline{\mathcal{CS}}(r)$, $i = 1, \ldots, n$. Using in the above arguments Theorem 2.4 instead of Theorem 2.3, in complete similarity one obtains that part (i) of Theorem 2.4 continues to hold for all $q \in (0, 1/2)$ and i.i.d. risks $X_i \sim \underline{\mathcal{CS}}(r)$, $i = 1, \ldots, n$. Parts (ii) of Theorems 2.3 and 2.4 for the classes $\overline{\mathcal{CS}}(r)$ and $\underline{\mathcal{CS}}(r)$ follow in the same way as in the proof of these theorems in the case of independent stable risks $X_i \sim S_\alpha(\sigma_i, \beta, 0)$.

Chapter 3

From Independence to Dependence
via Copulas and U-statistics

In this chapter, we show how to go from independent risks to risks
with an arbitrary dependence structure. In other words, we represent
any joint distribution and any copula function using certain functions
of independent random variables known as U-statistics. We also discuss various dependence measures that are useful in this context,
including a new measure suitable for measuring dependence between
vectors rather than between scalars. An interesting application of
this chapter's theory is to pricing options.

3.1. Introduction

The results we provide in this chapter may look abstract and theoretical at first. However, they appear to be quite useful in econometrics,
finance, statistics and probability theory. We start with discussing
why they are useful.

The results on characterizations of joint distributions provide a
unified approach to modelling multivariate dependence using simple functions of independent random variables. Moreover, they help
derive new results concerning convergence of multidimensional statistics of time series, useful in financial econometrics. Furthermore, they
lead us to the discussion of new measures of dependence that allow
one to uncover new patterns in data. Finally, the characterizations
of joint distributions provide tools that allow one to study important

properties of dependent time series, including Markov processes of an arbitrary order.

The results on characterizations of Markov processes provide a copula-based approach to construction of higher-order Markov processes — an alternative to the conventional method based on transition probabilities. The advantage of the new approach is that it allows one to separate the study of dependence properties (e.g., r-independence, m-dependence or conditional symmetry to be defined in Section 3.2.4) of a stochastic process from the analysis of marginal distributions (say, unconditional heavy-tailedness or skewness). In particular, the results provide methods for construction of higher-order Markov processes with arbitrary one-dimensional margins and additional dependence properties. These processes can be used, e.g., in the analysis of robustness of statistical and econometric methods to weak dependence. In addition, they serve as examples of non-Markovian processes that nevertheless satisfy Chapman-Kolmogorov stochastic equations.

As we show in the section on dependence measures, all multivariate statistics including various dependence coefficients widely used in practice are calculated as functionals of joint distributions. So in effect, this chapter also provides a device for the analysis of convergence of such multidimensional statistics. For example, the characterizations of joint distributions and copulas which we discuss help associate with each set of arbitrarily dependent r.v.'s a sum of U-statistics in independent r.v.'s with canonical kernels.

Along the way, a general methodology (of intrinsic interest within and outside economics and finance) is developed for analyzing key measures of dependence among r.v.'s. Using the methodology, we obtain sharp decoupling inequalities for comparing the expectations of arbitrary (integrable) functions of dependent variables to their corresponding counterparts with independent variables through the inclusion of multivariate dependence measures.

The chapter is organized as follows. Section 3.2 contains the results on general characterizations of copulas and joint distributions of dependent r.v.'s using U-statistics. Section 3.3 presents the results on characterizations of Markov processes using copulas. Section 3.4

uses the results of Sections 3.2-3.3 to study various dependence measures, while Section 3.5 uses them to give price bounds for options.

Appendix provides selected proofs of this chapter's theorems. For proofs of other theorems, we refer the interested reader to the appendix of de la Peña *et al.* (2004), de la Peña *et al.* (2006), Ibragimov (2009a) and Medovikov and Prokhorov (2016).

3.2. Characterizations of dependence

3.2.1. *Characterizations of joint distributions*

In the present section, we obtain explicit general representations of joint distributions of arbitrarily dependent r.v.'s with given one-dimensional marginal cdf's.

Theorem 3.1. *A function* $F : \mathbf{R}^n \to [0,1]$ *is a joint cdf with one-dimensional marginal cdf's* $F_k(x_k)$, $x_k \in \mathbf{R}$, $k = 1, \ldots, n$, *absolutely continuous with respect to the product of marginal cdf's* $\prod_{k=1}^{n} F_k(x_k)$, *if and only if there exist functions* $g_{i_1,\ldots,i_c} : \mathbf{R}^c \to \mathbf{R}$, $1 \leq i_1 < \ldots < i_c \leq n$, $c = 2, \ldots, n$, *satisfying the following conditions*

A1 (*integrability*):

$$\mathbb{E}|g_{i_1,\ldots,i_c}(\xi_{i_1}, \ldots, \xi_{i_c})| < \infty,$$

A2 (*degeneracy*):

$$\mathbb{E}(g_{i_1,\ldots,i_c}(\xi_{i_1}, \ldots, \xi_{i_{k-1}}, \xi_{i_k}, \xi_{i_{k+1}}, \ldots, \xi_{i_c}) | \xi_{i_1}, \ldots, \xi_{i_{k-1}}, \xi_{i_{k+1}}, \ldots, \xi_{i_c}) =$$
$$\int_{-\infty}^{\infty} g_{i_1,\ldots,i_c}(\xi_{i_1}, \ldots, \xi_{i_{k-1}}, x_{i_k}, \xi_{i_{k+1}}, \ldots, \xi_{i_c}) dF_{i_k}(x_{i_k}) = 0, \quad (a.s.)$$
$$1 \leq i_1 < \ldots < i_c \leq n, \ k = 1, 2, \ldots, c, \ c = 2, \ldots, n,$$

A3 (*positive definiteness*):

$$\sum_{c=2}^{n} \sum_{1 \leq i_1 < \ldots < i_c \leq n} g_{i_1,\ldots,i_c}(\xi_{i_1}, \ldots, \xi_{i_c}) \geq -1 \ (a.s.)$$

and the following representation holds for F :

$$F(x_1, \ldots, x_n) =$$

$$\int_{-\infty}^{x_1} \cdots \int_{-\infty}^{x_n} \left(1 + \sum_{c=2}^{n} \sum_{1 \le i_1 < \ldots < i_c \le n} g_{i_1, \ldots, i_c}(t_{i_1}, \ldots, t_{i_c}) \right) \prod_{i=1}^{n} dF_i(t_i).$$

$$(3.1)$$

Moreover, $g_{i_1, \ldots, i_c}(\xi_{i_1}, \ldots, \xi_{i_c}) = f_{i_1, \ldots, i_c}(\xi_{i_1}, \ldots, \xi_{i_c})$ *(a.s.),* $1 \le i_1 < \ldots < i_c \le n,$ $c = 2, \ldots, n,$ *where*

$$f_{i_1, \ldots, i_c}(x_{i_1}, \ldots, x_{i_c}) =$$

$$\sum_{k=2}^{c} (-1)^{c-k} \sum_{1 \le j_1 < \ldots < j_k \in \{i_1, \ldots, i_c\}} \left(\frac{dF(x_{j_1}, \ldots, x_{j_k})}{dF_{j_1} \ldots dF_{j_k}} - 1 \right).$$

As a corollary, it is not difficult to see that if r.v.'s $X_1, \ldots, X_n$ have a joint cdf given by (3.1) then the r.v.'s $X_{j_1}, \ldots, X_{j_k}$, $1 \le j_1 < \ldots < j_k \le n,$ $k = 2, \ldots, n,$ have the joint cdf

$$F(x_{j_1}, \ldots, x_{j_k}) = F(x_1, \ldots, x_n)|_{x_s = \infty, s \ne j_1, \ldots, j_k} =$$

$$\int_{-\infty}^{x_{j_1}} \cdots \int_{-\infty}^{x_{j_k}} \left(1 + \sum_{c=2}^{k} \sum_{i_1 < \ldots < i_c \in \{j_1, \ldots, j_k\}} g_{i_1, \ldots, i_c}(t_{i_1}, \ldots, t_{i_c}) \right) \prod_{i=1}^{k} dF_{j_i}(t_{j_i})$$

with the same functions $g_{i_1, \ldots, i_c}$.

Theorem 3.1 can be equivalently formulated as follows.

Theorem 3.2. *A function* $F : \mathbf{R}^n \to [0, 1]$ *is a joint cdf with one-dimensional marginal cdf's* $F_k(x_k),$ $x_k \in \mathbf{R},$ $k = 1, \ldots, n,$ *absolutely continuous with respect to the product of marginal cdf's* $\prod_{k=1}^{n} F_k(x_k),$

if and only if there exist functions $g_{i_1,\ldots,i_c} : \mathbf{R}^c \to \mathbf{R}$, $1 \le i_1 < \ldots <$ $i_c \le n$, $c = 2, \ldots, n$, *satisfying conditions A1–A3 and such that the measure element* $dF(x_1, \ldots, x_n)$ *can be expressed in the form*

$$dF(x_1, \ldots, x_n) =$$

$$\prod_{i=1}^{n} dF_i(x_i) \left(1 + \sum_{c=2}^{n} \sum_{1 \le i_1 < \ldots < i_c \le n} g_{i_1,\ldots,i_c}(x_{i_1}, \ldots, x_{i_c}) \right). \quad (3.2)$$

Sharakhmetov (2001) provides a proof of Theorem 3.2 in the case of density functions (of r.v.'s absolutely continuous with respect to the *Lebesgue* measure), with a mention that a similar representation holds for distributions of discrete r.v.'s. The setup considered in this chapter includes (among others) the class of vectors of dependent absolutely continuous and discrete r.v.'s as well as vectors of mixtures of absolutely continuous and discrete r.v.'s. Furthermore, our proof easily extends to the case of general Banach spaces, in particular, the spaces $\mathbf{R}^k$.

In the case of probability density functions and probability distributions of discrete r.v.'s, Theorems 3.1 and 3.2 have the following forms.

Theorem 3.3. *A function* $f(x_1, \ldots, x_n)$ *is a joint pdf with one-dimensional marginal pdf's* $f_k(x_k)$, $x_k \in \mathbf{R}$, $k = 1, \ldots, n$, *if and only if there exist functions* $g_{i_1,\ldots,i_c} : \mathbf{R}^c \to \mathbf{R}$, $1 \le i_1 < \ldots < i_c \le n$, $c = 2, \ldots, n$, *satisfying conditions A1–A3 and such that the following representation holds for* f:

$$f(x_1, \ldots, x_n) =$$

$$\prod_{i=1}^{n} f_i(x_i) \left(1 + \sum_{c=2}^{n} \sum_{1 \le i_1 < \ldots < i_c \le n} g_{i_1,\ldots,i_c}(x_{i_1}, \ldots, x_{i_c}) \right).$$

$$(3.3)$$

Theorem 3.4. *A function* $p(x_1, \ldots, x_n)$ *is a joint probability distribution of some discrete r.v.'s* $X_1, \ldots, X_n$: $p(x_1, \ldots, x_n)$ $= \mathbb{P} (X_1 = x_1, \ldots, X_n = x_n)$, *with one-dimensional marginal* *probability distributions* $p_k(x_k) = \mathbb{P}(X_k = x_k)$, $x_k \in \mathbf{R}$, $k = 1, \ldots, n$, *if and only if there exist functions* $g_{i_1, \ldots, i_c} : \mathbf{R}^c \to \mathbf{R}$, $1 \leq i_1 < \ldots < i_c \leq n$, $c = 2, \ldots, n$, *satisfying conditions A1–A3 and such that the following representation holds for* p:

$$p(x_1, \ldots, x_n) =$$

$$\prod_{i=1}^{n} p_i(x_i) \left(1 + \sum_{c=2}^{n} \sum_{1 \leq i_1 < \ldots < i_c \leq n} g_{i_1, \ldots, i_c}(x_{i_1}, \ldots, x_{i_c}) \right).$$

(3.4)

3.2.2. Characterizations of copulas

The following theorems give copula analogues of the representations in the previous section. Let $V_1, \ldots, V_n$ denote independent r.v.'s uniformly distributed on $[0, 1]$.

Theorem 3.5. *A function* $C : [0, 1]^n \to [0, 1]$ *is an absolutely continuous n-dimensional copula if and only if there exist functions* $\tilde{g}_{i_1, \ldots, i_c} : \mathbf{R}^c \to \mathbf{R}$, $1 \leq i_1 < \ldots < i_c \leq n$, $c = 2, \ldots, n$, *satisfying the conditions*

A4 (integrability):

$$\int_0^1 \cdots \int_0^1 |\tilde{g}_{i_1, \ldots, i_c}(t_{i_1}, \ldots, t_{i_c})| dt_{i_1} \ldots dt_{i_c} < \infty,$$

A5 (degeneracy):

$$\mathbb{E}(\tilde{g}_{i_1, \ldots, i_c}(V_{i_1}, \ldots, V_{i_{k-1}}, V_{i_k}, V_{i_{k+1}}, \ldots, V_{i_c}) | V_{i_1}, \ldots,$$

$$V_{i_{k-1}}, V_{i_{k+1}}, \ldots, V_{i_c}) =$$

$$\int_0^1 \tilde{g}_{i_1, \ldots, i_c}(V_{i_1}, \ldots, V_{i_{k-1}}, t_{i_k}, V_{i_{k+1}}, \ldots, V_{i_c}) dt_{i_k} = 0 \ (a.s.),$$

$1 \leq i_1 < \ldots < i_c \leq n$, $k = 1, 2, \ldots, c$, $c = 2, \ldots, n$,

A6 (positive definiteness):

$$\sum_{c=2}^{n} \sum_{1 \leq i_1 < \ldots < i_c \leq n} \tilde{g}_{i_1,\ldots,i_c}(V_{i_1}, \ldots, V_{i_c}) \geq -1 \; (a.s.)$$

and such that

$$C(u_1, \ldots, u_n) =$$

$$\int_0^{u_1} \ldots \int_0^{u_n} \left(1 + \sum_{c=2}^{n} \sum_{1 \leq i_1 < \ldots < i_c \leq n} \tilde{g}_{i_1,\ldots,i_c}(t_{i_1}, \ldots, t_{i_c}) \right) \prod_{i=1}^{n} dt_i.$$

$$(3.5)$$

Theorem 3.5 and Sklar's theorem given by Proposition 1.1 imply the following representation for a joint distribution of r.v.'s.

Theorem 3.6. *A function* $F : \mathbf{R}^n \to [0, 1]$ *is a joint cdf with the one-dimensional marginal cdf's* $F_k(x_k)$, $x_k \in \mathbf{R}$, $k = 1, \ldots, n$, *absolutely continuous with respect to the product of marginal cdf's* $\prod_{k=1}^{n} F_k(x_k)$ *if and only if there exist functions* $\tilde{g}_{i_1,\ldots,i_c} : [0, 1]^c \to \mathbf{R}$, $1 \leq i_1 < \ldots < i_c \leq n$, $c = 2, \ldots, n$, *satisfying the conditions A4–A6 and such that the following representation holds for* F:

$$F(x_1, \ldots, x_n) = \int_0^{F_1(x_1)} \ldots \int_0^{F_n(x_n)} \left(1 \right.$$

$$\left. + \sum_{c=2}^{n} \sum_{1 \leq i_1 < \ldots < i_c \leq n} \tilde{g}_{i_1,\ldots,i_c}(t_{i_1}, \ldots, t_{i_c}) \right) \prod_{i=1}^{n} dt_i, \qquad (3.6)$$

or, equivalently, if and only if the measure element dF *can be expressed in the form*

$$dF(x_1, \ldots, x_n) = \prod_{i=1}^{n} dF_i(x_i) \left(1 \right.$$

$$\left. + \sum_{c=2}^{n} \sum_{1 \leq i_1 < \ldots < i_c \leq n} \tilde{g}_{i_1,\ldots,i_c}(F_{i_1}(x_{i_1}), \ldots, F_{i_c}(x_{i_c})) \right).$$

It is easy to see that the functions g and $\tilde{g}$ in Theorems 3.1–3.6 are related in the following way: $g_{i_1,\ldots,i_c}(x_{i_1},\ldots,x_{i_c}) = \tilde{g}_{i_1,\ldots,i_c}(F_{i_1}(x_{i_1}),\ldots,F_{i_c}(x_{i_c}))$.

Following the terminology in the literature on U-statistics, it is natural to refer to functions g satisfying conditions A1–A3 in Theorem 3.1 and functions $\tilde{g}$ satisfying conditions A4–A6 in Theorem 3.5 as canonical kernels of the U-statistics in the representations for joint distributions and copulas.

Theorems 3.1–3.6 provide a general device for constructing multivariate copulas and joint distributions. For example, taking in (3.5) and (3.6) $n = 2$, $\tilde{g}_{1,2}(t_1, t_2) = \theta(1 - 2t_1)(1 - 2t_2)$, $\theta \in [-1, 1]$, we get the family of bivariate Eyraud-Farlie-Gumbel-Morgenstern (EFGM) copulas

$$C_\theta(u_1, u_2) = u_1 u_2 \left(1 + \theta(1 - u_1)(1 - u_2)\right) \qquad (3.7)$$

and the corresponding bivariate distributions $F_\theta(x_1, x_2) = F_1(x_1)\,F_2(x_2)(1 + \theta(1 - F_1(x_1))(1 - F_2(x_2))$.

More generally, taking $\tilde{g}_{i_1,\ldots,i_c}(t_{i_1},\ldots,t_{i_c}) = 0$, $1 \le i_1 < \ldots < i_c \le n$, $c = 2,\ldots,n - 1$, $\tilde{g}_{1,2,\ldots,n}(t_1, t_2,\ldots,t_n) = \theta(1 - 2t_1)(1 - 2t_2)\ldots(1 - 2t_n)$, we obtain the following multivariate EFGM copulas

$$C_\theta(u_1, u_2, \ldots, u_n) = \prod_{i=1}^{n} u_i \left(1 + \theta \prod_{i=1}^{n}(1 - u_i)\right) \qquad (3.8)$$

and corresponding multivariate cdf's $F_\theta(x_1, x_2, \ldots, x_n) = \prod_{i=1}^{n} F_i$ $(x_i)(1 + \theta \prod_{i=1}^{n}(1 - F_i(x_i)))$.

Moreover, if we let $\theta_{i_1,\ldots,i_c} \in \mathbf{R}$ be constants such that $\sum_{c=2}^{n} \sum_{1 \le i_1 < \ldots < i_c \le n} \theta_{i_1,\ldots,i_c}\delta_{i_1}\ldots\delta_{i_c} \ge -1$ for all $\delta_i \in \{0, 1\}$, $i = 1,\ldots,n$, then the choice $\tilde{g}_{i_1,\ldots,i_c}(t_{i_1},\ldots,t_{i_c}) = \theta_{i_1,\ldots,i_c}(1 - 2t_{i_1})(1 - 2t_{i_2})\ldots(1 - 2t_{i_c})$, $1 \le i_1 < \ldots < i_c \le n$, $c = 2,\ldots,n$, gives the following generalized multivariate EFGM copulas (see, e.g., Johnson and Kotz (1975); Cambanis (1977)):

$$C(u_1, \ldots, u_n) =$$

$$\prod_{k=1}^{n} u_k \left(1 + \sum_{c=2}^{n} \sum_{1 \le i_1 < \ldots < i_c \le n} \theta_{i_1,\ldots,i_c}(1 - u_{i_1})\ldots(1 - u_{i_c})\right)$$

$$(3.9)$$

and the corresponding cdf's

$$F(x_1, \ldots, x_n) = \prod_{i=1}^{n} F_i(x_i) \left(1 \right.$$

$$\left. + \sum_{c=2}^{n} \sum_{1 \leq i_1 < \ldots < i_c \leq n} \theta_{i_1, \ldots, i_c} (1 - F_{i_1}(x_{i_1})) \ldots (1 - F_{i_c}(x_{i_c})) \right).$$

The importance of the generalized EFGM copulas and cdf's stems, in particular, from the fact that, as shown by Sharakhmetov and Ibragimov (2002), they completely characterize joint distributions of two-valued r.v.'s.

Let now $\theta_{i_1, \ldots, i_c} \in \mathbf{R}$ be such that $\sum_{c=2}^{n} \sum_{1 \leq i_1 < \ldots < i_c \leq n} |\theta_{i_1, \ldots, i_c}| \leq 1$ (it is easy to see that this condition is satisfied if $\theta_{i_1, \ldots, i_c} = \lambda_{i_1} \ldots \lambda_{i_c}$, where $\sum_{i=1}^{n} |\lambda_i| \leq 1$). Taking

$$g_{i_1, \ldots, i_c}(t_{i_1}, \ldots, t_{i_c}) =$$

$$\theta_{i_1, \ldots, i_c}((l+1)t_{i_1}^l - (l+2)t_{i_1}^{l+1}) \ldots ((l+1)t_{i_c}^l - (l+2)t_{i_c}^{l+1}),$$

where $l \in \{0, 1, 2, \ldots\}$, we get the following extensions of EFGM copulas (3.9) that are natural to call power copulas:

$$C(u_1, \ldots, u_n) =$$

$$\prod_{i=1}^{n} u_i \left(1 + \sum_{c=2}^{n} \sum_{1 \leq i_1 < \ldots < i_c \leq n} \theta_{i_1, \ldots, i_c} (u_{i_1}^l - u_{i_1}^{l+1}) \ldots (u_{i_c}^l - u_{i_c}^{l+1}) \right).$$

$$(3.10)$$

We return to this class of copulas in Chapter 4.

Taking $n = 2$, $\tilde{g}_{1,2}(t_1, t_2) = \theta c(t_1, t_2)$, where c is a continuous function on the unit square $[0, 1]^2$ satisfying the properties $\int_0^1 c(t_1, t_2)dt_1 = \int_0^1 c(t_1, t_2)dt_2 = 0$, $1 + \theta c(t_1, t_2) \geq 0$ for all $0 \leq t_1, t_2 \leq 1$, one obtains the class of bivariate densities studied by Rüschendorf (1985) and Long and Krzysztofowicz (1995) (see also

Mari and Kotz (2001), pp. 73–78)

$$f(x_1, x_2) = f_1(x_1)f_2(x_2)(1 + \theta c(F_1(x_1), F_2(x_2)))$$

with covariance characteristic c and covariance scalar θ. Furthermore, from Theorems 3.1–3.6 it follows that this representation in fact holds for an arbitrary density function and the function $\theta c(t_1, t_2)$ is unique.

3.2.3. *Characterizations of expectations*

Denote by $\mathcal{G}_n$ the class of sums of U-statistics of the form

$$U_n(\xi_1, \ldots, \xi_n) = \sum_{c=2}^{n} \sum_{1 \le i_1 < \ldots < i_c \le n} g_{i_1, \ldots, i_c}(\xi_{i_1}, \ldots, \xi_{i_c}),$$

where the functions $g_{i_1, \ldots, i_c}$, $1 \le i_1 < \ldots < i_c \le n$, $c = 2, \ldots, n$, satisfy conditions A1–A3, and, as before, $\xi_1, \ldots, \xi_n$ are independent r.v.'s with cdf's $F_k(x_k)$, $x_k \in \mathbf{R}$, $k = 1, \ldots, n$. The following theorem puts into correspondence to any set of arbitrarily dependent r.v.'s a sum of U-statistics in independent r.v.'s with canonical kernels. This allows one to reduce problems for dependent r.v.'s to well-studied objects and to transfer results known for independent r.v.'s and U-statistics to arbitrary dependence. In what follows, the joint distributions considered are assumed to be absolutely continuous with respect to the product of the marginal distributions $\prod_{k=1}^{n} F_k(x_k)$.

Theorem 3.7. *R.v.'s $X_1, \ldots, X_n$ have one-dimensional cdf's $F_k(x_k)$, $x_k \in \mathbf{R}$, $k = 1, \ldots, n$, if and only if there exists a statistic $U_n \in \mathcal{G}_n$ such that for any Borel measurable function $f : \mathbf{R}^n \to \mathbf{R}$ for which the expectations exist*

$$\mathbb{E}f(X_1, \ldots, X_n) = \mathbb{E}f(\xi_1, \ldots, \xi_n)(1 + U_n(\xi_1, \ldots, \xi_n)). \quad (3.11)$$

Note that Theorem 3.7 holds for complex-valued functions f as well as for the real-valued ones. That is, choosing $f(x_1, \ldots, x_n) = \exp(i\sum_{k=1}^{n} t_k x_k)$, $t_k \in \mathbf{R}$, $k = 1, \ldots, n$, one gets the following representation for the joint characteristic function of the r.v.'s

$X_1, \ldots, X_n$:

$$\mathbb{E} \exp \left(i \sum_{k=1}^{n} t_k X_k \right) = \mathbb{E} \exp \left(i \sum_{k=1}^{n} t_k \xi_k \right)$$

$$+ \mathbb{E} \exp \left(i \sum_{k=1}^{n} t_k \xi_k \right) U_n(\xi_1, \ldots, \xi_n).$$

3.2.4. *Characterizations of certain dependence classes*

The following theorems give characterizations of different classes of dependent r.v.'s in terms of functions g that appear in the representations for joint distributions obtained in Section 3.2.1. Analogous results hold for functions $\tilde{g}$ that enter corresponding representations for copulas in Section 3.2.2.

Theorem 3.8. *The r.v.'s* $X_1, \ldots, X_n$ *with the one-dimensional cdf's* $F_k(x_k)$, $x_k \in \mathbf{R}$, $k = 1, \ldots, n$, *are independent if and only if the functions* $g_{i_1, \ldots, i_c}$ *in representations* (3.1) *and* (3.2) *satisfy the conditions*

$$g_{i_1, \ldots, i_c}(\xi_{i_1}, \ldots, \xi_{i_c}) = 0 \; (a.s.), \; 1 \le i_1 < \ldots < i_c \le n, c = 2, \ldots, n.$$

Theorem 3.9. *A sequence of r.v.'s* $\{X_n\}$ *is strictly stationary if and only if the functions* $g_{i_1, \ldots, i_c}$ *in representations* (3.1) *and* (3.2) *for any finite-dimensional distribution satisfy the conditions*

$$g_{i_1+h, \ldots, i_c+h}(\xi_{i_1}, \ldots, \xi_{i_c}) = g_{i_1, \ldots, i_c}(\xi_{i_1}, \ldots, \xi_{i_c}) \; (a.s.)$$
$$1 \le i_1 < \ldots < i_c \le n, \; c = 2, 3, \ldots, \; h = 0, 1, \ldots$$

Theorem 3.10. *A sequence of r.v.'s* $\{X_n\}$ *with* $\mathbb{E}X_k = 0$, $\mathbb{E}X_k^2 < \infty$, $k = 1, 2, \ldots$, *is weakly stationary if and only if the functions* g *in representations* (3.1) *and* (3.2) *for any finite-dimensional distribution have the property that the function* $h(s, t) = \mathbb{E}\xi_s \xi_t g_{st}(\xi_s, \xi_t)$, *depends only on* $|t - s|$, $t, s = 1, 2, \ldots$

Definition 3.1. R.v.'s $X_1, \ldots, X_n$ with $\mathbb{E}X_i = 0$, $i = 1, \ldots, n$, are called **orthogonal** if $\mathbb{E}X_i X_j = 0$ for all $1 \le i < j \le n$.

Theorem 3.11. *The r.v.'s* $X_1, \ldots, X_n$ *with* $\mathbb{E}X_k = 0$, $k = 1, \ldots, n$, *are orthogonal if and only if the functions* g *in representations* (3.1) *and* (3.2) *satisfy the conditions* $\mathbb{E}\xi_i\xi_j g_{ij}(\xi_i, \xi_j) = 0$, $1 \leq i < j \leq n$.

Definition 3.2. R.v.'s $X_1, \ldots, X_n$ are called **exchangeable** if all $n!$ permutations $(X_{\pi(1)}, \ldots, X_{\pi(n)})$ of the r.v.'s have the same joint distributions.

Theorem 3.12. *Identically distributed r.v.'s* $X_1, \ldots, X_n$ *are exchangeable if and only if the functions* $g_{i_1,\ldots,i_c}$ *in representations* (3.1) *and* (3.2) *satisfy the conditions* $g_{i_1,\ldots,i_c}(\xi_{i_1}, \ldots, \xi_{i_c}) = g_{i_{\pi(1)},\ldots,i_{\pi(c)}}(\xi_{i_{\pi(1)}}, \ldots, \xi_{i_{\pi(c)}})$ *(a.s.) for all* $1 \leq i_1 < \ldots < i_c \leq n$, $c = 2, \ldots, n$, *and all permutations* π *of the set* $\{1, \ldots, n\}$.

Definition 3.3. R.v.'s $X_1, \ldots, X_n$ are called m-**dependent** $(1 \leq m \leq n)$ if any two vectors $(X_{j_1}, X_{j_2}, \ldots, X_{j_{a-1}}, X_{j_a})$ and $(X_{j_{a+1}}, X_{j_{a+2}}, \ldots, X_{j_{l-1}}, X_{j_l})$, where $1 \leq j_1 < \ldots < j_a < \ldots < j_l \leq n$, $a = 1, 2, \ldots, l - 1$, $l = 2, \ldots, n$, $j_{a+1} - j_a \geq m$, are independent.

Theorem 3.13. *R.v.'s* $X_1, \ldots, X_n$ *are* m-*dependent if and only if the functions* g *in representations* (3.1) *and* (3.2) *satisfy the conditions*

$$g_{i_1,\ldots,i_k,i_{k+1},\ldots,i_c}(\xi_{i_1}, \ldots, \xi_{i_k}, \xi_{i_{k+1}}, \ldots, \xi_{i_c}) =$$
$$g_{i_1,\ldots,i_k}(\xi_{i_1}, \ldots, \xi_{i_k}) g_{i_{k+1},\ldots,i_c}(\xi_{i_{k+1}}, \ldots, \xi_{i_c})$$

for all $1 \leq i_1 < \ldots < i_k < i_{k+1} \ldots < i_c \leq n$, $i_{k+1} - i_k \geq m$, $k = 1, \ldots, c - 1$, $c = 2, \ldots, n$.

Definition 3.4. R.v.'s $X_1, \ldots, X_n$ form a **multiplicative system** of order $\alpha \in \mathbf{N}$ (shortly, $MS(\alpha)$) if $\mathbb{E}|X_j|^\alpha < \infty$, $j = 1, \ldots, n$, and $\mathbb{E}\prod_{j=1}^n X_j^{\alpha_j} = \prod_{j=1}^n \mathbb{E}X_j^{\alpha_j}$ for any $\alpha_j \in \{0, 1, \ldots, \alpha\}$, $j = 1, \ldots, n$.

The systems $MS(1)$ and $MS(2)$ under the names multiplicative and strongly multiplicative systems, respectively, were introduced by Alexits (1961). Multiplicative systems of an arbitrary order were considered, e.g., by Kwapien (1987) and Sharakhmetov (1993). Examples of the multiplicative systems $MS(1)$ are given, besides independent r.v.'s, by the lacunary trigonometric systems $\{\cos 2\pi n_k x, \sin 2\pi n_k x, k = 1, 2, \ldots\}$ on the interval $[0, 1]$ with the Lebesgue measure for $n_{k+1}/n_k \geq 2$ — important in Fourier analysis of time series — and also by such important classes of dependent r.v.'s as martingale-difference sequences. Examples of strongly multiplicative systems (that is, of systems $MS(2)$) are given by the lacunary trigonometric systems for $n_{k+1}/n_k \geq 3$ and martingale-difference sequences $X_1, \ldots, X_n$ satisfying the conditions $\mathbb{E}(X_n^2|X_1, \ldots, X_{n-1}) = b_n^2 \in \mathbf{R}$, $n = 1, 2, \ldots$ Examples of systems $MS(\alpha)$ include the lacunary trigonometric systems with large lacunas, that is, with $n_{k+1}/n_k \geq \alpha+1$ and also ϵ-independent and asymptotically independent r.v.'s introduced by Zolotarev (1991).

Theorem 3.14. *R.v.'s* $X_1, \ldots, X_n$ *form a multiplicative system of order* α *if and only if the functions* $g_{i_1,\ldots,i_c}$ *in representations* (3.1) *and* (3.2) *satisfy the conditions* $\mathbb{E}\xi_{i_1}^{\alpha_{i_1}} \ldots \xi_{i_c}^{\alpha_{i_c}} g_{i_1,\ldots,i_c}(\xi_{i_1}, \ldots, \xi_{i_c}) = 0$, $1 \leq i_1 < \ldots < i_c \leq n$, $c = 2, \ldots, n$, $\alpha_j \in \{0, 1, \ldots, \alpha\}$, $j = 1, \ldots, n$.

Definition 3.5. R.v.'s $X_1, \ldots, X_n$ are called r-**independent** $(2 \leq r < n)$ if any r of them of are jointly independent.

Theorem 3.15. *The r.v.'s* $X_1, \ldots, X_n$ *are* r-*independent if and only if the functions* $g_{i_1,\ldots,i_c}$ *in representations* (3.1) *and* (3.2) *satisfy the conditions* $g_{i_1,\ldots,i_c}(\xi_{i_1}, \ldots, \xi_{i_c}) = 0$ *(a.s.)*, $1 \leq i_1 < \ldots < i_c \leq n$, $c = 2, \ldots, r$.

Note the following special case. Let $F_1(x), \ldots, F_n(x)$ be arbitrary one-dimensional distribution functions, $\alpha_1, \ldots, \alpha_n \in (-1, 1)\backslash\{0\}$, $\sum_{i=1}^n |\alpha_i| \leq 1$. Taking in Theorem 3.6 $g_{i_1,\ldots,i_c}(t_{i_1}, \ldots, t_{i_c}) = 0$, $1 \leq i_1 < \ldots < i_c \leq n, c = 2, \ldots, n, c \neq r+1$, $g_{i_1,\ldots,i_{r+1}}(t_{i_1}, \ldots, t_{i_{r+1}}) = \frac{\alpha_1 \ldots \alpha_n}{\alpha_{i_1} \ldots \alpha_{i_r+1}}((k+1)t_{i_1}^k - (k+2)t_{i_1}^{k+1}) \ldots ((k+1)t_{i_c}^k - (k+2)t_{i_c}^{k+1})$, $k = 0, 1, 2, \ldots$, we obtain the following extensions of the examples of r-independent r.v.'s obtained by Wang (1990):

$$F(x_1, \ldots, x_n) = \prod_{i=1}^n F_i(x_i) \left(1 + \right.$$

$$\left. \sum_{1 \leq i_1 < \ldots < i_{r+1} \leq n} \frac{\alpha_1 \ldots \alpha_n}{\alpha_{i_1} \ldots \alpha_{i_r+1}} \prod_{m=1}^{r+1} (F_{i_m}^k(x_{i_m}) - F_{i_m}^{k+1}(x_{i_m})) \right)$$

$k = 0, 1, 2, \ldots$ (Wang's examples are with $k = 0$).

3.2.5. *Reduction property for multiplicative systems*

Using the U-characterizations for joint distributions and copulas, we can establish an important structural property of multiplicative systems that can be referred as a reduction. It shows that r.v.'s forming a multiplicative system of order α and taking not more than $\alpha + 1$ values are in fact jointly independent. Below, $card(A_i)$ denotes the number of elements in (finite) sets A_i.

Theorem 3.16. *Let $\alpha \in \mathbf{N}$, and let A_i, $i = 1, \ldots, n$, be sets of real numbers such that $card(A_i) \le \alpha + 1$, $i = 1, \ldots, n$. The r.v.'s $X_1, \ldots, X_n$ taking values in $A_1, \ldots, A_n$, respectively, form a multiplicative system of order α if and only if they are jointly independent.*

A result due to Sharakhmetov and Ibragimov (2002) follows from Theorem 3.16 with $\alpha = 1$. A sequence of r.v.'s $\{X_n\}$ on a probability space $(\Omega, \Im, P)$ assuming two values is a martingale-difference with respect to an increasing sequence of σ-algebras $\Im_0 = (\Omega, \emptyset) \subseteq \Im_1 \subseteq \ldots \subseteq \Im$ if and only the r.v.'s $\{X_n\}$ are jointly independent.

In addition, if a sequence of r.v.'s $\{X_n\}$ assuming three values is a martingale-difference with respect to $(\Im_n)$ such that $\mathbb{E}(X_n^2 | \Im_{n-1}) = b_n^2 \in \mathbf{R}$, then the r.v.'s are jointly independent.

Characterizations of joint distributions and copulas presented in Sections 3.2.1 and 3.2.2 will be used in the next sections to establish further reduction properties of different dependence structures and time series, including Markov processes of an arbitrary order satisfying additional dependence assumptions.

3.3. Characterizations of Markov processes

Darsow *et al.* (1992) obtained important characterizations of first-order Markov processes in terms of copula functions corresponding to their two-dimensional distributions. Chen and Fan (2004, 2006b) consider parametric copula estimation procedures for time series based on bivariate copulas and applied the results to evaluating density forecasts. Fermanian *et al.* (2004) establish weak convergence of empirical copula processes in the case of independently observed vectors with dependent components. Doukhan *et al.* (2004) focus on the analysis of the asymptotics of empirical copula processes for weakly dependent sequences of random vectors.

In this section, we present characterizations of Markov processes of an arbitrary order in terms of copulas corresponding to their finite-dimensional distributions — Section 3.3.1. These results extend the characterizations of first-order Markov processes in terms of bivariate copulas obtained by Darsow *et al.* (1992). The results show that a Markov process of order k is fully determined by its $(k + 1)$-dimensional copulas and one-dimensional marginal cdf's. The characterizations thus provide a justification for estimation of finite-dimensional copulas of time series with higher-order Markovian dependence structure.

Using these results and the U-statistics representations for joint distributions and copulas from the previous section, we provide necessary and sufficient conditions for higher-order Markov processes to exhibit several additional dependence properties, such as m-dependence, r-independence or conditional symmetry — Section 3.3.2). These conditions show, in particular, that dependence properties of copula-based time series provide additional non-trivial restrictions on the U-statistics characterizations of copulas for the processes in consideration that can be used in inference on their properties.

Using copula-based characterizations of higher-order Markov processes, in Sections 3.3.3 and 3.3.4, we present an analysis of applicability and limitations of different classes of copulas in constructing higher-order Markov processes with prescribed dependence properties. In Section 3.3.3, we focus on processes based on expansions by linear functions (bivariate and multivariate EFGM copulas) as well as on more general copulas that involve products of nonlinear functions of the arguments, such as power copulas. We provide impossibility/reduction-type results similar in spirit to those in Section 3.2.5 that show that time series based on such copulas that simultaneously exhibit Markovness and m-dependence or r-independence properties are, in fact, sequences of independent r.v.'s.

Motivated by the reduction and impossibility results, in Section 3.3.4, we introduce a new class of copulas based on expansions by

Fourier polynomials (Fourier copulas) that, in contrast to the copula families considered in Section 3.3.3, allow one to combine higher-order Markovness with m-dependence or r-independence.

3.3.1. *Copula-based characterizations of Markovness*

Darsow *et al.* (1992) obtained the following necessary and sufficient conditions for a time series process based on bivariate copulas to be first-order Markov. For copulas $A, B : [0,1]^2 \to [0,1]$, set

$$(A * B)(x, y) = \int_0^1 \frac{\partial A(x,t)}{\partial t} \cdot \frac{\partial B(t,y)}{\partial t} dt.$$

Further, for copulas $A : [0,1]^m \to [0,1]$ and $B : [0,1]^n \to [0,1]$, define their $\star$−product $A \star B : [0,1]^{m+n-1} \to [0,1]$ via

$$A \star B(x_1, \ldots, x_{m+n-1}) =$$
$$\int_0^{x_m} \frac{\partial A(x_1, \ldots, x_{m-1}, \xi)}{\partial \xi} \cdot \frac{\partial B(\xi, x_{m+1}, \ldots, x_{m+n-1})}{\partial \xi} d\xi.$$

As shown in Darsow *et al.* (1992), the operators $*$ and $\star$ on copulas are distributive over convex combinations, associative and continuous in each place, but not jointly continuous.

Darsow *et al.* (1992) proved that the transition probabilities $\mathbb{P}(s, x, t, A) = \mathbb{P}(X_t \in A | X_s = x)$ of a real-valued stochastic process $\{X_t\}_{t \in T}$, $T \subseteq \mathbf{R}$, satisfy Chapman-Kolmogorov equations (3.20) if and only if the copulas corresponding to bivariate distributions of X_t are such that

$$C_{st} = C_{su} * C_{ut} \tag{3.12}$$

for all $s, u, t \in T$ such that $s < u < t$. Darsow *et al.* (1992) also show that a real-valued stochastic process $\{X_t\}_{t \in T}$ is a first-order Markov process if and only if the copulas corresponding to the finite-dimensional distributions of $\{X_t\}$ satisfy the conditions

$$C_{t_1, \ldots, t_n} = C_{t_1 t_2} \star C_{t_2 t_3} \star \ldots \star C_{t_{n-1} t_n}$$

for all $t_1, \ldots, t_n \in T$ such that $t_k < t_{k+1}$, $k = 1, \ldots, n-1$.

Let $m, n \geq k \geq 1$. Let A and B be, respectively, m- and n-dimensional copulas such that

$$A(u_1, \ldots, u_{m-k}, \xi_1, \ldots, \xi_k)\Big|_{u_i=1, i=1, \ldots, m-k}$$

$$= B(\xi_1, \ldots, \xi_k, u_{k+1}, \ldots, u_n)\Big|_{u_i=1, i=k+1, \ldots, n}$$

$$= C(\xi_1, \ldots, \xi_k), \tag{3.13}$$

$\xi_i \in [0, 1]$, $i = 1, \ldots, k$, where C is a k-dimensional copula (relation (3.13) means that a k-dimensional margin of the copula A is the same as a k-dimensional margin of the copula B).

Let $V_1, \ldots, V_m$ and $W_1, \ldots, W_n$ be r.v.'s with joint cdf's A and B (see Definition 1.1 in the Introduction). Denote by $A_{1, \ldots, m|m-k+1, \ldots, m}$ $(u_1, \ldots, u_{m-k}, \xi_1, \ldots, \xi_k) = \mathbb{P}(V_1 \leq u_1, \ldots, V_{m-k} \leq u_{m-k} | V_{m-k+1} = \xi_1, \ldots, V_m = \xi_k)$ and $B_{1, \ldots, n|1, \ldots, k}(\xi_1, \ldots, \xi_k, u_{m+1}, \ldots, u_{m+n-k}) = \mathbb{P}(W_{k+1} \leq u_{m+1}, \ldots, W_n \leq u_{m+n-k} | W_1 = \xi_1, \ldots, W_k = \xi_k)$ the conditional analogues of the copulas A and B. One has

$$A_{1, \ldots, m|m-k+1, \ldots, m}(u_1, \ldots, u_{m-k}, \xi_1, \ldots, \xi_k)$$

$$= \frac{\partial^k A(u_1, \ldots, u_{m-k}, \xi_1, \ldots, \xi_k)}{\partial v_{m-k+1} \ldots \partial v_m} \Big/ \frac{\partial^k C(\xi_1, \ldots, \xi_k)}{\partial v_1 \ldots \partial v_k},$$

$$B_{1, \ldots, n|1, \ldots, k}(\xi_1, \ldots, \xi_k, u_{m+1}, \ldots, u_{m+n-k})$$

$$= \frac{\partial^k B(\xi_1, \ldots, \xi_k, u_{m+1}, \ldots, u_{m+n-k})}{\partial v_1 \ldots \partial v_k} \Big/ \frac{\partial^k C(\xi_1, \ldots, \xi_k)}{\partial v_1 \ldots \partial v_k},$$

$$\tag{3.14}$$

where $\partial^k A(v_1, \ldots, v_m)/\partial v_{m-k+1} \ldots \partial v_m$, $\partial^k B(v_1, \ldots, v_n)/\partial v_1 \ldots \partial v_k$ and $\partial^k C(v_1, \ldots, v_k)/\partial v_1 \ldots \partial v_k$ denote the partial derivatives of the copulas A, B and C.

Further, define the $\star^k$−product of the copulas A and B, $D = A \star^k B : [0, 1]^{m+n-k} \to [0, 1]$ via the relation

$$D(u_1, \ldots, u_{m+n-k}) =$$

$$\int_0^{u_{m-k+1}} \ldots \int_0^{u_m} A_{1, \ldots, m|m-k+1, \ldots, m}(u_1, \ldots, u_{m-k}, \xi_1, \ldots, \xi_k)$$

$$\times B_{1, \ldots, n|1, \ldots, k}(\xi_1, \ldots, \xi_k, u_{m+1}, \ldots, u_{m+n-k}) C(d\xi_1, \ldots, d\xi_k).$$

$$\tag{3.15}$$

The $\star^k$—operator is a generalization of the star $\star$—operator in (3.12) considered in Darsow *et al.* (1992); the $\star$—operator in (3.12) is a particular case of its above $\star^k$— analogue with $k = 1$. Similar to the case of $k = 1$ in Darsow *et al.* (1992), one can show that the operator $\star^k$ is associative, distributive over convex combinations and continuous in each place (but not jointly continuous).

In terms of the densities $\frac{\partial^m A(v_1,...,v_m)}{\partial v_1...\partial v_m}$, $\frac{\partial^n B(v_1,...,v_n)}{\partial v_1...\partial v_n}$ and $\frac{\partial^{m+n-k} D(v_1,...,v_{m+n-k})}{\partial v_1...\partial v_{m+n-k}}$ of the copulas A, B and $D = A \star^k B$, relation (3.15) is equivalent to the following:

$$
\frac{\partial^{m+n-k} D(u_1, \ldots, u_{m+n-k})}{\partial v_1 \ldots \partial v_{m+n-k}} =
$$
$$
\frac{\partial^m A(u_1, \ldots, u_{m-k}, u_{m-k+1}, \ldots, u_m)}{\partial v_1 \ldots \partial v_m} \times
$$
$$
\frac{\partial^n B(u_{m-k+1}, \ldots, u_m, u_{m+1}, \ldots, u_{m+n-k})}{\partial v_1 \ldots \partial v_n} \Big/
$$
$$
\frac{\partial^k C(u_{m-k+1}, \ldots, u_m)}{\partial v_1 \ldots \partial v_k},
$$

or, equivalently,

$$
\frac{\partial^{m+n-k} D(u_1, \ldots, u_{m+n-k})}{\partial v_1 \ldots \partial v_{m+n-k}} \cdot \frac{\partial^k C(u_{m-k+1}, \ldots, u_m)}{\partial v_1 \ldots \partial v_k} =
$$
$$
\frac{\partial^m A(u_1, \ldots, u_{m-k}, u_{m-k+1}, \ldots, u_m)}{\partial v_1 \ldots \partial v_m} \times
$$
$$
\frac{\partial^n B(u_{m-k+1}, \ldots, u_m, u_{m+1}, \ldots, u_{m+n-k})}{\partial v_1 \ldots \partial v_n}.
$$

Let $T \subseteq \mathbf{R}$. The processes considered throughout the chapter are assumed to be real-valued and continuous and to be defined on the same probability space $(\Omega, \mathfrak{S}, P)$. Stationarity refers to strict stationary. For a r.v. X on $(\Omega, \mathfrak{S}, P)$ and $x \in \mathbf{R}$, $\mathbb{I}_{X<x}$ denotes the indicator of the event $\{X < x\}$. In addition, as usual, for r.v.'s $Y_1, \ldots, Y_s, X_1, \ldots, X_l$ on $(\Omega, \mathfrak{S}, P)$ and $x_1, \ldots, x_l \in \mathbf{R}$, $\mathbb{P}(X_1 < x_1, \ldots, X_l < x_l | Y_1, \ldots, Y_s)$ stand for $\mathbb{E}(\mathbb{I}_{X_1<x_1} \ldots \mathbb{I}_{X_l<x_l} | Y_1, \ldots, Y_s)$. For two r.v.'s X and Y on $(\Omega, \mathfrak{S}, P)$, we write $X = Y$ if $X = Y$ (a.s.).

Definition 3.6. A process $\{X_t\}_{t\in T}$ is called a **Markov process of order** $k \geq 1$ if, for all $t, t_i \in T$, $i = 1, \ldots, n$, such that $t_1 < \ldots < t_{n-k} < t_{n-k+1} < \ldots < t_n < t$ and all $x \in \mathbf{R}$,

$$\mathbb{P}\big(X_t < x \big| X_{t_1}, \ldots, X_{t_{n-k}}, X_{t_{n-k+1}}, \ldots, X_{t_n}\big) =$$

$$\mathbb{P}\big(X_t < x \big| X_{t_{n-k+1}}, \ldots, X_{t_n}\big). \tag{3.16}$$

Throughout the rest of the section, $C_{t_1,\ldots,t_k}$, $t_i \in T$, $i = 1, \ldots, k$, $t_1 < \ldots < t_k$, stand for copulas corresponding to the joint distribution of the r.v.'s $X_{t_1}, \ldots, X_{t_k}$ in the process $\{X_t\}_{t\in T}$ in consideration. In addition, throughout the chapter, formulated equalities and inequalities for two functions f and g defined on $[a,b]^n \subseteq \mathbf{R}^n$ are understood to hold almost everywhere on $[a,b]^n$. That is, we write $f = g$ (or $f(u) = g(u)$) if f and g coincide almost everywhere on $[a,b]^n$: $f(u) = g(u)$ for all $u \in [a,b]^n \backslash \mathcal{A}$, where $\mathcal{A}$ is a subset of $[a,b]^n$ with the Lebesgue measure zero. The meaning of the inequalities $f \geq g$ and $f \leq g$ (or $f(u) \geq g(u)$ and $f(u) \leq g(u)$) is similar.

The following theorem provides a characterization of Markov processes of an arbitrary order in terms of their $(k + 1)$-dimensional copulas.

Theorem 3.17. *A real-valued stochastic process* $\{X_t\}_{t\in T}$, *is a Markov process of order* k, $k \geq 1$, *if and only if for all* $t_i \in T$, $i = 1, \ldots, n$, $n \geq k + 1$, *such that* $t_1 < \ldots < t_n$,

$$C_{t_1,\ldots,t_n} = C_{t_1,\ldots,t_{k+1}} \star^k C_{t_2,\ldots,t_{k+2}} \star^k \ldots \star^k C_{t_{n-k},\ldots,t_n}. \tag{3.17}$$

Let $n \geq k + 1$ and $s \geq 1$. For an n-dimensional copula C denote by C^s the s-fold product $\star^k$ of C with itself.

Corollary 3.1. *A sequence of identically distributed r.v.'s* $\{X_t\}_{t=1}^{\infty}$ *is a stationary Markov process of order* k, $k \geq 1$, *if and only if for all* $n \geq k + 1$,

$$C_{1,\ldots,n}(u_1, \ldots, u_n) = \underbrace{C \star^k C \star^k \ldots \star^k C}_{n-k+1}(u_1, \ldots, u_n)$$

$$= C^{n-k+1}(u_1, \ldots, u_n), \tag{3.18}$$

where C *is a* $k + 1-$*dimensional copula such that* $C_{i_1+h,\ldots,i_l+h} = C_{i_1,\ldots,i_l}$, $1 \leq h \leq k + 1 - i_l$, $1 \leq i_1 < \ldots < i_l \leq k + 1$, $l = 2, \ldots, k$, and $C_{j_1,\ldots,j_l}$, $1 \le j_1 < \ldots < j_l \le k+1$, denote the corresponding marginals of C: $C_{j_1,\ldots,j_l} = C|_{u_i=1, i \ne j_1,\ldots,j_l}$.

Let, as above, $\{X_t\}_{t \in T}$ be a Markov process of order k with finite-dimensional copulas $C_{t_1,\ldots,t_n}(u_1, \ldots, u_n)$, $t_i \in T$, $i = 1,\ldots,n$, $t_1 < \ldots < t_n$. For $t \in T$, denote by F_t the cdf of X_t. Let, for $t_1 < \ldots < t_k$, $x_1,\ldots,x_k \in \mathbf{R}$ and Borel sets $A \in \mathcal{B}(\mathbf{R})$, $p(t_1,\ldots,t_k,t_{k+1},x_1,\ldots,x_k,A) = \mathbb{P}(X_{t_{k+1}} \in A | X_{t_1} = x_1,\ldots,X_{t_k} = x_k)$ stand for the transition probabilities of $\{X_t\}_{t \in T}$. Using (3.14), it is not difficult to see, similar to the case of first-order Markov processes in the proof of Darsow *et al.* (1992, Theorem 3.2), that, for all Borel sets $A \in \mathcal{B}(\mathbf{R})$ of the form $A = (-\infty, x_{k+1})$, $x_{k+1} \in \mathbf{R}$, one has

$$p(t_1,\ldots,t_k,t_{k+1},x_1,\ldots,x_k,A) =$$

$$\frac{\dfrac{\partial^k C_{t_1,\ldots,t_k,t_{k+1}}\left(F_{t_1}(x_1),\ldots,F_{t_k}(x_k),F_{t_{k+1}}(x_{k+1})\right)}{\partial u_1 \ldots \partial u_k}}{\dfrac{\partial^k C_{t_1,\ldots,t_k}\left(F_{t_1}(x_1),\ldots,F_{t_k}(x_k)\right)}{\partial u_1 \ldots \partial u_k}}, \tag{3.19}$$

where, as before, $\partial^k C_{t_1,\ldots,t_k,t_{k+1}}(u_1,\ldots,u_k,u_{k+1})/\partial u_1 \ldots \partial u_k$ and $\partial^k C_{t_1,\ldots,t_k}(u_1,\ldots,u_k)/\partial u_1 \ldots \partial u_k$ denote partial derivatives of the copulas $C_{t_1,\ldots,t_k,t_{k+1}}(u_1,\ldots,u_k,u_{k+1})$ and $C_{t_1,\ldots,t_k}(u_1,\ldots,u_k)$.

Theorem 3.17 and Corollary 3.1 provide copula-based characterizations of higher-order Markovian processes that are alternatives to the conventional characterization using their transition probabilities (and the initial distribution). One can specify a Markov process of order k by prescribing all one-dimensional marginal distributions and a family of $(k+1)$-dimensional copulas. Then one can generate the copulas of higher order and, thus, the finite-dimensional cdf's using (3.17). As discussed in Section 1.2.1, the advantage of the approach based on copulas is that it allows one to separate in the analysis the properties of time series determined by one-dimensional distributions from their dependence characteristics. Relations (3.19), on the other hand, allow one to recover one characterization of higher-order Markov processes given the other.

Corollary 3.1, together with the inversion method for constructing copulas described in Section 1.2.1 (see relations (1.1) in Proposition 1.1), provide a device for obtaining new Markov processes
of an arbitrary order that exhibit dependence properties similar to
those of a given Markov process of the same order but with different marginals. Namely, let $\{X_t\}_{t=1}^{\infty}$ be a stationary Markov process
of order $k \geq 1$ with $(k + 1)$-dimensional cdf $\tilde{F}(x_1, \ldots, x_{k+1})$ and
the one-dimensional marginal cdf F. Then the $(k + 1)$-dimensional
copula generating the process $\{X_t\}_{t=1}^{\infty}$ is, via formula (1.1) in Proposition 1.1, $C(u_1, \ldots, u_{k+1}) = \tilde{F}(F^{-1}(u_1), \ldots, F^{-1}(u_{k+1}))$. Given an
arbitrary one-dimensional cdf G, the stationary k-th order Markov
process that has the dependence structure similar to that of $\{X_t\}_{t=1}^{\infty}$
but a different one-dimensional marginal cdf G can be constructed
via (3.18) by generating its copulas of an arbitrary order and substituting the new one-dimensional cdf to obtain its finite-dimensional
cdf's. For instance, taking $\tilde{F}$ to be the $(k + 1)$-dimensional normal
cdf with a linear correlation matrix R : $\tilde{F} = \Phi_R^{k+1}(x_1, \ldots, x_{k+1})$ as
in (1.3) with $n = k + 1$, one obtains a stationary k-th order Markov
process based on the normal copula $C_R^{k+1}(u_1, \ldots, u_{k+1})$ in (1.4) with
$n = k + 1$. In the case $k = 1$ and $T = \mathbf{R}$, the construction provides a
first-order Markov process of Darsow *et al.* (1992, Example 4.3) that
is referred to as a Brownian motion with non-Gaussian marginal
distributions therein.

More generally, in the case $k \geq 1$, one obtains a process with an
arbitrary one-dimensional marginal cdf's whose dependence structure is similar to that of a Gaussian autoregressive process of order
k (see the discussion at the beginning of Section 3.3.3). In addition, using examples of (possibly higher-order) Markov processes
that satisfy additional dependence assumptions available in the literature, such as k-independent k-th order Markov processes or m-
dependent Markov processes of the first order constructed by Levy
(1949), Rosenblatt and Slepian (1962), Aaronson *et al.* (1992) and
Matúš (1998), one can use the above inversion procedure to construct Markov processes that exhibit similar dependence properties but have one-dimensional marginals different from those in the
examples.

In what follows, we refer to the processes $\{X_t\}_{t=1}^{\infty}$ constructed via (3.18) as stationary k-th order Markov processes based on the $((k+1)$-dimensional) copula C or as stationary C-based k-th order Markov processes for short.

One interesting remark is worth making. Let $f : \mathbf{R} \to \mathbf{R}$ be a strictly increasing function and let $\{X_t\}_{t=1}^{\infty}$ be a stationary C-based k-th order Markov process. Denote $Y_t = f(X_t)$, $t \geq 1$. Since copulas are invariant under strictly increasing transformations of r.v.'s (see Proposition 1.2 in Section 1.2.1) and Markov property is preserved by strictly monotone (hence one-to-one) functions of Markov processes, $\{Y_t\}_{t=1}^{\infty}$ is a stationary k-th order Markov process based on the same copula C.[1] Then, using the results from Section 1.2.1 we can derive the expressions relating the copulas of a k-th order Markov process $\{X_t\}_{t=1}^{\infty}$ to those of a k-th order Markov process determined by $Y_t = f_t(X_t)$, $t \geq 1$, where $f_t : \mathbf{R} \to \mathbf{R}$ are strictly monotone functions. For instance, in the case of the first-order Markov processes X_t with bivariate copulas $C_{t_1 t_2}$, the copulas $\tilde{C}_{t_1 t_2}$ of the process $Y_t = f_t(X_t)$ are related to $C_{t_1 t_2}$ as follows (see Nelsen (1999), Theorem 2.4.4.):

(1) $\tilde{C}_{t_1,t_2}(u,v) = u - C_{t_1,t_2}(u, 1-v)$, if f_{t_1} is strictly increasing and f_{t_2} is strictly decreasing;

(2) $\tilde{C}_{t_1,t_2}(u,v) = v - C_{t_1,t_2}(1-u, v)$, if f_{t_1} is strictly decreasing and f_{t_2} is strictly increasing;

(3) $\tilde{C}_{t_1,t_2}(u,v) = u + v - 1 + C_{t_1,t_2}(1-u, 1-v)$, if both f_{t_1} and f_{t_2} are strictly decreasing.

3.3.2. *Combining Markovness with other dependence properties*

From copula-based characterizations of Markov processes of an arbitrary order in Section 3.3.1, it follows, in particular, that any copulas and, thus, any U-statistics-based representation of a copula function

---

[1] In general, Markovness is not preserved by many-to-one transformations. Rosenblatt (1971, Ch. III) provides conditions under which general functions of Markov processes are still Markovian.

in Section 3.2.2 can be used to construct higher-order Markov processes. The results in this and the next section show that additional dependence properties of such time series impose further restrictions on the U-statistics employed in the representations. These restrictions allow one to obtain characterizations of Markov processes of an arbitrary order that satisfy additional assumptions of r-independence or m-dependence.

The following definitions are time series analogues of Definitions 3.3 and 3.5 in Section 3.2.4.

Definition 3.7. Let $r \geq 2$ and let $T \subseteq \mathbf{R}$ be an index set that contains at least $r + 1$ elements. A process $\{X_t\}_{t \in T}$ is called r-**independent** if any r r.v.'s among X_t, $t \in T$, are jointly independent.

Definition 3.8. Let $m \geq 1$ and let $T \subseteq \mathbf{R}$ be an index set that contains at least $m + 2$ elements. A process $\{X_t\}_{t \in T}$ is called m-**dependent** if, for all $1 \leq a \leq l - 1$ and any indices $j_s \in T$, $s = 1, \ldots, l$, such that $1 \leq j_1 < \cdots < j_a < \cdots < j_l$ and $j_{a+1} - j_a \geq m + 1$, the vectors $(X_{j_1}, X_{j_2}, \ldots, X_{j_{a-1}}, X_{j_a})$ and $(X_{j_{a+1}}, X_{j_{a+2}}, \ldots, X_{j_{l-1}}, X_{j_l})$ are independent.

A number of studies have focused on problems of combining Markovian structures with other types of dependence. Levy (1949) constructed a 2-nd order Markov process consisting of pairwise independent uniformly distributed r.v.'s (a 2-nd order pairwise independent Markov process). Motivated by applications in the study of the mechanism of human vision, Rosenblatt and Slepian (1962) constructed stationary N-th order Markov processes consisting of discrete r.v.'s such that every N variables of the process are independent while $N + 1$ adjacent variables of the process are not independent (stationary N-th order N-independent Markov process). Rosenblatt and Slepian (1962) also obtained a result that is natural to refer to as an impossibility or a reduction property for Markov processes. This result shows that all N-th order N-independent Markov processes with two-valued X_t's are trivial in that they are processes of independent r.v.'s.

Higher order Markov r-independent processes are important in testing empirically the sensitivity of statistical procedures developed on the independence assumption to weak dependence in the data generating process (see, Rosenblatt and Slepian (1962)). In addition, such processes are of interest since they provide examples which are not Markovian of first order but whose first order transition probabilities $\mathbb{P}(s, x, t, A) = \mathbb{P}(X_t \in A | X_s = x)$ nevertheless satisfy the Chapman-Kolmogorov stochastic equation

$$\mathbb{P}(s, x, t, A) = \int_{-\infty}^{\infty} \mathbb{P}(u, \xi, t, A) \mathbb{P}(s, x, u, d\xi) \qquad (3.20)$$

for all Borel sets A, all $s < t$ in T, $u \in (s, t) \cap T$ and for almost all $x \in \mathbf{R}$.[2]

Markov processes with 1-dependence appeared for the first time in Aaronson *et al.* (1992) and were considered, e.g., by Burton *et al.* (1993) and Matúš (1996), where the focus was on 1-dependent Markov shifts and on the structure of block-factors. Matúš (1998) studied m-dependent Markov sequences consisting of discrete r.v.'s and showed, in particular, that generally no stationary sequence of r.v.'s which is Markov of order n but not of order $n - 1$ and m-dependent but not $(m - 1)$-dependent exists if the state space of the sequence has small cardinality (another type of an impossibility/reduction result for Markov processes). Matúš (1998) also showed that to ensure the existence of Markov processes of order $n = 1$ that are m-dependent but not $(m - 1)$-dependent, the number of attainable states must be at least $m + 2$ and this bound is tight.

The following result gives a characterization of stationary k-independent k-th order Markov processes in terms of the canonical functions g in U-statistic-based representations for their copulas presented in the previous chapter. Below, $[x]$ stands for the integer part of $x \in \mathbf{R}$.

Theorem 3.18. *Let C be a $(k + 1)$-dimensional copula. A sequence of r.v.'s $\{X_t\}_{t=1}^{\infty}$, is a stationary k-independent C-based k-th order*

---

[2]Examples of non-Markovian processes for which the Chapman-Kolmogorov equation is satisfied were also given, e.g., by Feller (1959) and Rosenblatt (1960).

Markov process if and only if the density of C *has the form*

$$\frac{\partial^{k+1} C(u_1, \ldots, u_{k+1})}{\partial u_1 \ldots \partial u_{k+1}} = 1 + g(u_1, \ldots, u_{k+1}), \qquad (3.21)$$

where $g : [0,1]^{k+1} \to \mathbf{R}$ *is a function satisfying the conditions*

$$\int_0^1 \ldots \int_0^1 |g(u_1, \ldots, u_{k+1})| du_1 \ldots du_{k+1} < \infty, \qquad (3.22)$$

$$\int_0^1 \ldots \int_0^1 \prod_{j=1}^{s} g(u_j, \ldots, u_{k+j}) du_{i_1} \ldots du_{i_s}$$

$$= \int_0^1 \ldots \int_0^1 g(u_1, \ldots, u_{k+1}) g(u_2, \ldots, u_{k+2}) \ldots$$

$$g(u_s, \ldots, u_{k+s}) du_{i_1} \ldots du_{i_s}$$

$$= 0 \qquad (3.23)$$

for all $s \le i_1 < \ldots < i_s \le k+1$, $s = 1, 2, \ldots, \left[\frac{k+1}{2}\right]$, *and*

$$g(u_1, \ldots, u_{k+1}) \ge -1. \qquad (3.24)$$

Integration in condition (3.23) is with respect to all combinations of s variables among the arguments $u_s, u_{s+1}, \ldots, u_{k+1}$ that are common to all functions $g(u_1, \ldots, u_{k+1})$, $g(u_2, \ldots, u_{k+2})$, $\ldots$, $g(u_s, \ldots, u_{k+s})$ appearing in the integrand. These conditions ensure that all k-dimensional marginals of the copula of $X_1, \ldots, X_{k+s}$, $s \ge 1$, are product copulas (1.2) with $n = k$ and thus the k-independence property is satisfied for the stationary k-th order Markov process in consideration.

The following theorem provides a characterization of Markov processes satisfying m-dependence properties in terms of U-statistic representations of their copulas using canonical functions g defined in previous section.

Theorem 3.19. *Let C be a bivariate copula. A sequence of r.v.'s $\{X_t\}_{t=1}^{\infty}$ is a stationary m-dependent C-based first-order Markov process if and only if the density of C satisfies*

$$\frac{\partial^2 C(u_1, u_2)}{\partial u_1 \partial u_2} = 1 + g(u_1, u_2), \tag{3.25}$$

where $g : [0,1]^2 \to \mathbf{R}$ is a function satisfying the conditions

$$\int_0^1 \int_0^1 |g(u_1, u_2)| du_1 du_2 < \infty, \tag{3.26}$$

$$\int_0^1 g(u_1, u_2) du_i = 0, \; i = 1, 2, \tag{3.27}$$

$$g(u_1, u_2) \geq -1 \tag{3.28}$$

and such that

$$\int_0^1 \cdots \int_0^1 \prod_{i=1}^{m+1} g(u_i, u_{i+1}) du_2 du_3 \ldots du_{m+1}$$

$$= \int_0^1 \cdots \int_0^1 g(u_1, u_2) g(u_2, u_3) \cdots$$

$$g(u_{m+1}, u_{m+2}) du_2 du_3 \ldots du_{m+1}$$

$$= 0. \tag{3.29}$$

Similar to (3.23), integration in condition (3.29) is with respect to the variables $u_2, u_3, \ldots, u_{m+1}$ that appear more than once among the arguments of the functions $g(u_1, u_2)$, $g(u_2, u_3)$, $\ldots$, $g(u_{m+1}, u_{m+2})$. This condition ensures that the r.v.'s X_1 and X_{m+2} are independent and, more generally, independence holds between the vectors $(X_1, \ldots, X_n)$ and $(X_{m+n+1}, \ldots, X_{m+n+j})$, $n, j \geq 1$.

Similar to the proof of Theorems 3.18 and 3.19, one can obtain necessary and sufficient conditions for a Markov process of an arbitrary order to be m-dependent and also exhibit certain additional dependence properties. For instance, let C be a $(k + 1)$-dimensional copula and suppose that $\{X_t\}_{t=1}^{\infty}$ is a stationary C-based k-th order Markov process such that any k r.v.'s among $X_1, \ldots, X_{k+1}$ are independent (clearly, this assumption is weaker than k-independence of the process $\{X_t\}_{t=1}^{\infty}$ in Theorem 3.19). Using Theorems 3.5 and 3.15 and Corollary 3.1 as in the proof of Theorems 3.18 and 3.19, one can show that the process $\{X_t\}_{t=1}^{\infty}$ is m-dependent for some $m \geq k+1$ if and only if the density of C has form (3.21) with a function $g : [0,1]^{k+1} \to \mathbf{R}$ that satisfies conditions (3.22) and (3.24) and is such that

$$\int_0^1 \cdots \int_0^1 \prod_{j=1}^{m-k+2} g(u_j, \ldots, u_{k+j}) du_2 \ldots du_{m+1}$$

$$= \int_0^1 \cdots \int_0^1 g(u_1, \ldots, u_{k+1}) g(u_2, \ldots, u_{k+2}) \cdots$$

$$g(u_{m-k+2}, \ldots, u_{m+2}) du_2 \ldots du_{m+1}$$

$$= 0.$$

In a number of applications, including those in finance, Markov and martingale properties hold simultaneously. The martingale property, in contrast to the Markov (first and higher order) properties, is not determined by finite-dimensional copulas only and can be affected by changes in one-dimensional marginal distributions. Indeed, using (3.14) with $m = 2$ and $k = 1$ (or Theorem 3.1 in Darsow *et al.* (1992)), it is not difficult to see that a stationary process $\{X_t\}_{t=1}^{\infty}$ with bivariate copulas $C(u, v)$ and the univariate cdf $F(x)$ is a martingale difference sequence with respect to the natural filtration $\mathfrak{S}_t = \sigma(X_1, \ldots, X_t)$ if and only if $\int_{-\infty}^{\infty} x \frac{\partial C(F(x), F(y))}{\partial u \partial v} dF(x) = 0$. The martingale property is determined by copulas alone for the class of martingale differences that satisfy conditional symmetry assumptions.

Definition 3.9. A sequence $\{X_t\}_{t=1}^{\infty}$ on a probability space $(\Omega, \mathfrak{S}, P)$ is **a conditionally symmetric martingale difference** with respect to an increasing sequence of σ−algebras $\mathfrak{S}_0 = (\Omega, \emptyset) \subseteq \mathfrak{S}_1 \subseteq \mathfrak{S}_2 \subseteq \ldots \subseteq \mathfrak{S}_n \subseteq \mathfrak{S}$ if, for all $t \geq 1$, the r.v. X_t is $\mathfrak{S}_t$-measurable and conditionally symmetric given $\mathfrak{S}_{t-1}$, that is, $\mathbb{P}(X_t > x | \mathfrak{S}_{t-1}) = \mathbb{P}(X_t < -x | \mathfrak{S}_{t-1})$, $x \geq 0$.

Theorem 3.20. *Let C be a bivariate copula. A stationary C-based first-order Markov process $\{X_t\}_{t=1}^{\infty}$ consisting of symmetric r.v.'s is a conditionally symmetric martingale difference with respect to the natural filtration $\Im_0 = (\Omega, \emptyset)$, $\Im_t = \sigma(X_1, \ldots, X_t)$, $t \geq 1$, if and only if*

$$\frac{\partial C(u_1,\, 1/2 - u)}{\partial u_1} + \frac{\partial C(u_1,\, 1/2 + u)}{\partial u_1} = 1 \qquad (3.30)$$

for all $u_1 \in [0,1]$, $u \in [0, 1/2)$, or, equivalently, if the density of C satisfies

$$\frac{\partial C(u_1,\, 1/2 - u)}{\partial u_1 \partial u_2} = \frac{\partial C(u_1,\, 1/2 + u)}{\partial u_1 \partial u_2} \qquad (3.31)$$

for all $u_1 \in [0,1]$, $u \in [0, 1/2)$.

3.3.3. *Reduction property for Markov processes*

For a Gaussian process, pairwise independence coincides with joint independence. Therefore, it is not surprising that a stationary higher-order Markov process based on a normal copula exhibits r-independence if and only if it is an i.i.d. sequence. Formally, let $C(u_1, \ldots, u_{k+1}) = C_R(u_1, \ldots, u_{k+1})$ be the normal copula with linear correlation matrix R defined in (1.4) with $n = k + 1$ and let $\{X_t\}_{t=1}^{\infty}$ be a stationary C-based k-th order Markov process. Let $Y_t = \Phi^{-1}[F(X_t)]$, $t \geq 1$, where $\Phi(x)$ is the standard normal univariate cdf and $F(x)$ is the cdf of X_t. Since the random vector $(Y_1, \ldots, Y_{k+1})$ has the multivariate normal distribution $\mathcal{N}(0, R)$ with correlation matrix R:

$$(Y_1, \ldots, Y_{k+1}) \sim \mathcal{N}(0, R), \qquad (3.32)$$

we conclude that the process $\{X_t\}_{t=1}^{\infty}$ is r-independent for some $r \geq 2$ if and only if $R = I$, that is, if and only if $\{X_t\}_{t=1}^{\infty}$ is a sequence of i.i.d. r.v.'s.

Let now $k = 1$ and let $C(u_1, u_2) = C_\rho(u_1, u_2)$ be the bivariate normal copula with correlation coefficient ρ (see (1.5)). Then (as in, e.g., Example 1 of Chen and Fan (2006b)) we conclude that $\{Y_t\}_{t=1}^{\infty}$

is a Gaussian process and, thus, for $t \geq m + 2$,

$$Y_t = \rho Y_{t-1} + \epsilon_t = \rho^{m+1} Y_{t-m-1} + \sum_{k=0}^{m} \rho^k \epsilon_{t-k}, \qquad (3.33)$$

where ϵ_t has a normal distribution: $\epsilon_t \sim \mathcal{N}(0, 1 - \rho^2)$. From (3.33) we conclude that Y_t and Y_{t-m-1} (and, thus, X_t and X_{t-m-1}) are independent if and only if $\rho = 0$. Thus, $\{X_t\}_{t=1}^{\infty}$ is a stationary m-dependent C_ρ-based Markov process (of the first order) if and only if $\rho = 0$, that is, if and only if $\{X_t\}_{t=1}^{\infty}$ is a sequence of i.i.d. r.v.'s.

Theorems 3.18 and 3.19 further imply several reduction and impossibility results for Markov processes satisfying m-dependence and r-independence conditions that are similar in spirit to those in the case of normal copulas. The results show that a number of copula-based time series that simultaneously exhibit Markovness and m-dependence or r-independence properties are, in fact, sequences of independent r.v.'s.

These results for Markov processes are similar in spirit to reduction results for multiplicative systems in Section 3.2.5, which show that if r.v.'s $X_1, \ldots, X_n$ form an $MS(\alpha)$ and each of them takes not more than $\alpha + 1$ values, then the r.v.'s X_i, $i = 1, \ldots, n$, are, in fact, jointly independent.

Theorem 3.21 shows that a construction of non-trivial Markov processes of higher order that exhibit r-independence is impossible on the base of copulas whose densities in Theorem 3.18 have functions g with a separable product form.

Theorem 3.21. *Let $k \geq 2$ and let C be a $(k + 1)$-dimensional copula that has density* (3.21), *where $g(u_1, u_2, \ldots, u_{k+1}) = \alpha f(u_1) f(u_2) \ldots f(u_{k+1})$ for some $\alpha \in \mathbf{R}$ and some continuous function $f : [0, 1] \to \mathbf{R}$. A sequence of r.v.'s $\{X_t\}_{t=1}^{\infty}$ is a stationary k-independent C-based k-th order Markov process if and only if $\{X_t\}_{t=1}^{\infty}$ is a sequence of i.i.d. r.v.'s.*

An example of copulas C in the separable product form of Theorem 3.21 is given by the special case of $(k + 1)$-dimensional EFGM copulas (3.8):

$$C(u_1, u_2, \ldots, u_{k+1}) = \prod_{i=1}^{k+1} u_i \Big(1 + \alpha(1 - u_1)(1 - u_2) \ldots (1 - u_{k+1}) \Big),$$

(3.34)

where $-1 \le \alpha \le 1$. These copulas have densities (3.21) with

$$g(u_1, u_2, \ldots, u_{k+1}) = \alpha(1 - 2u_1)(1 - 2u_2) \ldots (1 - 2u_{k+1}).$$

(3.35)

Corollary 3.2. *Let* $k \ge 2$ *and let* C *be a* $(k+1)$-*dimensional EFGM copula* (3.34) *with density* (3.21) *where* g *is given by* (3.35). *A sequence of r.v.'s* $\{X_t\}_{t=1}^{\infty}$ *is a stationary* k-*independent* C-*based* k-*th order Markov process if and only if it is a sequence of i.i.d. r.v.'s.*

Corollary 3.3 is a generalization of Corollary 3.2 to $(k+1)$-dimensional power copulas (3.10) given by

$$C(u_1, u_2, \ldots, u_{k+1})$$

$$= \prod_{i=1}^{k+1} u_i \Big(1 + \alpha(u_1^l - u_1^{l+1})(u_2^l - u_2^{l+1}) \ldots (u_{k+1}^l - u_{k+1}^{l+1}) \Big),$$

(3.36)

$-1 \le \alpha \le 1$, where $l \in \{0, 1, 2, \ldots\}$ (copulas (3.36) reduce to those in (3.34) if $l = 0$). These copulas have density (3.21) in which

$$g(u_1, u_2, \ldots, u_{k+1})$$

$$= \alpha \Big((l+1)u_1^l - (l+2)u_1^{l+1} \Big) \Big((l+1)u_2^l - (l+2)u_2^{l+1} \Big) \cdot \ldots \cdot$$

$$\Big((l+1)u_{k+1}^l - (l+2)u_{k+1}^{l+1} \Big).$$

(3.37)

Corollary 3.3. *Let* $k \ge 2$ *and let* C *be a* $(k+1)$-*dimensional power copula* (3.36) *with density* (3.21), *where* g *is given by* (3.37). *A sequence of r.v.'s* $\{X_t\}_{t=1}^{\infty}$ *is a stationary* k-*independent* C-*based* k-*th order Markov process if and only if* $\{X_t\}_{t=1}^{\infty}$ *is a sequence of i.i.d. r.v.'s.*

Theorem 3.22 is an analogue of Theorem 3.21 that provides impossibility/reduction results for m-dependent Markov processes. Theorem 3.22 shows that construction of non-trivial examples (that is, those more general than sequences of i.i.d. r.v.'s) of stationary Markov processes exhibiting m-dependence is impossible on the base of bivariate copulas that have, similar to Theorem 3.21, the function g in a separable product form.

Theorem 3.22. *Suppose that C is a bivariate copula that has the density $\partial^2 C(u_1, u_2)/\partial u_1 \partial u_2 = 1 + \alpha f(u_1) f(u_2)$ for some $\alpha \in \mathbf{R}$ and some continuous function $f : [0,1] \to \mathbf{R}$. A sequence of r.v.'s $\{X_t\}_{t=1}^{\infty}$ is a stationary m-dependent C-based Markov process (of the first order) if and only if $\{X_t\}_{t=1}^{\infty}$ is a sequence of i.i.d. r.v.'s.*

The following corollary is a specialization of Theorem 3.22 to the special case of bivariate EFGM copulas (3.7), that is copulas (3.34) with $k = 1$:

$$C(u_1, u_2) = u_1 u_2 \Big(1 + \alpha(1 - u_1)(1 - u_2) \Big), \quad -1 \le \alpha \le 1, \quad (3.38)$$

that have density (3.25) with

$$g(u_1, u_2) = \alpha(1 - 2u_1)(1 - 2u_2). \quad (3.39)$$

Corollary 3.4. *Let C be a bivariate copula EFGM copula (3.38) with density (3.25), where g is given by (3.39). A sequence of r.v.'s $\{X_t\}_{t=1}^{\infty}$ is a stationary m-dependent C-based Markov process (of the first order) if and only if $\{X_t\}_{t=1}^{\infty}$ is a sequence of i.i.d. r.v.'s.*

The results in this section that demonstrate that Markov processes with m-dependence and r-independent Markov processes of higher order cannot be constructed from EFGM copulas and other separable copulas complement and substantially generalize the results of Cambanis (1991). Cambanis (1991) show that the most common dependence structures such as constant, exponential and m-dependence cannot be exhibited by stationary processes $\{X_t\}$ whose finite-dimensional copulas are the following multivariate analogues of the bivariate EFGM copulas (3.7) that are a particular case of the generalized multivariate EFGM copulas (3.9):

$$C_{j_1,\ldots,j_n}(u_{j_1},\ldots,u_{j_n})$$

$$= \prod_{s=1}^{n} u_{j_k}\left(1 + \sum_{1\leq l<m\leq n} \alpha_{lm}(1-u_{j_l})(1-u_{j_m})\right).$$

The results also complement the above-mentioned results by Rosenblatt and Slepian (1962) on non-existence of non-trivial N-th order N-independent Markov processes consisting of two-valued r.v.'s since, as follows from Sharakhmetov and Ibragimov (2002), the finite-dimensional copulas of sequences of r.v.'s concentrated on two points have multivariate EFGM structure (3.9).

Interestingly, in contrast to normal copulas, t-copulas cannot be used to construct (possibly higher-order) Markov processes that exhibit r-independence or m-dependence even if their correlation matrices are identity matrices corresponding to the case of uncorrelatedness. For instance, let $C(u_1,\ldots,u_{k+1}) = C_{\nu,R}^t(u_1,\ldots,u_{k+1})$ be a t-copula with correlation matrix R defined in (1.6) with $n = k+1$ and let $\{X_t\}_{t=1}^{\infty}$ be a stationary C-based k-th order Markov processes. Then relation (3.32) holds for $Y_t = (\sqrt{S}/\sqrt{\nu}) \cdot t_\nu^{-1}[F(X_t)]$, where $S \sim \chi_\nu^2$ is a r.v. with chi-square distribution with ν degrees of freedom that is independent of $\{X_t\}_{t=1}^{\infty}$, $t_\nu(x)$ is the cdf of the univariate Student t-distribution with ν degrees of freedom and $F(x)$ is the cdf of X_t. As above, we obtain that $R = I$ and $\{Y_t\}_{t=1}^{\infty}$ is a sequence of i.i.d. standard normal r.v.'s if $\{X_t\}_{t=1}^{\infty}$ is r-independent for some $r \geq 2$. Since the components of the random vector

$$(F^{-1}[t_\nu(\sqrt{\nu}Y_{t-1}/\sqrt{S})], F^{-1}[t_\nu(\sqrt{\nu}Y_t/\sqrt{S})]) \qquad (3.40)$$

are dependent if Y_{t-1} and Y_t are i.i.d. standard normal r.v.'s independent of $S \sim \chi_\nu^2$, we thus conclude that there does not exist a stationary r-independent $C_{\nu,R}^t$-based k-th order Markov process for any correlation matrix R.

Let $k = 1$ and let $C(u_1, u_2) = C_{\nu,\rho}^t(u_1, u_2)$ be a bivariate t-copula in (1.7). Then the process $Y_t = (\sqrt{S}/\sqrt{\nu}) \cdot t_\nu^{-1}[F(X_t)]$ satisfies (3.33). We thus conclude that if $\{X_t\}_{t=1}^{\infty}$ is a stationary m-dependent $C_{\nu,\rho}^t$-based Markov process (of the first order), then $\rho = 0$ and $\{Y_t\}_{t=1}^{\infty}$ is a sequence of i.i.d. standard normal r.v.'s. As before, this, together with dependence of the components of random vector (3.40), implies that there does not exist a stationary m-dependent $C_{\nu,\rho}^t$-based Markov process of the first order for any value of ρ.

3.3.4. *Fourier copulas*

The results on limitations of Eyraud-Farlie-Gumbel-Mongenstern, separable, normal and t-copulas presented in the previous section emphasize the substantial technical difficulty in constructing copula-based time series with flexible dependence structures. A class of copulas based on expansions by Fourier polynomials we introduce in this section allows one to overcome this difficulty.

It is not difficult to check that the conditions of Theorem 3.18 are satisfied for the following functions g:

$$
\begin{aligned}
&g(u_1,\ldots,u_{k+1}) \\
&= \sum_{j=1}^{N} \left[\alpha_j \sin\left(2\pi \sum_{i=1}^{k+1} \beta_i^j u_i\right) + \gamma_j \cos\left(2\pi \sum_{i=1}^{k+1} \beta_i^j u_i\right) \right],
\end{aligned}
\tag{3.41}
$$

where $N \geq 1$, and $\alpha_j, \gamma_j \in \mathbf{R}$, and $\beta_i^j \in \mathbf{Z}$, $i = 1,\ldots,k+1$, $j = 1,\ldots,N$, are arbitrary numbers such that

$$
\beta_1^{j_1} + \sum_{l=2}^{s} \delta_{l-1} \beta_l^{j_l} \neq 0,
$$

for $j_1,\ldots,j_s \in \{1,\ldots,N\}$, $\delta_1,\ldots,\delta_{s-1} \in \{-1,1\}$, $s = 2,\ldots,k+1$, and

$$
1 + \sum_{j=1}^{N} [\alpha_j \delta_j + \gamma_j \delta_{j+N}] \geq 0
$$

for $\delta_1,\ldots,\delta_{2N} \in \{-1,1\}$. We refer to the copulas C corresponding to the functions g,

$$
C(u_1,\ldots,u_{k+1}) = \int_0^{u_1} \cdots \int_0^{u_{k+1}} (1 + g(u_1,\ldots,u_{k+1})) du_1 \ldots du_{k+1},
$$

as $(k + 1)$-dimensional *Fourier* copulas. Each such copula can thus be used to construct a stationary k-independent k-th order Markov process via (3.18).

Similarly, conditions of Theorem 3.19 are satisfied with $m = 1$ for the bivariate Fourier copulas corresponding to the functions g defined in (3.41) with $k = 1$, that is, for the Fourier copulas

$$C(u_1, u_2) = \int_0^{u_1} \int_0^{u_2} (1 + g(u_1, u_2)) du_1 du_2, \qquad (3.42)$$

where

$$g(u_1, u_2) = \sum_{j=1}^{N} [\alpha_j \sin(2\pi(\beta_1^j u_1 + \beta_2^j u_2))$$

$$+ \gamma_j \cos(2\pi(\beta_1^j u_1 + \beta_2^j u_2))], \qquad (3.43)$$

$N \geq 1$, $\alpha_j, \gamma_j \in \mathbf{R}$, and $\beta_1^j, \beta_2^j \in \mathbf{Z}$, $j = 1, \ldots, N$, are arbitrary numbers such that $\beta_1^{j_1} + \beta_2^{j_2} \neq 0$ for $j_1, j_2 \in \{1, \ldots, N\}$ and $\beta_1^{j_1} - \beta_2^{j_2} \neq 0$, $1 + \sum_{j=1}^{N} [\alpha_j \delta_j + \gamma_j \delta_{j+N}] \geq 0$ for $\delta_1, \ldots, \delta_{2N} \in \{-1, 1\}$. The processes constructed from copulas (3.42) via (3.18) thus give examples of stationary 1-dependent first-order Markov processes.

3.4. Measures of dependence

In this section, we discuss various dependence measures. We introduce some new measures and also apply the results from Section 3.2 to study properties of some well-known measures and convergence of multidimensional statistics of time series.

In particular, we discuss conditions for convergence in distribution of m-dimensional statistics $h(X_t, X_{t+1}, \ldots, X_{t+m-1})$ of time series $\{X_t\}$ in terms of weak convergence of $h(\xi_t, \xi_{t+1}, \ldots, \xi_{t+m-1})$, where $\{\xi_t\}$ is a sequence of independent copies of X_t's, and convergence to zero of measures of intertemporal dependence in $\{X_t\}$. The tools used include new sharp estimates for the distance between the distribution function of an arbitrary statistic in dependent random variables and the distribution function of the statistic in independent copies of the random variables in terms of the measures of dependence of the random variables. Furthermore, we obtain new sharp complete decoupling moment and probability inequalities for dependent random variables in terms of their dependence characteristics.

3.4.1. *Problems with correlation*

One of the central problems in statistics, economics, finance and risk management is to develop appropriate measures of dependence, particularly measures of serial dependence in time series and measure of vector rather than scalar dependence. Appropriate dependence measures permit early detection of contagion, correct pricing and portfolio optimization decisions, among other things.

The most widely applied dependence measure is, by all means, the correlation defined, for r.v.'s X, Y, with finite second moments $\mathbb{E}X^2, \mathbb{E}Y^2 < \infty$, as follows

$$\operatorname{corr}(X, Y) = \frac{\operatorname{cov}(X, Y)}{\sqrt{\mathbb{V}(X)\mathbb{V}(Y)}},$$

where $\operatorname{cov}(X, Y) = \mathbb{E}(X - \mathbb{E}X)(Y - \mathbb{E}Y)$ denotes covariance between X and Y and $\mathbb{V}(X) = \mathbb{E}(X - \mathbb{E}X)^2$, $\mathbb{V}(Y) = \mathbb{E}(Y - \mathbb{E}Y)^2$ denote variances of X and Y, respectively.

However, many studies have observed that the use of correlation as a measure of dependence is problematic in many settings. For example, Boyer *et al.* (1997) reported that correlations provide very little information about the underlying dependence structure in cases of asymmetric dependence. More generally, Blyth (1996) and Shaw (1997) point out that the linear correlation fails to capture nonlinear dependencies in data. Embrechts *et al.* (2002) present a rigorous study concerning the problems related to the use of correlation as measure of dependence in risk management and finance. A notable case when the use of correlation as a measure of dependence becomes problematic is when we have departures from multivariate normality and, more generally, from elliptic distributions — a setting of particular interest to us (see, among others, Embrechts *et al.* (2002); Chapter 5 of McNeil *et al.* (2005); de la Peña *et al.* (2006)).

Other problems with using correlation include the fact that it is a bivariate measure and even using its time varying versions, at best, leads to only capturing pairwise dependencies, rather than more complicated multivariate dependence structures. In fact, the same critique applies to other bivariate measures of dependence such as bivariate Pearson's (1894) ϕ^2, Kendall's (1938) τ, Spearman's (1904) ρ, Hoeffding's (1940) Φ^2, as well as Kullback-Leibler and Shannon mutual information.

Perhaps even more importantly, correlations are defined for variable with finite second order moments

$$\mathbb{V}(X) < \infty, \quad \mathbb{V}(Y) < \infty, \quad \text{cov}(X, Y) < \infty$$

and, as discussed in Section 1.2.2, a bulk of financial and commodity market data exhibit heavy-tailed behavior with higher moments failing to exist, including the second order moments (see, e.g., Loretan and Phillips (1994), Cont (2001), and references therein). Even if second moments are finite, reliable estimation of correlations and autocorrelations especially under dependence is problematic as it requires finite fourth-order moments (see, e.g., Davis and Mikosch (1998); Mikosch and Stărică (2000); Ibragimov *et al.* (2015); Hill and Prokhorov (2016)).

3.4.2. *Some alternative measures*

By now, alternative measures of association between $n \geq 2$ scalar components of a random vector $\mathbf{X} = (X_1, \ldots, X_n)$ have received substantial attention in the literature. Such measures include multivariate extensions of Kendall's τ, Spearman's ρ, Pearson's ϕ^2, Hoeffding's Φ^2, various multivariate divergence measures such as relative entropy measures of Joe (1987, 1989) (see, e.g., Nelsen (1996); Joe (1997, 2014), for surveys). Most desirable of them are invariant to increasing transformations and therefore capture dependence regardless of specific marginals. As will be clear shortly, this means they are functionals of the copula.

Substantial work has focused on statistical and econometric applications of mutual information and other dependence measures and concepts (see, among others, Lehmann (1966); Golan (2002); Golan and Perloff (2002); Massoumi and Racine (2002); Miller and

Liu (2002); Soofi and Retzer (2002); Ullah (2002), and references therein), including works on estimating entropy measures of serial dependence in time series and related problems (see, e.g., Robinson (1991); Granger and Lin (1994); Hong and White (2005)), the study of multifractals and Boltzmann-Gibbs statistics (Tsallis (1988)), and the problem of testing for conditional dependence and non-causality (Fernandes and Flôres (2002); Medovikov (2015)).

Less attention has been paid to constructing measures of multivariate association between several *vectors* $(\mathbf{X}_1, \ldots, \mathbf{X}_p)$ of dimension $n_1, \ldots, n_p$, invariant to dependence between the *within*-vector components, even though such measures are key to multivariate modelling in many fields. This is certainly the case for the study of contagion in financial markets, where we look for a measure of dependence between entire markets, which is robust to co-movements within them (see, e.g., Medovikov and Prokhorov (2016)). The reason why such measures are of interest in their own right is that mutual independence between elements of vectors is not in general implied by pairwise independence.

These alternative dependence measures — both scalar and vector versions — are rank-based, which makes them functional of the copula. In fact, many conventional generalizations of correlation such as Spearman's ρ or Kendall's τ are known as rank *correlations* and can be expressed in terms of the copula functions. Moreover, various divergence and entropy measures mentioned above are functionals of the multivariate distribution and thus of the implied copula. We now define some of these measures.

Let $X_1, \ldots, X_n$ be r.v.'s with one-dimensional cdf's $F_k(x_k)$, $k = 1, \ldots, n$, joint cdf $F(x_1, \ldots, x_n)$ and copula function $C(u_1, \ldots, u_n)$ and copula density $c(u_1, \ldots, u_n)$.

Definition 3.10.

(1) Multivariate analog of Pearson's ϕ^2 coefficient

$$\phi^2_{X_1,\ldots,X_n} = \int_{-\infty}^{\infty} \cdots \int_{-\infty}^{\infty} \frac{(dF(x_1, \ldots, x_n))^2}{dF_1(x_1) \ldots dF_n(x_n)} - 1$$

$$= \int_{-\infty}^{\infty} \cdots \int_{-\infty}^{\infty} \left(\frac{dF(x_1, \ldots, x_n)}{dF_1 \ldots dF_n} \right)^2 dF_1(x_1) \ldots dF_n(x_n) - 1$$

$$= \int_{-\infty}^{\infty} \cdots \int_{-\infty}^{\infty} (c(F_1(x_1), \ldots, F_n(x_n)))^2 dF_1(x_1) \ldots dF_n(x_n) - 1$$

$$= \mathbb{E}c(F_1(X_1), \ldots, F_n(X_n)) - 1$$

(2) Relative entropy

$$\delta_{X_1, \ldots, X_n} = \int_{-\infty}^{\infty} \cdots \int_{-\infty}^{\infty} \log \left(\frac{dF(x_1, \ldots, x_n)}{dF_1 \ldots dF_n} \right) dF(x_1, \ldots, x_n)$$

$$= \mathbb{E} \log c(F_1(X_1), \ldots, F_n(X_n))$$

(3) Multivariate analogue of Hoeffding's Φ^2

$$\Phi^2_{X_1, \ldots, X_n} = \mathbb{K} \int_{-\infty}^{\infty} \cdots \int_{-\infty}^{\infty} [F(x_1, \ldots, x_n)$$

$$- F_1(x_1) \ldots F_n(x_n)]^2 dF_1(x_1) \ldots dF_n(x_n)$$

$$= \mathbb{K} \int_{-\infty}^{\infty} \cdots \int_{-\infty}^{\infty} [C(F_1(x_1), \ldots, F_n(x_n))$$

$$- F_1(x_1) \ldots F_n(x_n)]^2 dF_1(x_1) \ldots dF_n(x_n)$$

(4) Kendall's τ

$$\tau_{X_1, X_2} = 4 \int_{-\infty}^{\infty} \int_{-\infty}^{\infty} C(F_1(x_1), F_2(x_2)) dC(F_1(x_1), F_2(x_2)) - 1$$

$$= 4\mathbb{E}C(F_1(X_1), F_2(X_2)) - 1$$

(5) Spearman's ρ

$$\rho_{X_1, X_2} = 12 \int_{-\infty}^{\infty} \int_{-\infty}^{\infty} F_1(x_1) F_2(x_2) dC(F_1(x_1), F_2(x_2)) - 3$$

$$= 12\mathbb{E}F_1(X_1) F_2(X_2) - 3$$

(6) Upper and lower tail dependence coefficients λ

$$\lambda_u = \lim_{u \to 1} \mathbb{P}(X_1 > F_1^{-1}(u) | X_2 > F_2^{-1}(u)) = \lim_{u \to 1} \frac{1 - 2u + C(u, u)}{1 - u}$$

$$\lambda_l = \lim_{u \to 0} \mathbb{P}(X_1 \leq F_1^{-1}(u) | X_2 \leq F_2^{-1}(u)) = \lim_{u \to 0} \frac{C(u, u)}{u}$$

The multivariate Pearson's ϕ^2 coefficient and the relative entropy are particular cases of multivariate divergence measures

$$D^{\psi}_{X_1,\ldots,X_n} = \int_{-\infty}^{\infty} \cdots \int_{-\infty}^{\infty} \psi\left(\frac{dF(x_1,\ldots,x_n)}{dF_1\ldots dF_n}\right) dF_1(x_1)\ldots dF_n(x_n)$$

$$= \mathbb{E}\frac{\psi(c(F_1(X_1),\ldots,F_n(X_n)))}{c(F_1(X_1),\ldots,F_n(X_n))}$$

where ψ is a strictly convex function on $\mathbf{R}$ satisfying $\psi(1) = 0$ and $\frac{dF(x_1,\ldots,x_n)}{dF_1(x_1)\ldots dF_n(x_n)}$ is to be taken to be 0 if $x_1,\ldots,x_n$ are not points of increase of $F_1,\ldots,F_n$, respectively. Bivariate divergence measures were considered, e.g., by Ali and Silvey (1966) and Joe (1989). The multivariate Pearson's ϕ^2 corresponds to $\psi(x) = x^2 - 1$ and the relative entropy is obtained with $\psi(x) = x \log x$.

In the case of absolutely continuous r.v.'s $X_1,\ldots,X_n$ the measures $\delta_{X_1,\ldots,X_n}$ and $\phi^2_{X_1,\ldots,X_n}$ were introduced by Joe (1987, 1989). In the case of two r.v.'s X_1 and X_2 the measure $\phi^2_{X_1,X_2}$ was introduced by Pearson (1894) and was studied, among others, by Lancaster (1958). In the bivariate case, the measure δ_{X_1,X_2} is commonly known as Shannon or Kullback-Leibler mutual information between X_1 and X_2 (Kullback and Leibler (1948, 1951)).

A special case considered by Joe (1989) is $(X_1,\ldots,X_n)' \sim N(\mu,\Sigma)$. In this case,

$$\phi^2_{X_1,\ldots,X_n} = |R(2\mathbb{I}_n - R)|^{-1/2} - 1,$$

where $\mathbb{I}_n$ is the $n \times n$ identity matrix, provided that the correlation matrix R corresponding to Σ has the maximum eigenvalue of less than 2 and is infinite otherwise ($|A|$ denotes the determinant of a matrix A). In addition to that, if $\text{diag}(\Sigma) = (\sigma_1^2,\ldots,\sigma_n^2)$, then $\delta_{X_1,\ldots,X_n} = -.5\log(|\Sigma|/\prod_{i=1}^n \sigma_i^2)$. In the bivariate case with correlation coefficient ρ,

$$(\phi^2_{X_1,X_2}/(1 + \phi^2_{X_1,X_2}))^{1/2} = (1 - \exp(-2\delta_{X_1,X_2}))^{1/2} = |\rho|.$$

The measure $\Phi^2_{X_1,\ldots,X_n}$ is a multivariate extension of the bivariate measure of association initially proposed by Hoeffding (1940) (also see Fisher and Sen (1994)). The extension was proposed by Gaißer *et al.* (2010).

The other measures, τ, ρ and λ, are common in copula literature but are defined only for $n = 2$.

A class of measures of dependence closely related to the multivariate divergence measures ϕ^2, δ and Φ^2 is the class of generalized entropies introduced by Tsallis (1988) in the study of multifractals and generalizations of Boltzmann-Gibbs statistics (also see Golan (2002); Golan and Perloff (2002); Fernandes and Flôres (2002)). It is defined as follows:

Definition 3.11.

(1) Generalized entropy measure of dependence

$$\rho^{(q)}_{X_1,\ldots,X_n}$$

$$= (1/(1-q))\Bigg(1 - \int_{-\infty}^{\infty}\cdots\int_{-\infty}^{\infty}\left(\frac{dF(x_1,\ldots,x_n)}{dF_1\ldots dF_n}\right)^{1-q} dF_1(x_1)\ldots dF_n(x_n)\Bigg),$$

where $q > 0$, $q \neq 1$, is known as the entropic index. In the limiting case $q \to 1$, the dependence measure $\rho^{(q)}$ becomes the relative entropy $\delta_{X_1,\ldots,X_n}$ and in the case $q = 1/2$ it becomes the scaled squared Hellinger distance between dF and $dF_1\ldots dF_n$:

(2) Scaled squared Hellinger distance

$$\rho^{(1/2)}_{X_1,\ldots,X_n}$$

$$= 1/2\Bigg(1 - \int_{-\infty}^{\infty}\cdots\int_{-\infty}^{\infty}\left(\frac{dF(x_1,\ldots,x_n)}{dF_1\ldots dF_n}\right)^{1/2} dF_1(x_1)\ldots dF_n(x_n)\Bigg)$$

This is equal to $2H^2_{X_1,\ldots,X_n}$, where $H_{X_1,\ldots,X_n}$ stands for the Hellinger distance.

The generalized entropy has the form of the multivariate divergence measures $D^\psi_{X_1,\ldots,X_n}$ with $\psi(x) = (1/(1-q))(1-x^{1-q})$.

In information theory terminology (see, e.g., Akaike (1973)), the various multivariate dependence measures in Definition 3.10 represent the mean amount of information available for discriminating between a density f of dependent r.v.'s and the density of their independent copies, where as before independent copies are independent r.v.'s with the same marginals. Let the joint density of the independent copies be $f_0 = \prod_{k=1}^{n} f_k(x_k)$. Then, the mean amount of information can be written as $I(f_0, f; \zeta) = \int \zeta(f(x)/f_0(x))f(x)dx$, where ζ is a properly chosen function. For example, $\phi^2 = I(f_0, f; \zeta_1)$, where $\zeta_1(x) = x$; $\delta = I(f_0, f; \zeta_2)$, where $\zeta_2(x) = \log(x)$; and $D^\psi = I(f_0, f, \zeta_3)$, where $\zeta_3(x) = \psi(x)/x$.

3.4.3. *Sharp moment and probability inequalities*

This section derives sharp inequalities for dependent r.v.'s in terms of their dependence characteristics, including the dependence measures discussed in the previous section. These are the so called complete decoupling inequalities — the interested reader may want to consult de la Peña (1990); de la Peña and Giné (1999) and de la Peña and Lai (2001) for a discussion of decoupling inequalities and their applications.

Let $\xi_1, \ldots, \xi_n$ denote independent r.v.'s and let $g_{i_1,\ldots,i_c}$ $(x_{i_1}, \ldots, x_{i_c})$ be functions defined in Theorems 3.1 and 3.2, then we can use Theorem 3.7 to derive the following expressions for $\delta_{X_1,\ldots,X_n}$, $\phi^2_{X_1,\ldots,X_n}$, $D^\psi_{X_1,\ldots,X_n}$, $\rho^{(q)}_{X_1,\ldots,X_n}$, including $2H^2_{X_1,\ldots,X_n}$ for $q = 1/2$, and $I(f_0, f; \zeta)$:

$$
\begin{aligned}
\delta_{X_1,\ldots,X_n} &= \mathbb{E}\log\left(1 + U_n(X_1, \ldots, X_n)\right) \\
&= \mathbb{E}\left(1 + U_n(\xi_1, \ldots, \xi_n)\right)\log(1 + U_n(\xi_1, \ldots, \xi_n)),
\end{aligned}
$$

$$(3.44)$$

$$
\begin{aligned}
\phi^2_{X_1,\ldots,X_n} &= \mathbb{E}\left(1 + U_n(\xi_1, \ldots, \xi_n)\right)^2 - 1 \\
&= \mathbb{E}U_n^2(\xi_1, \ldots, \xi_n) \\
&= \mathbb{E}U_n(X_1, \ldots, X_n),
\end{aligned}
$$

$$(3.45)$$

$$
D^\psi_{X_1,\ldots,X_n} = \mathbb{E}\psi\left(1 + U_n(\xi_1, \ldots, \xi_n)\right),
$$

$$(3.46)$$

$$
\rho^{(q)}_{X_1,\ldots,X_n} = (1/(1-q))(1 - \mathbb{E}(1 + U_n(\xi_1, \ldots, \xi_n))^q),
$$

$$(3.47)$$

$$2H^2_{X_1,\dots,X_n} = 1/2(1 - \mathbb{E}(1 + U_n(\xi_1,\dots,\xi_n))^{1/2}), \qquad (3.48)$$

$$I(f_0, f; \zeta) = \mathbb{E}\zeta(1 + U_n(\xi_1,\dots,\xi_n))(1 + U_n(\xi_1,\dots,\xi_n)), \qquad (3.49)$$

where

$$U_n(x_1,\dots,x_n) = \sum_{c=2}^{n} \sum_{1 \le i_1 < \dots < i_c \le n} g_{i_1,\dots,i_c}(x_{i_1},\dots,x_{i_c})$$

is the sum of U-statistics corresponding to r.v.'s $X_1,\dots,X_n$.

It follows from display (3.45) that $\phi^2_{X_1,\dots,X_n}$ can be expanded in terms of the "canonical" functions g as follows:

$$\phi^2_{X_1,\dots,X_n} = \sum_{c=2}^{n} \sum_{1 \le i_1 < \dots < i_c \le n} \mathbb{E}g^2_{i_1,\dots,i_c}(\xi_{i_1},\dots,\xi_{i_c}).$$

In particular, if $X_1,\dots,X_n$ have the generalized multivariate EFGM copula (3.9) then $\phi^2_{X_1,\dots,X_n}$ can be written as follows:

$$\phi^2_{X_1,\dots,X_n} = \sum_{c=2}^{n} \sum_{1 \le i_1 < \dots < i_c \le n} \alpha^2_{i_1,\dots,i_c}.$$

It is well known that the mutual information between two r.v.'s X_1 and X_2 is nonnegative (see, e.g., Cover and Thomas (1991), p. 27). The multivariate analog of this property for $\delta_{X_1,\dots,X_n}$ follows from the results obtained by Joe (1989). It is interesting to note that nonnegativity of $\delta_{X_1,\dots,X_n}$ can be easily obtained from (3.44): since the function $(1+x)\log(1+x)$ is convex in $x \ge 0$, then, by Jensen's inequality, $\delta_{X_1,\dots,X_n} \ge (1 + \mathbb{E}U_n(\xi_1,\dots,\xi_n))\log(1 + \mathbb{E}U_n(\xi_1,\dots,\xi_n)) = 0$.

The following theorem gives an inequality between $\delta_{X_1,\dots,X_n}$ and $\phi^2_{X_1,\dots,X_n}$ that generalizes and improves the results obtained, in the bivariate case, by Dragomir (2000) (see also Mond and Pečarić (2001)).

Theorem 3.23. *The following inequalities hold:*

$$\delta_{X_1,\dots,X_n} \le \log(1 + \phi^2_{X_1,\dots,X_n}) \le \phi^2_{X_1,\dots,X_n}.$$

The characterization results obtained in previous sections, in particular Theorem 3.7, allow us to simplify the analysis of convergence in distribution of dependent variables to the study of convergence in distribution of independent counterparts plus convergence of some dependence measure.

Let Y be some r.v. and let ψ be a convex function increasing on $[1, \infty)$ and decreasing on $(-\infty, 1)$ with $\psi(1) = 0$.

Theorem 3.24. *For a double array* $\{X_i^n\}, i = 1, \ldots, n, \ n = 0, 1, \ldots, \infty$ *let functionals* $\phi_{n,n}^2 = \phi_{X_1^n, X_2^n, \ldots, X_n^n}^2$, $\delta_{n,n} = \delta_{X_1^n, X_2^n, \ldots, X_n^n}$, $D_{n,n}^\psi = D_{X_1^n, X_2^n, \ldots, X_n^n}^\psi$, $\rho_{n,n}^{(q)} = \rho_{X_1^n, X_2^n, \ldots, X_n^n}^{(q)}$, $q \in (0,1)$, $H_{n,n} = (1/2\rho_{n,n}^{(1/2)})^{1/2}$, $n = 0, 1, 2, \ldots$ *denote the corresponding dependence measures. Then, if*

$$\sum_{i=1}^{n} \xi_i^n \xrightarrow{d} Y$$

as $n \to \infty$, *and either* $\phi_{n,n}^2 \to 0$, $\delta_{n,n} \to 0$, $D_{n,n}^\psi \to 0$, $\rho_{n,n}^{(q)} \to 0$ *or* $H_{n,n} \to 0$ *as* $n \to \infty$, *then*

$$\sum_{i=1}^{n} X_i^n \xrightarrow{d} Y$$

as $n \to \infty$.

In the following theorem, we apply this idea to reduce the analysis of convergence of multidimensional statistics of a time series to the analysis of convergence of certain measures of intertemporal dependence, such as ϕ^2, δ, D^ψ and $I(f_0, f; \zeta)$.

Let $h : \mathbf{R}^m \to \mathbf{R}$ be an arbitrary function of m arguments; we can think of h as an m-dimensional statistic based on a time series.

Theorem 3.25. *For a time series* $\{X_t\}_{t=0}^{\infty}$ *define the functionals*

$$\phi_t^2 = \phi_{X_t, X_{t+1}, \ldots, X_{t+m-1}}^2,$$

$$\delta_t = \delta_{X_t, X_{t+1}, \ldots, X_{t+m-1}},$$

$$D_t^\psi = D_{X_t, X_{t+1}, \ldots, X_{t+m-1}}^\psi,$$

$$\rho_t^{(q)} = \rho_{X_t, X_{t+1}, \ldots, X_{t+m-1}}^{(q)}, \quad q \in (0,1),$$

$$H_t = (1/2\rho_t^{(1/2)})^{1/2}, \quad t = 0, 1, 2, \ldots$$

Then, if

$$h(\xi_t, \xi_{t+1}, \ldots, \xi_{t+m-1}) \xrightarrow{d} Y$$

as $t \to \infty$, *and either* $\phi_t^2 \to 0$, $\delta_t \to 0$, $D_t^\psi \to 0$, $\rho_t^{(q)} \to 0$ *or* $H_t \to 0$
as $t \to \infty$, *then*

$$h(X_t, X_{t+1}, \ldots, X_{t+m-1}) \xrightarrow{d} Y$$

as $t \to \infty$.

Examples can illustrate the point. Suppose $\{X_t\}_{t=0}^\infty$ are Gaussian processes with $(X_t, X_{t+1}, \ldots, X_{t+m-1}) \sim N(\mu_{t,m}, \Sigma_{t,m})$. Then, the conditions of Theorem 3.25 apply if $|R_{t,m}(2I_m - R_{t,m})| \to 1$ or $|\Sigma_{t,m}|/\sum_{i=0}^{m-1} \sigma_{t+i}^2 \to 1$, as $t \to \infty$, where $R_{t,m}$ denote correlation matrices corresponding to $\Sigma_{t,m}$ and $(\sigma_t^2, \ldots, \sigma_{t+m-1}^2) = \text{diag}(\Sigma_{t,m})$. As another example, let $\{X_t\}_{t=1}^\infty$ have the generalized EFGM copulas (3.9). Then, the conditions of the theorem are satisfied if $\phi_t^2 = \sum_{c=2}^m \sum_{i_1 < \ldots < i_c \in \{t,t+1,\ldots,t+m-1\}} \alpha_{i_1,\ldots,i_c}^2 \to 0$ as $t \to \infty$. As indicated before, according to Sharakhmetov and Ibragimov (2002), the class of time series with generalized EFGM copulas includes all time series for which the underlying r.v. is two-valued.

It is important to emphasize here that since Theorem 3.24 holds for Tsallis entropy and multivariate divergence measures, this allows us to study convergence of statistics of time series in the case when only lower moments of the above-mentioned U-statistics underlying the dependence generating process for the time series exist. Therefore, they provide a unifying approach to studying convergence in "heavy-tailed" and "standard" situations, where "standard" means convergence of Pearson's coefficient and of mutual information and entropy, which correspond, respectively, to the cases of second moments of the U-statistics and the first moments multiplied by the logarithm.

The following theorem provides an estimate for the distance between the distribution function of an arbitrary statistic in dependent r.v.'s from the distribution function of the statistic in independent copies of the r.v.'s. The inequality complements (and can be better than) the well-known Pinsker inequality for total variation between the densities of dependent and independent r.v.'s in terms of the relative entropy (see, e.g., Miller and Liu (2002)).

Theorem 3.26. *The following inequality holds for an arbitrary statistic* $h(X_1, \ldots, X_n)$:

$$|\mathbb{P}(h(X_1, \ldots, X_n) \leq x) - \mathbb{P}(h(\xi_1, \ldots, \xi_n) \leq x)|$$
$$\leq \phi_{X_1, \ldots, X_n} \max[(\mathbb{P}(h(\xi_1, \ldots, \xi_n) \leq x))^{1/2},$$
$$\times (\mathbb{P}(h(\xi_1, \ldots, \xi_n) > x))^{1/2}], \; x \in \mathbf{R}. \tag{3.50}$$

The following theorems allow one to reduce the problems of evaluating expectations of general statistics in dependent r.v.'s $X_1, \ldots, X_n$ to the case of independence. The theorems contain complete decoupling results for statistics in dependent r.v.'s using the relative entropy and the multivariate Pearson's ϕ^2 coefficient. The results provide generalizations of earlier known results on complete decoupling of r.v.'s from particular dependence classes, such as martingales and adapted sequences of r.v.'s to the case of arbitrary dependence.

Theorem 3.27. *If* $f : \mathbf{R}^n \to \mathbf{R}$ *is a nonnegative function, then the following sharp inequalities hold for* $x \in \mathbf{R}$:

$$\mathbb{E}f(X_1, \ldots, X_n) \leq \mathbb{E}f(\xi_1, \ldots, \xi_n) + \phi_{X_1, \ldots, X_n}(\mathbb{E}f^2(\xi_1, \ldots, \xi_n))^{1/2},$$
$$\tag{3.51}$$
$$\mathbb{E}f(X_1, \ldots, X_n) \leq (1 + \phi^2_{X_1, \ldots, X_n})^{1/q}(\mathbb{E}f^q(\xi_1, \ldots, \xi_n))^{1/q}, \; q \geq 2,$$
$$\tag{3.52}$$
$$\mathbb{E}f(X_1, \ldots, X_n) \leq \mathbb{E}\exp(f(\xi_1, \ldots, \xi_n)) - 1 + \delta_{X_1, \ldots, X_n}, \tag{3.53}$$
$$\mathbb{E}f(X_1, \ldots, X_n) \leq \left(1 + D^{\psi}_{X_1, \ldots, X_n}\right)^{(q-1)/q}(\mathbb{E}f^q(\xi_1, \ldots, \xi_n))^{1/q},$$
$$\tag{3.54}$$

$q > 1$, *where* $\psi(x) = |x|^{q/(q-1)} - 1$.

It is interesting to note that from relation (3.45) and inequality (3.51), it follows that the following representation holds for the multivariate Pearson coefficient $\phi_{X_1,\ldots,X_n}$:

$$\phi_{X_1,\ldots,X_n} =$$

$$\max_{f:\mathbb{E}f(\xi_1,\ldots,\xi_n)=0,\mathbb{E}f^2(\xi_1,\ldots,\xi_n)<\infty} \frac{(\mathbb{E}f(X_1,\ldots,X_n) - \mathbb{E}f(\xi_1,\ldots,\xi_n))}{(\mathbb{E}f^2(\xi_1,\ldots,\xi_n))^{1/2}}.$$

$$(3.55)$$

The following result gives complete decoupling inequalities for the tail probabilities of arbitrary statistics in dependent r.v.'s.

Theorem 3.28. *The following inequalities hold:*

$$\mathbb{P}(h(X_1,\ldots,X_n) > x) \leq \mathbb{P}(h(\xi_1,\ldots,\xi_n) > x)$$

$$+ \phi_{X_1,\ldots,X_n}(\mathbb{P}(h(\xi_1,\ldots,\xi_n) > x))^{1/2},$$

$$\mathbb{P}(h(X_1,\ldots,X_n) > x) \leq (1 + \phi^2_{X_1,\ldots,X_n})^{1/2}(\mathbb{P}(h(\xi_1,\ldots,\xi_n) > x))^{1/2},$$

$$\mathbb{P}(h(X_1,\ldots,X_n) > x) \leq (e - 1)\mathbb{P}(h(\xi_1,\ldots,\xi_n) > x) + \delta_{X_1,\ldots,X_n},$$

$$\mathbb{P}(h(X_1,\ldots,X_n) > x) \leq (1 + D^{\psi}_{X_1,\ldots,X_n})^{(q-1)/q}$$

$$\times (\mathbb{P}(h(\xi_1,\ldots,\xi_n) > x))^{1/q},$$

$q > 1$, $x \in \mathbf{R}$, *where* $\psi(x) = |x|^{q/(q-1)} - 1$.

3.4.4. *Vector version of Hoeffding's* Φ^2

In this section, we consider a measure that captures dependence between vectors rather than scalars.

Mutual independence between n scalar components of $\mathbf{X}$ is characterized by the independence, or product, copula (1.2), which can be written as $C^{\perp}(\mathbf{u}) = \prod_{j=1}^n u_j$, $\mathbf{u} \in [0,1]^n$. It is therefore natural to develop measures of association based on the distance $C(\mathbf{u}) - C^{\perp}(\mathbf{u})$ similar to the multivariate version of Hoeffding's Φ^2 defined in Definition 3.10.

Gaißer *et al.* (2010) write the measure as

$$\Phi^2 = \frac{||C - C^{\perp}||_2^2}{||M - C^{\perp}||_2^2},$$

$$(3.56)$$

where $||\cdot||_2$ denotes the L_2-norm, and $||M - C^{\perp}||_2^2$ is the normalization factor where the function $M(\mathbf{u}) = \min(u_1,\ldots,u_n)$, $\mathbf{u} \in [0,1]^n$, is the so-called *comonotone copula* representing an almost-sure strictly-increasing functional relationship between all components of $\mathbf{X}$. Here, M is the upper Frechet–Hoeffding bound, that is, for any valid copula C, $C(\mathbf{u}) \leq M(\mathbf{u})$, $\forall \mathbf{u} \in [0,1]^n$. It is not difficult to see that this definition is equivalent to that given in Definition 3.10 if we let $\mathbb{K} = \|M - C^{\perp}\|_2^2$.

It is well known that Φ^2 is constrained to the $[0,1]$ interval where the case $\Phi^2 = 1$ occurs when there is co-monotonicity between all components of $\mathbf{X}$, while $\Phi^2 = 0$ corresponds to mutual independence. However, Φ^2 is not well-suited for the measurement of dependence between random *vectors* since even under mutual independence *between* vectors, Φ^2 may be non-zero due to dependence *within* them. Medovikov and Prokhorov (2016) propose a measure of multivariate association that is invariant to dependence *within* vectors.

Let $\mathbf{X}$ be partitioned into p sub-vectors $(\mathbf{X}_1, \ldots, \mathbf{X}_p)$ of dimension $n_1, \ldots, n_p$, respectively. Define $b_0 = 0$ and, for any integer $k \in \{1, \ldots, p\}$, set $b_k = \sum_{\ell=1}^{k} n_\ell$. Moreover, for $k = 1, \ldots, n$ and $\mathbf{u} \in [0,1]^n$, let

$$\mathbf{u}^{\{k\}} = (u_{b_{k-1}+1}, \ldots, u_{b_k}) \in \mathbb{R}^{n_k},$$
$$\mathbf{u}^{[k]} = (\mathbf{1}_{n_1+\cdots+n_{k-1}}, \mathbf{u}^{\{k\}}, \mathbf{1}_{n_{k+1}+\cdots+n_p}) \in \mathbb{R}^n.$$

Here, for $m \in \mathbb{N}$, $\mathbf{1}_m$ denotes an m-dimensional vector of ones. For $k = 1, \ldots, p$, let

$$C^{\{k\}} : [0,1]^{n_k} \to [0,1], \quad C^{\{k\}}(\mathbf{v}) = C(\mathbf{1}_{n_1+\cdots+n_{k-1}}, \mathbf{v}, \mathbf{1}_{n_{k+1}+\cdots+n_p})$$

denote the copula of $\mathbf{X}_k$. Also, define a copula $C^{\Pi} : [0,1]^n \to [0,1]$ through

$$C^{\Pi}(\mathbf{u}) = \prod_{k=1}^{p} C^{\{k\}}(\mathbf{u}^{\{k\}}) = \prod_{k=1}^{p} C(\mathbf{u}^{[k]}).$$

This copula serves as a vector analogue of the scalar independence copula $C^{\perp}$.

First we note that $\mathbf{X}_1, \ldots, \mathbf{X}_p$ are mutually independent if and only if $C = C^{\Pi}$. In this case, the joint distribution of p vectors is a product of p multivariate marginals, corresponding to the vector dimensions. This observation suggests that we can define a measure of dependence between the vectors $\mathbf{X}_1, \ldots, \mathbf{X}_p$ by considering the following distance between C and C^Π.

Definition 3.12. A vector version of Hoeffding Φ^2 is defined as follows

$$\bar{\Phi}^2 := \bar{\Phi}^2(C, n_1, \ldots, n_p) := \frac{\|C - C^\Pi\|_2^2}{\|M - C^\Pi\|_2^2},$$

where as before $\|\cdot\|_2$ denotes the L_2-norm.

Now, $C^{\{k\}}$, $k \in \{1, \ldots, p\}$ is used in the distance and in the normalizing factor. The scalar version is obtained as a special case when $n_j = 1, j = 1, \ldots, p$.

Clearly, other bounds may be used instead of M. There may exist a vector equivalent of scalar Frecher-Hoeffding bound. For example, we may consider using $\min(C^{\{1\}}, \ldots, C^{\{p\}})$ instead. However, little is known about the properties of this object, specifically we cannot guarantee that $C \leq \min(C^{\{1\}}, \ldots, C^{\{p\}})$ for any copula C. What is known is that this would be the correct Frechet-Hoeffding bound for some copula $C_c(C^{\{1\}}, \ldots, C^{\{p\}})$ but, unless all margins are scalar distributions, the resulting distribution is a n-variate copula only if $C_c(u_1, \ldots, u_p) = C_\perp(u_1, \ldots, u_p)$ (see, e.g., Quesada-Molina and Rodriguez-Lallena (1994); Genest *et al.* (1995b)). So this would not be the right bound to use.

$\bar{\Phi}^2$ is bounded to the interval $[0, 1]$ and $\bar{\Phi}^2 = 1$ when $C = M$ and $\bar{\Phi}^2 = 0$ when $C = C^\Pi$. Given partition sizes $n_1, \ldots, n_p$, for every permutation π of partition order $\{1, \ldots, p\}$, we have that $\bar{\Phi}^2(C, n_1, \ldots, n_p) = \bar{\Phi}^2(C, n_{\pi(1)}, \ldots, n_{\pi(p)})$, as long as the composition of each partition is maintained. Given partition sizes $n_1, \ldots, n_p$ and partition composition, $\bar{\Phi}^2$ is invariant with respect to permutations of components within each partition and with respect to strictly-increasing transformations of one or many components of $\mathbf{X}$.

For a sample $\mathbf{X}^{(1)}, \ldots, \mathbf{X}^{(N)}$, $\mathbf{X}^{(i)} = (X_1^{(i)}, \ldots, X_n^{(i)}) \sim F = C(F_1, \ldots, F_n)$ and continuous marginal cdf's $F_1, \ldots, F_n$, let $\hat{\mathbf{U}}^{(i)} = (\hat{U}_1^{(i)}, \ldots, \hat{U}_n^{(i)})$ denote *pseudo-observations* from the copula C

defined through

$$\hat{U}_j^{(i)} = \frac{1}{N}\left(\text{rank of } X_j^{(i)} \text{ among } X_j^{(1)},\ldots,X_j^{(N)}\right),$$

for $i = 1,\ldots,N$ and $j = 1,\ldots,n$. Let $\hat{C}_N : [0,1]^n \to [0,1]$ denote the associated empirical copula, defined for $\mathbf{u} = (u_1,\ldots,u_n) \in \mathbb{R}^n$ as

$$\hat{C}_N(\mathbf{u}) = \frac{1}{N}\sum_{i=1}^{N}\mathbb{I}(\hat{\mathbf{U}}^{(i)} \leq \mathbf{u}) = \frac{1}{N}\sum_{i=1}^{N}\prod_{j=1}^{n}\mathbb{I}(\hat{U}_j^{(i)} \leq u_j),$$

where $\mathbb{I}(\cdot)$ is the indicator function. The empirical copula easily allows to define a sample version of $\bar{\Phi}^2$ through

$$\hat{\bar{\Phi}}_N^2 := \bar{\Phi}^2(\hat{C}_N, n_1,\ldots,n_p) = \frac{\|\hat{C}_N - \hat{C}_N^{\Pi}\|_2^2}{\|M - \hat{C}_N^{\Pi}\|_2^2},$$

where $\hat{C}_N^{\Pi}(\mathbf{u}) = \prod_{k=1}^{p}\hat{C}_N(\mathbf{u}^{[k]})$.

Medovikov and Prokhorov (2016) use results obtained by Kojadi-novic and Holmes (2009) to show that the estimator $\hat{\bar{\Phi}}_N^2$ can be calculated directly from the pseudo-observations.

The following theorem provides a weak convergence result regarding $\hat{\bar{\Phi}}_N^2$ using functional weak convergence of the empirical copula process $\mathbb{C}_N = \sqrt{N}(\hat{C}_N - C)$ and of the process $\mathbb{C}_N^{\Pi}$, defined on $\mathbf{u} \in [0,1]^n$ through

$$\mathbb{C}_N^{\Pi}(\mathbf{u}) = \sqrt{N}\{\hat{C}_N^{\Pi}(\mathbf{u}) - C^{\Pi}(\mathbf{u})\}.$$

Naturally, both $\mathbb{C}_N$ and $\mathbb{C}_N^{\Pi}$ can be considered as elements of the space of real-valued, bounded functions on $[0,1]^n$, denoted by $\ell^{\infty}([0,1]^n)$, equipped with the uniform metric induced by the sup-norm $\|f\|_{\infty} = \sup_{\mathbf{u}\in[0,1]^n}|f(\mathbf{u})|$. Weak convergence of the empirical copula process $\mathbb{C}_N$ has been investigated by various authors under slightly different assumptions (see, e.g., Rüschendorf (1976); Gaenssler and Stute (1987); Fermanian *et al.* (2004); Segers (2012); Bücher and Volgushev (2013)). Under appropriate smoothness conditions and under $C = C^{\Pi}$, Kojadinovic and Holmes (2009, Theorem 3) give a weak convergence result for $\mathbb{H}_N = \mathbb{C}_N - \mathbb{C}_N^{\Pi}$. The following theorem can be regarded as a generalization of their result to multivariate marginals.

Theorem 3.29. *Let C be a copula such that, for any $j = 1, \ldots, n$, the jth first order partial derivative $\dot{C}_j = \partial C / \partial u_j$ exists and is continuous on the set $\{\mathbf{u} \in [0,1]^n : u_j \in (0,1)\}$. Let $\mathbb{B}_C$ denote a C-Brownian bridge on $[0,1]^n$, i.e., a centered Gaussian process with continuous sample paths and covariance*

$$\operatorname{cov}\{\mathbb{B}_C(\mathbf{u}), \mathbb{B}_C(\mathbf{v})\} = C(\mathbf{u} \wedge \mathbf{v}) - C(\mathbf{u})C(\mathbf{v}).$$

Then, $(\mathbb{C}_N, \mathbb{C}_N^\Pi) \xrightarrow{d} (\mathbb{C}_C, \mathbb{C}_C^\Pi)$ in $\{\ell^\infty([0,1]^n)\}^2$, where

$$\mathbb{C}_C(\mathbf{u}) = \mathbb{B}_C(\mathbf{u}) - \sum_{j=1}^n \dot{C}_j(\mathbf{u})\mathbb{B}_C(1, \ldots, 1, u_j, 1, \ldots, 1),$$

with $\dot{C}_j$ defined as 0 wherever it does not exist, and with

$$\mathbb{C}_C^\Pi(\mathbf{u}) = \sum_{k=1}^p \mathbb{C}_C(\mathbf{u}^{[k]}) \prod_{\substack{k'=1 \\ k' \neq k}}^p C(\mathbf{u}^{[k']}).$$

Theorem 3.29 also holds under many serial dependence scenarios for time series (e.g., under α-mixing), see Bücher and Volgushev (2013). More precisely, provided the weak limit $\tilde{\mathbb{B}}_C$ of the empirical process

$$\mathbf{u} \mapsto \sqrt{N}\left\{\frac{1}{N}\sum_{i=1}^N \mathbb{I}(\mathbf{U}^{(i)} \leq \mathbf{u}) - C(\mathbf{u})\right\}$$

exists and has continuous sample paths, the statement of Theorem 3.29 holds with the C-Brownian bridge $\mathbb{B}_C$ replaced by $\tilde{\mathbb{B}}_C$.

Theorem 3.30. *Under the assumptions of Theorem 3.29,*

(a) *if $\bar{\Phi}^2 \neq 0$, then*

$$\sqrt{N}(\hat{\bar{\Phi}}_N^2 - \bar{\Phi}^2) \xrightarrow{d} 2\frac{\int_{[0,1]^n}\{C(\mathbf{u}) - C^\Pi(\mathbf{u})\}\{\mathbb{C}_C(\mathbf{u}) - \mathbb{C}_C^\Pi(\mathbf{u})\}\, d\mathbf{u}}{\|M - C^\Pi\|_2^2}$$

$$+ 2\frac{\|C - C^\Pi\|_2^2 \int \mathbb{C}_C^\Pi(\mathbf{u})\{M(\mathbf{u}) - C^\Pi(\mathbf{u})\}\, d\mathbf{u}}{\|M - C^\Pi\|_2^4}.$$

(b) *if $\bar{\Phi}^2 = 0$, then*

$$N\hat{\bar{\Phi}}_N^2 \xrightarrow{d} \frac{\|\mathbb{C}_C - \mathbb{C}_C^\Pi\|_2^2}{\|M - C^\Pi\|_2^2}.$$

Note that in the case $\bar{\Phi}^2 \neq 0$ the limiting distribution of $\sqrt{N}(\hat{\bar{\Phi}}_N^2 - \bar{\Phi}^2)$ is Gaussian. This follows from the fact that $\mathbb{C}_C$ is a centered Gaussian process. The limiting variance depends in a complicated way on C and its partial derivatives. Moreover, in practice, the true copula is unknown; and even if it was known, analytical calculations, even approximate, of these quantities can be intractable. Bootstrap distributions will be used instead.

3.5. Bounds on options

It turns out that the results in Section 3.2 can be used in derivation of options bounds. This section discusses several such results obtained by de la Peña *et al.* (2004).

Section 3.5.1 presents the main results on semiparametric bounds for the expected payoffs and prices of contingent claims in the multi-period trinomial model with dependent log-returns. These results allow one to reduce the analysis to the i.i.d. multiperiod case of the binomial model or the two-period case of the derivative pricing model with log-normal returns. Section 3.5.2 shows how the approach of Section 3.5.1 can also be used to obtain bounds for path-dependent derivatives and, as an illustration, provides estimates for Asian options in the trinomial option pricing model.

3.5.1. *Bounds on European options*

Let $\{u_t\}_{t=1}^{\infty}$ be a sequence of nonnegative numbers and let $\Im_0 = (\Omega, \emptyset) \subseteq \Im_1 \subseteq \ldots \Im_t \subseteq \ldots \subseteq \Im$ be an increasing sequence of σ-algebras on a probability space $(\Omega, \Im, P)$. We consider a market consisting of two assets. The first asset is a risky asset with the trinomial price process

$$S_0 = s, \quad S_t = S_{t-1}X_t, \quad t \geq 1, \tag{3.57}$$

where the $(X_t)_{t=1}^{\infty}$ is an $(\Im_t)$-adapted sequence of nonnegative r.v.'s representing the asset's gross returns. The second asset is a money-market account with a risk-free rate of return r.

First, we assume that random log-returns $\log(X_t)$ form an $(\Im_t)$-martingale-difference sequence and take on three values u_t, $-u_t$

and 0:

$$\mathbb{P}(\log(X_t) = u_t) = \mathbb{P}(\log(X_t) = -u_t) = p_t, \qquad (3.58)$$

$$\mathbb{P}(\log(X_t) = 0) = 1 - 2p_t,$$

$0 \le p_t \le 1/2$, $t = 1, 2, \ldots$ (so that, in period t, the price of the asset increases to $S_t = \exp(u_t)S_{t-1}$ with probability p_t, decreases to $S_t = \exp(-u_t)S_{t-1}$ with the same probability or stays the same: $S_t = S_{t-1}$ with probability $1 - 2p_t$). As usual, in what follows, we denote by $\mathbb{E}_t$, $t \ge 0$, the conditional expectation operator $\mathbb{E}_t = \mathbb{E}(\cdot|\Im_t)$.

For $t \ge 0$, let $\tilde{S}_\tau$, $\tau \ge t$, be the price process with $\tilde{S}_\tau = S_t$ and $\tilde{S}_\tau = \tilde{S}_{\tau-1}\exp(u_\tau\epsilon_\tau)$, $\tau > t$, where, conditionally on $\Im_t$, $(\epsilon_\tau)_{\tau=t+1}^{\infty}$ is a sequence of i.i.d. symmetric Bernoulli r.v.'s:

$$\mathbb{P}(\epsilon_\tau = 1|\Im_t) = \mathbb{P}(\epsilon_\tau = -1|\Im_t) = 1/2 \quad \text{for } \tau > t.$$

The following theorem provides bounds for the time-t expected payoff of contingent claims in the trinomial model with an arbitrary increasing convex payoff functions $\phi : \mathbf{R}_+ \to \mathbf{R}$. These estimates reduce the problem of derivative pricing in the multiperiod trinomial financial market model to the case of the multiperiod binomial model with i.i.d. returns.

Theorem 3.31. *For any increasing convex function $\phi : \mathbf{R}_+ \to \mathbf{R}$, the following bound holds:* $\mathbb{E}_t\phi(S_T) \le \mathbb{E}_t\phi(\tilde{S}_T)$ *for all* $0 \le t < T$.

The choice of the function $\phi(x) = \max(x - K, 0)$, $x \ge 0$, in Theorem 3.31 immediately provides estimates for the time-t expected payoffs of a European call option with strike price $K \ge 0$ on the asset expiring at time T. Furthermore, using the results of Eaton (1974), we also obtain bounds for the expected payoff of European call options in the trinomial model in terms of the expected payoff of power options with the payoff function $\phi(x) = [\max(x - K, 0)]^3$ written on an asset with log-normally distributed price. These bounds are similar in spirit to the estimates for linear combinations of i.i.d. symmetric Bernoulli r.v.'s and t-statistics under symmetry obtained by Eaton (1970, 1974) (see also Edelman (1986, 1990)) and to the estimates in the binomial model in terms of Poisson r.v.'s obtained by de la Peña *et al.* (2004). These bounds essentially reduce the problem of option pricing in the trinomial multiperiod model to the problem of pricing in the case of two-periods and the standard assumption of log-normal returns.

Theorem 3.32. *The following bounds hold:*

$$\mathbb{E}_t \max(S_T - K, 0) \leq \mathbb{E}_t \max(\tilde{S}_T - K, 0)$$

$$\leq \left\{ \mathbb{E}_t \left[\max \left(S_t e^{Z \sqrt{\sum_{k=t+1}^{T} u_k^2}} - K, 0 \right) \right]^3 \right\}^{1/3},$$

$$(3.59)$$

where, conditionally on $\mathfrak{F}_t$, Z has the standard normal distribution.

Analogues of the bounds in Theorems 3.31 and 3.32 also hold for price processes with asymmetric trivariate distributions of the log-returns. For instance, evidently, the bounds continue to hold in the case of the log-returns X_t with the trivariate distributions $\mathbb{P}(\log(X_t) = u_t) = p_t$, $\mathbb{P}(\log(X_t) = -d_t) = q_t$, $\mathbb{P}(\log(X_t) = 0) = 1 - p_t - q_t$, where $0 < u_t \leq d_t$, $0 \leq p_t \leq q_t \leq 1/2$, $t = 1, 2, \ldots$

In addition, further generalizations of the bounds in the trinomial model to the asymmetric case may be obtained using symmetriza-tion inequalities for (generalized) moments of sums of r.v.'s. Bounds similar to those given by Theorems 3.31 and 3.32 hold as well for the expected stop-loss for a sum of three-value risks that form a martingale-difference sequence.

Theorem 3.33. *Suppose that the r.v.'s $\{X_t\}_{t=1}^{\infty}$, form an $(\mathfrak{F}_t)-$martingale-difference sequence and have distributions*

$$\mathbb{P}(X_t = u_t) = \mathbb{P}(X_t = -u_t) = p_t, \; \mathbb{P}(X_t = 0) = 1 - 2p_t, \quad (3.60)$$

$0 \leq p_t \leq 1/2$, $t = 1, 2, \ldots$ Further, let $\{\epsilon_t\}_{t=1}^{\infty}$ be a sequence of i.i.d. symmetric Bernoulli r.v.'s: $\mathbb{P}(\epsilon_t = 1) = \mathbb{P}(\epsilon_t = -1) = 1/2$, $t = 1, 2, \ldots$, and let Z denote a standard normal r.v. Then the following bound holds for the expectation of any convex function $\phi : \mathbf{R} \to \mathbf{R}$ of

the sum of the risks $\sum_{t=1}^{T} X_t$:

$$\mathbb{E}\phi\left(\sum_{t=1}^{T} X_t\right) \leq \mathbb{E}\phi\left(\sum_{t=1}^{T} u_t \epsilon_t\right).$$

In particular, the following bounds hold for the expected stop-loss $\mathbb{E}\max(\sum_{t=1}^{T} X_t - K, 0)$, $K \geq 0$:

$$\mathbb{E}\max\left(\sum_{t=1}^{T} X_t - K, 0\right) \leq \mathbb{E}\max\left(\sum_{t=1}^{T} u_t \epsilon_t - K, 0\right)$$

$$\leq \left\{\mathbb{E}\left[\max\left(Z\sqrt{\sum_{k=t+1}^{T} u_k^2} - K, 0\right)\right]^3\right\}^{1/3}.$$

Let us now turn to the problem of making inferences on the European call option price in the trinomial model. As is well-known, the trinomial option pricing model is incomplete and allows for an infinite number of equivalent probability measures under which the discounted asset price process is a martingale. The different risk-neutral measures lead to different prices for contingent claims in the model, all of which are consistent with market prices of the underlying assets. Therefore, the standard no-arbitrage pricing approach breaks down.

The inequalities for the expected payoffs of contingent claims in Theorems 3.31 and 3.32 (under the *true* probability measure), together with estimates for the call return over its lifetime, on the other hand, provide bounds for possible prices of the option. Let R_S and R_C denote, respectively, the required gross returns (over the periods $t+1, \ldots, T$) on the underlying asset with the trinomial price process (3.57), (3.58) and on the European call option with the strike price K on the asset expiring at time $T > t$. The price of the call option at time t is given by $C_t = R_C \mathbb{E}_t \max(S_T - K, 0)$ (where the expectation is taken with respect to the true probability measure) and, similarly, the price of the asset satisfies $S_t = R_S \mathbb{E}_t S_T$.

As follows from Rodriguez (2003) (see also Theorem 8 in Merton (1973)), risk-averse investors require a higher rate of return on the call option than on its underlying asset and, therefore, $R_C < R_S$.[3] Combining this with the bounds given by Theorem 3.32, we immediately obtain estimates for the prices of European call options in the trinomial model. Similar bounds also hold for other contingent claims with convex payoff function; they can be obtained using Theorem 3.31 and estimates for the gross return on the contingent claims over their lifetimes. For instance, in the case of independent returns X_t, $R_S = S_t/\mathbb{E}_t S_T = (\prod_{k=t+1}^{T} \mathbb{E}_t X_k)^{-1} = \{\prod_{k=t+1}^{T}[1 + 2(cosh(u_k) - 1)p_k]\}^{-1}$, where $cosh(x) = (\exp(x) + \exp(-x))/2$, $x \in \mathbf{R}$, is the hyperbolic cosine. We thus have the following estimates for the call option prices.

Theorem 3.34. *In the case of the trinomial option pricing model with risk-averse investors and independent returns X_t with distribution* (3.58), *the time-t prices of the European call option on the asset satisfy the following bounds:*

$$C_t \leq \left\{ \prod_{k=t+1}^{T} [1 + 2(cosh(u_k) - 1)p_k] \right\}^{-1} \mathbb{E}_t \max(\tilde{S}_T - K, 0)$$

$$\leq \left\{ \prod_{k=t+1}^{T} [1 + 2(cosh(u_k) - 1)p_k] \right\}^{-1}$$

$$\times \left\{ \mathbb{E}_t \left[\max \left(S_t e^{Z\sqrt{\sum_{k=t+1}^{T} u_k^2}} - K, 0 \right) \right]^3 \right\}^{1/3}, \qquad (3.61)$$

where, conditionally on $\Im_t$, Z has the standard normal distribution.

---

[3] As discussed in Jagannathan (1984) and Rodriguez (2003), this result depends critically on the Rothschild and Stiglitz (1970) definition of risk orderings. Grundy (1991) provides an example in which the expected return on the option is less than the risk-free rate; however the condition $dC(S)/dS < 0$ in his example conflicts with theoretical models and empirical findings, see Rodriguez (2003).

3.5.2. *Bounds for Asian options*

The approach presented in the previous section allows one to obtain semiparametric bounds for the expected payoffs and prices of path-dependent contingent claims in the trinomial model. As an illustration, we derive estimates for the expected payoffs of Asian options written on an asset with the trinomial price process. Similar bounds for other path-dependent contingent claims may also be derived.

Let $0 < t \le T - n$. Consider an Asian call option with strike price K expiring at time T written on the average of the past n prices of the asset with price process (3.57). The time-t expected payoff of the option is $\mathbb{E}_t(A_{n,T})$, where

$$A_{n,T} = \max \left[\left(\sum_{k=T-n+1}^{T} S_k \right) \Big/ n - K, 0 \right]$$

$$= \max \left[S_{T-n} \left(\sum_{k=T-n+1}^{T} X_{T-n+1} \ldots X_k \right) \Big/ n - K, 0 \right]$$

$$= \max \left[S_t \prod_{j=t+1}^{T-n+1} X_j \left(1 + \sum_{k=T-n+2}^{T} X_{T-n+2} \ldots X_k \right) \Big/ n - K, 0 \right].$$

$$\tag{3.62}$$

Using the convexity of the payoff function of the Asian option, we can state, similar to Theorems 3.31–3.33, the following bounds for the trinomial Asian option pricing model.

Theorem 3.35. *If the log-returns* $\log(X_t)$ *form an* $(\Im_t)$*-martingale-difference sequence and have the distribution* (3.58)*, then the expected payoff of the Asian option satisfies*

$$\mathbb{E}_t \max \left[\left(\sum_{k=T-n+1}^{T} S_k \right) \Big/ n - K, 0 \right]$$

$$\le \mathbb{E}_t \max \left[S_t \left(\sum_{k=T-n+1}^{T} \exp \left(\sum_{j=t+1}^{k} u_j \epsilon_j \right) \right) \Big/ n - K, 0 \right].$$

$$\tag{3.63}$$

Estimate (3.63) becomes even simpler in the case of an Asian option written on an asset whose gross returns form a three-valued martingale-difference sequence (so that the model represents the trinomial financial market with short-selling where the gross returns can take negative values):

Theorem 3.36. *If the returns* (X_t) *form an* $(\Im_t)-martingale-difference and have the distribution (3.60), then the expected payoff of the Asian option satisfies the inequality*

$$
\mathbb{E}_t \max \left[\left(\sum_{k=T-n+1}^{T} S_k \right) \middle/ n - K, 0 \right]
$$

$$
\leq \mathbb{E}_t \max \left[S_t \left(\sum_{k=T-n+1}^{T} \left(\prod_{j=t+1}^{k} u_j \right) \epsilon_k \right) \middle/ n - K, 0 \right].
$$

$$(3.64)$$

The advantage of the semiparametric bounds approach to option pricing, presented in this section, is that it does not require the knowledge of the entire distribution of the underlying asset's price. The bounds are easy to calculate and depend only on several parameters of the asset's price distribution, such as moments or the values taken by the asset's returns. Therefore, they provide an appealing alternative to exact option pricing and computationally expensive numerical pricing methods.

3.6. Concluding remarks

This chapter discussed mechanism of generating arbitrary dependence and its measures. We discussed characterizations of dependence in joint distributions, copulas, expectations, certain dependence classes and multiplicative systems.

The results are very general and have numerous implications. For example, they provide complete positive answers to the problems raised by Kotz and Seeger (1991) concerning characterizations of density weighting functions (d.w.f.) of dependent r.v.'s, existence of a method for constructing d.w.f.'s, and derivation of d.w.f.'s for a given model of dependence (see also Miller and Liu (2002) for a discussion of d.w.f.'s).

Furthermore, from the results presented in this chapter, it follows that weak convergence in the dependent case is implied by similar asymptotic results under independence together with convergence to zero of one of a series of dependence measures including the multivariate extension of Pearson's correlation, the relative entropy or other multivariate divergence measures. The results in this chapter provide a justification for estimation of $(k + 1)$-dimensional copulas for stationary time series with k-th order Markovian dependence structure. For instance, the results imply that all finite-dimensional copulas and, thus, all copula-based multivariate dependence measures and properties of such time series (e.g., multivariate analogues of Spearman's rho or relative entropy for finite-dimensional distributions; and β-mixing properties) are determined by and can be recovered from their $(k + 1)$-dimensional copulas.

The chapter also shows how dependence properties of time series place additional non-trivial restrictions on copulas of the processes in consideration that can be applied in inference on the properties of the processes. These restrictions allow one to construct time series with prescribed dependence structures that can be used, for instance, in the analysis of the robustness of statistical and econometric procedures to dependence.

The results suggest new approaches to obtaining approximations of higher-order Markov-processes and their functionals, including those that arise in contingent claim pricing and other applications in economics, finance and risk management (see, e.g., Cherubini *et al.* (2004, 2012); McNeil *et al.* (2005)), using approximations to their copulas of a given order. This can be accomplished, for instance, using expansions of the copulas by orthogonal functions or degenerate U-statistic kernels as in Theorem 3.5, Bernstein polynomials (see, e.g., Sancetta and Satchell (2004); Burda and Prokhorov (2014)), Hermite polynomials, or Fourier polynomials similar to the constructions discussed in Section 3.3.4. (see Lowin (2007) for an in-depth analysis of properties and economic and financial applications of Fourier copulas and their generalizations).

This chapter discussed issues with using second moment based measures such as correlations and proposed new and robust dependence measures. It also presented a number of new semiparametric bounds on the expected payoffs and prices of European call options and path-dependent contingent claims.

3.7. Appendix: Proofs

Proof of Theorem 3.1. Let us first prove the necessity part of the theorem. Denote

$$T(x_1, \ldots, x_n) = \int_{-\infty}^{x_1} \cdots \int_{-\infty}^{x_n} \left(1 + \right.$$

$$\left. \sum_{c=2}^{n} \sum_{1 \leq i_1 < \ldots < i_c \leq n} g_{i_1, \ldots, i_c}(t_{i_1}, \ldots, t_{i_c}) \right) \prod_{i=1}^{n} dF_i(t_i).$$

Let $k \in \{1, \ldots, n\}$, $x_k \in \mathbf{R}$. Let us show that

$$T(\infty, \ldots, \infty, x_k, \infty, \ldots, \infty) = F_k(x_k), \qquad (3.65)$$

$x_k \in \mathbf{R}$, $k = 1, \ldots, n$. It suffices to consider the case $k = 1$. We have

$$T(x_1, \infty, \ldots, \infty)$$

$$= \underbrace{\int_{-\infty}^{x_1} \int_{-\infty}^{\infty} \cdots \int_{-\infty}^{\infty}}_{n}$$

$$\left(1 + \sum_{c=2}^{n} \sum_{1 \leq i_1 < \ldots < i_c \leq n} g_{i_1, \ldots, i_c}(t_{i_1}, \ldots, t_{i_c}) \right) \prod_{i=1}^{n} dF_i(t_i)$$

$$= \underbrace{\int_{-\infty}^{x_1} \int_{-\infty}^{\infty} \cdots \int_{-\infty}^{\infty}}_{n} \prod_{i=1}^{n} dF_i(t_i)$$

$$+ \sum_{c=2}^{n} \sum_{1 \leq i_1 < \ldots < i_c \leq n} \underbrace{\int_{-\infty}^{x_1} \int_{-\infty}^{\infty} \cdots \int_{-\infty}^{\infty}}_{n} g_{i_1, \ldots, i_c}(t_{i_1}, \ldots, t_{i_c}) \prod_{i=1}^{n} dF_i(t_i)$$

$$= \Sigma' + \Sigma''.$$

Obviously, $\Sigma' = F_1(x_1)$. Furthermore, it is easy to see that there is at least one t_s of $t_2, \ldots, t_n$ among the arguments of each of the functions $g_{i_1,\ldots,i_c}(t_{i_1}, \ldots, t_{i_c})$, $1 \leq i_1 < \ldots < i_c \leq n$, $c = 2, \ldots, n$, in the latter summand. By condition A2 we get, therefore, that $\Sigma'' = 0$. Consequently, $T(x_1, \infty, \ldots, \infty) = F_1(x_1)$, $x_1 \in \mathbf{R}$, and (3.65) holds. It is evident that

$$\lim_{x_k \to -\infty} T(x_1, \ldots, x_k, \ldots, x_n) = 0 \qquad (3.66)$$

for all $x_j \in \mathbf{R}$, $j = 1, \ldots, n$, $j \neq k$, $k = 1, \ldots, n$. Since

$$T(x_1, \ldots, x_n)$$
$$= \prod_{i=1}^{n} F_i(x_i)$$
$$+ \mathbb{E}\left[\left(\sum_{c=2}^{n} \sum_{1 \leq i_1 < \ldots < i_c \leq n} g_{i_1,\ldots,i_c}(\xi_{i_1}, \ldots, \xi_{i_c}) \right) \right.$$
$$\left. \times \prod_{i=1}^{n} \mathbb{I}(\xi_i \leq x_i) \right],$$

from the monotone convergence theorem we obtain that $T(x_1, \ldots, x_n)$ is right-continuous in $(x_1, \ldots, x_n) \in \mathbf{R}^n$. Let $\delta_{[a,b)}^k T(x_1, \ldots, x_n) = T(x_1, \ldots, x_{k-1}, b, x_{k+1}, \ldots, x_n) - T(x_1, \ldots, x_{k-1}, a, x_{k+1}, \ldots, x_n)$, $a < b$. By integrability of the functions $g_{i_1,\ldots,i_c}$ and condition A3, we obtain

$$\delta_{(a_1,b_1]}^1 \delta_{(a_2,b_2]}^2 \cdots \delta_{(a_n,b_n]}^n T(x_1, \ldots, x_n)$$
$$= \prod_{i=1}^{n} \mathbb{P}(a_i < \xi_i \leq b_i)$$
$$+ \mathbb{E}\left[\left(\sum_{c=2}^{n} \sum_{1 \leq i_1 < \ldots < i_c \leq n} g_{i_1,\ldots,i_c}(\xi_{i_1}, \ldots, \xi_{i_c}) \right) \right.$$
$$\left. \times \prod_{i=1}^{n} \mathbb{I}(a_i < \xi_i \leq b_i) \right]$$
$$\geq 0 \qquad (3.67)$$

for all $a_i < b_i$, $i = 1, \ldots, n$.

Note that (3.66) and (3.67) are immediate if the probability space and the random variables are defined in the canonical way with $\Omega = \mathbf{R}^n$ and $X_i(\omega) = \omega_i$ for $\omega = (\omega_1, \ldots, \omega_n)$ and $\mathbb{P}(A) = \int_A (1 + \sum_{1 \leq i_1 < \ldots < i_c \leq n} g_{i_1, \ldots, i_c}(t_{i_1}, \ldots, t_{i_c})) \prod_{i=1}^n dF_i(t_i)$.

Right-continuity of $T(x_1, \ldots, x_n)$ and (3.65)–(3.67) imply that $T(x_1, \ldots, x_n)$ is a joint cdf of some r.v.'s $X_1, \ldots, X_n$ with the one-dimensional cdf's $F_k(x_k)$. Furthermore, the joint cdf $T(x_1, \ldots, x_n)$ satisfies (3.1).

Let us now prove the sufficiency part. Consider the functions

$$f_{i_1, \ldots, i_c}(x_{i_1}, \ldots, x_{i_c})$$

$$= \sum_{s=2}^c (-1)^{c-s} \sum_{j_1 < \ldots < j_s \in \{i_1, \ldots, i_c\}} \left(\frac{dF(x_{j_1}, \ldots, x_{j_s})}{dF_{j_1} \ldots dF_{j_s}} - 1 \right),$$

$1 \leq i_1 < \ldots < i_c \leq n$, $c = 2, \ldots, n$. Obviously, the functions $f_{i_1, \ldots, i_c}$ satisfy condition A1. Let us show that they satisfy condition A2. It suffices to consider the case $i_1 = 1, i_2 = 2, \ldots, i_c = c, k = 1$. We have

$$\mathbb{E} g_{1,2,\ldots,c}(\xi_1, x_2, \ldots, x_c)$$

$$= \int_{-\infty}^{\infty} g_{1,2,\ldots,c}(x_1, x_2, \ldots, x_c) dF_1(x_1)$$

$$= \int_{-\infty}^{\infty} \sum_{s=2}^c (-1)^{c-s} \left[\sum_{2 \leq i_2 < \ldots < i_s \leq c} \left(\frac{dF(x_1, x_{i_2} \ldots, x_{i_s})}{dF_1 dF_{i_2} \ldots dF_{i_s}} - 1 \right) \right.$$

$$\left. + \sum_{2 \leq i_1 < \ldots < i_s \leq c} \left(\frac{dF(x_{i_1}, x_{i_2} \ldots, x_{i_s})}{dF_{i_1} dF_{i_2} \ldots dF_{i_s}} - 1 \right) \right] dF_1(x_1)$$

$$= \sum_{s=2}^c (-1)^{c-s} \left[\sum_{2 \leq i_2 < \ldots < i_s \leq c} \left(\frac{dF(x_{i_2} \ldots, x_{i_s})}{dF_{i_2} \ldots dF_{i_s}} - 1 \right) \right.$$

$$\left. + \sum_{2 \leq i_1 < \ldots < i_s \leq c} \left(\frac{dF(x_{i_1} \ldots, x_{i_s})}{dF_{i_1} \ldots dF_{i_s}} - 1 \right) \right] = 0.$$

By the inversion formula (see, e.g., Borovskikh and Korolyuk (1997) pp. 177–178), it follows that if $a_{i_1, \ldots, i_c}$, $b_{i_1, \ldots, i_c}$, $1 \leq i_1 < \ldots <$

$i_c \leq n,\ c = 2, \ldots, n$, are arbitrary numbers then the relations

$$b_{i_1,\ldots,i_c} = \sum_{c=2}^{n} \sum_{j_1 < \ldots < j_s \in \{i_1,\ldots,i_c\}} a_{j_1,\ldots,j_s},$$

$$1 \leq i_1 < \ldots < i_c \leq n, \quad c = 2, \ldots, n,$$

and

$$a_{i_1,\ldots,i_c} = \sum_{s=2}^{c} (-1)^{c-s} \sum_{j_1 < \ldots < j_s \in \{i_1,\ldots,i_c\}} b_{i_1,\ldots,i_c},$$

$$1 \leq i_1 < \ldots < i_c \leq n, \quad c = 2, \ldots, n,$$

are equivalent. Taking here $a_{i_1,\ldots,i_c} = g_{i_1,\ldots,i_c}(x_{i_1},\ldots,x_{i_c})$, $b_{i_1,\ldots,i_c} = \frac{dF(x_{i_1},\ldots,x_{i_c})}{dF_{i_1}(x_{i_1})\ldots dF_{i_c}(x_{i_c})} - 1$, $1 \leq i_1 < \ldots < i_c \leq n, c = 2, \ldots, n$, we obtain, in particular, that

$$\frac{dF(x_1,\ldots,x_n)}{dF_1 \ldots dF_n} = 1 + \sum_{c=2}^{n} \sum_{1 \leq i_1 < \ldots < i_c \leq n} g_{i_1,\ldots,i_c}(x_{i_1},\ldots,x_{i_c}).$$

$$(3.68)$$

Therefore, representation (3.1) holds and the functions $f_{i_1,\ldots,i_c}$ satisfy condition A3. Suppose now that there exists another set of functions $g_{i_1,\ldots,i_c}$ satisfying conditions A1–A3 and such that (3.1) holds and, equivalently, (3.2) holds. Then, we have

$$\sum_{c=2}^{s} \sum_{i_1 < \ldots < i_c \in \{j_1,\ldots,j_s\}} f_{i_1,\ldots,i_c}(\xi_{i_1},\ldots,\xi_{i_c})$$

$$= \sum_{c=2}^{s} \sum_{i_1 < \ldots < i_c \in \{j_1,\ldots,j_s\}} g_{i_1,\ldots,i_c}(\xi_{i_1},\ldots,\xi_{i_c}) \qquad (3.69)$$

(a.s.) for $1 \leq j_1 < \ldots < j_s \leq n$, $s = 2, \ldots, n$. From (3.69) we subsequently obtain that $f_{i_1,i_2}(\xi_{i_1},\xi_{i_2}) = g_{i_1,i_2}(\xi_{i_1},\xi_{i_2})$ (a.s.), $1 \leq i_1 < i_2 \leq n$; $f_{i_1,i_2,i_3}(\xi_{i_1},\xi_{i_2},\xi_{i_3}) = g_{i_1,i_2,i_3}(\xi_{i_1},\xi_{i_2},\xi_{i_3})$ (a.s.), $1 \leq i_1 < i_2 < i_3 \leq n$; $\ldots$, $f_{1,2,\ldots,n}(\xi_1,\xi_2,\ldots,\xi_n) = g_{1,2,\ldots,n}(\xi_1,\xi_2,\ldots,\xi_n)$ (a.s.), that is $g_{i_1,\ldots,i_c}(\xi_{i_1},\ldots,\xi_{i_c}) = f_{i_1,\ldots,i_c}(\xi_{i_1},\ldots,\xi_{i_c})$ (a.s.), $1 \leq i_1 < \ldots < i_c \leq n, c = 2, \ldots, n$. The proof is complete.

Proof of Theorem 3.18. Let C be a $(k + 1)$-dimensional copula and let $\{X_t\}_{t=1}^{\infty}$ be a stationary C-based k-th order Markov process. Using Theorems 3.5 and 3.15, we obtain that if the process $\{X_t\}_{t=1}^{\infty}$ is k-independent, then the density of the copula C has formed (3.5) with $n = k+1$ and the functions g such that $g_{i_1,\ldots,i_c}(u_{i_1}, \ldots, u_{i_c}) = 0$, $1 \leq i_1 < \ldots < i_c \leq n$, $c = 2, \ldots, k$, that is, (3.21) holds with $g(u_1, \ldots, u_{k+1}) = g_{1,\ldots,k+1}(u_1, \ldots, u_{k+1})$. In addition, by the same theorems, the above function g satisfies conditions (3.22) and (3.24) and is such that

$$\int_0^1 g(u_1, \ldots, u_{k+1})du_j = 0, \quad j = 1, 2, \ldots, k + 1. \qquad (3.70)$$

Further, from Corollary 3.1, it follows that the density of the copula $C_{1,2,\ldots,k+1,k+2}$ of the r.v.'s $X_1, X_2, \ldots, X_{k+1}, X_{k+2}$ is given by

$$\frac{\partial^{k+2} C_{1,2,\ldots,k+1,k+2}(u_1, u_2, \ldots, u_{k+1}, u_{k+2})}{\partial u_1 \partial u_2 \ldots \partial u_{k+1} \partial u_{k+2}}$$

$$= (1 + g(u_1, \ldots, u_{k+1}))(1 + g(u_2, \ldots, u_{k+2}))$$

$$= 1 + g(u_1, \ldots, u_{k+1}) + g(u_2, \ldots, u_{k+2})$$

$$+ g(u_1, \ldots, u_{k+1})g(u_2, \ldots, u_{k+2}). \qquad (3.71)$$

Using (3.70) and (3.71), we get that, for $2 \leq i_1 < i_2 \leq k + 1$, the density of the copula of the r.v.'s X_j, $j \in \{1, 2, \ldots, k + 2\}\backslash\{i_1, i_2\}$ is given by

$$1 + \int_0^1 \int_0^1 g(u_1, \ldots, u_{k+1})g(u_2, \ldots, u_{k+2})du_{i_1}du_{i_2}.$$

This and k-independence of $\{X_t\}$ imply that

$$\int_0^1 \int_0^1 g(u_1, \ldots, u_{k+1})g(u_2, \ldots, u_{k+2})du_{i_1}du_{i_2} = 0, \qquad (3.72)$$

$2 \leq i_1 < i_2 \leq k + 1$. In complete similarity, by considering the k-dimensional marginal copulas of the r.v.'s $X_1, X_2, \ldots, X_{k+2}, X_{k+3}$ and using (3.70), (3.71) and (3.72), we obtain

$$\int_0^1 \int_0^1 \int_0^1 g(u_1, \ldots, u_{k+1})g(u_2, \ldots, u_{k+2})g(u_3, \ldots, u_{k+3})$$

$$\times du_{i_1}du_{i_2}du_{i_3} = 0,$$

$3 \leq i_1 < i_2 < i_3 \leq k + 1$.

Continuing in the same fashion, we get that the property that $\{X_t\}_{t=1}^{\infty}$ is a stationary k-independent C-based k-th order Markov process implies that (3.23) holds for all $s \le u_{i_1} < \ldots < u_{i_s} \le k + 1$, $s = 1, 2, \ldots, [\frac{k+1}{2}]$.

Suppose now that relation (3.23) holds for all $s \le u_{i_1} < \ldots < u_{i_s} \le k + 1$, $s = 1, 2, \ldots, [\frac{k+1}{2}]$. One then easily gets that (3.23) also holds for all $1 \le u_{i_1} < \ldots < u_{i_s} \le k + s$, $s \ge 1$, and the product $g(u_1, \ldots, u_{k+1}) g(u_2, \ldots, u_{k+2}) \ldots g(u_s, \ldots, u_{k+s})$ that appears in the density $\frac{\partial^{k+s} C_{1,\ldots,k+s}(u_1,\ldots,u_{k+s})}{\partial u_1 \ldots \partial u_{k+s}} = \prod_{j=1}^{s}(1 + g(u_j, \ldots, u_{k+j}))$ of the copula of $X_1, X_2, \ldots, X_{k+s}$. It is not difficult to see, similar to the above analysis, that this implies that the copula $C_{1,2,\ldots,k+s}$ has k-dimensional marginal copulas in product form (1.2) with $n = k$. Thus, the r.v.'s $X_1, X_2, \ldots, X_{k+s}$ are k-independent for all $s \ge 1$.

Proof of Theorem 3.19. Let C be a bivariate copula and let $\{X_t\}_{t=1}^{\infty}$ be a stationary C-based first-order Markov process. By Theorem 3.5, the density of the copula C is given by (3.25) with the function $g(u_1, u_2)$ satisfying conditions (3.26)–(3.28). In addition, from Corollary 3.1, it follows that the density of the copula $C_{1,2,\ldots,m+1,m+2}$ of the r.v.'s $X_1, X_2, \ldots, X_{m+1}, X_{m+2}$ has the form

$$\frac{\partial^{m+2} C_{1,2,\ldots,m+1,m+2}(u_1, u_2, \ldots, u_{m+1}, u_{m+2})}{\partial u_1 \partial u_2 \ldots \partial u_{m+1} \partial u_{m+2}}$$

$$= (1 + g(u_1, u_2))(1 + g(u_2, u_3)) \ldots (1 + g(u_{m+1}, u_{m+2}))$$

$$= \prod_{s=1}^{m+1} (1 + g(u_s, u_{s+1})).$$

Using relations (3.27), we thus get that the copula $C_{1,m+2}$ of the r.v.'s X_1 and X_{m+2} is given by

$$C_{1,m+2}(u_1, u_{m+2}) = \int_0^1 \ldots \int_0^1 \prod_{s=1}^{m+1} (1 + g(u_s, u_{s+1})) du_2 \ldots du_{m+1}$$

$$= 1 + \int_0^1 \ldots \int_0^1 \prod_{s=1}^{m+1} g(u_s, u_{s+1}) du_2 \ldots du_{m+1}.$$

Thus, the copula $C_{1,m+2}$ is the product copula: $C_{1,m+2}(u_1, u_{m+2}) = u_1 u_{m+2}$ if and only if condition (3.29) is satisfied. It is not difficult to see that if (3.29) holds, then one also has

$$\int_0^1 \cdots \int_0^1 \prod_{s=1}^{m+n+j} g(u_s, u_{s+1}) du_{n+1} \cdots du_{m+n} = 0$$

for all $n \geq 1$, $j \geq 0$. Similar to the above analysis, this relation, together with (3.27), implies that the random vectors $(X_1, \ldots, X_n)$ and $(X_{m+n+1}, \ldots, X_{m+n+j+1})$ in the stationary C-based Markov process $\{X_t\}_{t=1}^\infty$ are independent for all $n \geq 1$, $j \geq 0$.

Proof of Theorem 3.20. By Markov property, $\mathbb{P}(X_t > x | \Im_{t-1}) = \mathbb{P}(X_t < -x | \Im_{t-1})$, $x \geq 0$, if and only if $\mathbb{P}(X_t > x | X_{t-1}) = \mathbb{P}(X_t < -x | X_{t-1})$, $x \geq 0$. The latter inequality, in turn, is equivalent to $\mathbb{P}(V_n > 1/2 + u | V_{n-1}) = \mathbb{P}(V_n < 1/2 - u | V_{n-1})$, $u \in [0, 1/2)$, where $V_t = F(X_t)$ and, by stationarity, to $\mathbb{P}(V_2 > 1/2 + u | V_1) = \mathbb{P}(V_2 < 1/2 - u | V_1)$, $u \in [0, 1/2)$. We have therefore, that $\{X_t\}$ is a conditionally symmetric martingale difference if and only if $\frac{\partial C(V_1, 1/2-u)}{\partial u_1} = 1 - \frac{\partial C(V_1, 1/2+u)}{\partial u_1}$, or, equivalently, if and only if (3.30) and (3.31) hold.

Chapter 4

Limits of Diversification under Fat Tails and Dependence

This chapter revisits the question of when portfolio diversification does and does not work raised in Chapter 2. What is different is that now we let the portfolio components have certain more or less arbitrary forms of dependence. The types of dependence we consider include general power-type copula families, including those discussed in Section 3.3, as well as dependence structures induced by convolutions of α-symmetric distributions and models with common multiplicative and additive shocks.

4.1. Introduction

Let, as before, VaR_q denote level-q Value-at-Risk, i.e., the $(1 - q)$-quantile of the distribution of losses — we interpret positive values as losses. For example, assuming that the cdf of a portfolio risk Z is strictly increasing, its 5%-VaR is the value of loss x for which the probability $\mathbb{P}(Z > x)$ is 0.05. More generally, the $q\%$ Value-at-Risk (VaR) of a portfolio risk Z is

$$\text{VaR}_q(z) = \inf\{z \in \mathbb{R} : \mathbb{P}(Z > z) \leq q\}.$$

For simplicity, consider the problem of optimal portfolio allocation in the VaR framework with two risks that are possibly extremely heavy tailed. Let $w = (w_1, w_2) \in \mathbb{R}^2$ be the portfolio weights such that $w_1 + w_2 = 1$. Let X_j be a loss, $j = 1, 2$ which has a power law distribution (1.8) with the tail index α_j. The tail of the aggregate loss distribution can be written as $\mathbb{P}(w_1X_1 + w_2X_2 > x)$, where the weighted average loss $w_1X_1 + w_2X_2$ corresponds to a portfolio of two risks with weights w_1 and w_2. Unless one of the weights is zero, the portfolio is diversified.

The problem of interest is to minimize $\text{VaR}_q(w_1X_1 + w_2X_2)$ over the weights w for a given $q \in (0, 1/2)$. When X_1 and X_2 are i.i.d. risks with stable distributions, Chapter 2 showed that, for all non-zero w's, $\mathbb{P}(w_1X_1 + w_2X_2 > x) \leq \mathbb{P}(X_1 > x)$ if $\alpha_j > 1, j = 1, 2$. Equivalently, the VaR of a diversified portfolio $\text{VaR}_q(w_1X_1 + w_2X_2 > x)$ is not greater than that of a not diversified one, $\text{VaR}_q(X_1 > x)$, if $\alpha_j > 1$. Put differently, diversification helps to lower the VaR for moderately, but not extremely, heavy tailed risks. If $\alpha_j < 1$, $j = 1, 2$, then $\mathbb{P}(w_1X_1 + w_2X_2 > x) \geq \mathbb{P}(X_1 > x)$; that is, for extremely heavy-tailed risks the benefits of diversification disappear and the least risky portfolio is the undiversified one that consists of only one risk. For example, if X_j's are i.i.d. stable with $\alpha = 1/2$, that is, if they are Levy distributed, the aggregate loss of an equally weighted portfolio $\frac{X_1+X_2}{2}$ has the same distribution as $2X_1$ and thus $\text{VaR}_q\left(\frac{X_1+X_2}{2}\right) = 2\text{VaR}_q(X_1) > \text{VaR}_q(X_1)$. Section 2.4 discussed this in a general context with any number of i.i.d. risks, using majorization theory.

VaR is a widely used risk measure, whose popularity in a wide range of areas in finance is attributed to the recommendations of the Basel Committee on Banking Supervision. A series of recent works have studied the problem of portfolio optimization in the VaR framework, focusing on the situation when the portfolio components are independent and have a heavy tailed distribution — see Chapter 2. An interesting conclusion from that work is that if tails are extremely heavy then diversification increases riskiness in terms of VaR.

This theoretical property of VaR known as non-subadditivity or non-coherence is often weighted against the practical considerations of the ease of calculation and backtesting and smaller data requirements, compared to subadditive and coherent risk measures such as Expected Shortfall (ES) (see, e.g., Danielsson *et al.* (2013); Garcia *et al.* (2007); Ibragimov *et al.* (2015)). Naturally, the use of the coherent ES is in principle possible only for moderately heavy-tailed risks as it is defined only under the assumption of finite first moments. At the same time, VaR is defined for any risk and can be used under extreme heavy-tailedness with tail indices smaller than one and infinite first moments.

It is well established in today's finance that in practice risks are dependent in some usually unknown fashion and that the tail behavior of other risk measures including ES is closely related to the tail behavior of sums of dependent risks used in VaR analysis (see, e.g., Alink *et al.* (2005)). Therefore, a better understanding of when VaR is non-subadditive in non-i.i.d. settings is key to continued use of VaR as a robust risk measure.

As discussed in Chapter 2, the literature on the properties of VaR for independent risks is very wide and has a long tradition. More recently, Ibragimov and Walden (2007) and Ibragimov (2009b) focused on the case of convolutions of i.i.d. stable random variable with infinite variance and showed that VaR is subadditive provided the mean is finite. Similar results were also obtained for asymptotically large losses without the assumption of a stable distribution (see, e.g., McNeil *et al.* (2005)).

However, extensions to non-independence have been limited to specific cases. For example, Ibragimov and Walden (2007) and Ibragimov (2009b) consider dependence arising from common multiplicative shocks, Embrechts *et al.* (2009) and Chen *et al.* (2012) consider Archimedean copulas, Asmussen and Rojas-Nandayapa (2008) consider the normal copula, Albrecher *et al.* (2006) consider several copula classes permitting explicit solutions such as Archimedean copulas (see also Kortschak and Albrecher (2009)). An interesting result arising from these studies is that the subadditivity property of VaR is generally affected by both the strength of dependence and the tail behavior of the marginals, however in some cases only heavy tails of the marginals matter.

This chapter covers selected results on subadditivity of VaR in non-i.i.d. settings. Section 4.2 discusses the complications arising from allowing for non-i.i.d.-ness. Section 4.3 considers classes of dependence structures motivated by several widely used copula families and their approximations. In Section 4.4, we consider dependence implied by models with common shocks, both multiplicative and additive. Section 4.5 provides noteworthy theoretical applications of the results in Section 4.4 to VaR of portfolio components, to variances of financial indices and to random effects estimation. It also discusses some extensions. Section 4.6 concludes. Section 4.7 contains proofs of selected results presented in this chapter.

4.2. Dependence vs independence

It follows from Section 2.4 that the limits of diversification results hold for i.i.d. losses regardless of the weights w_j used in construction of a diversified portfolio. Therefore, in this section we consider an equally weighted portfolio $w_1 = w_2 = 1/2$. To state the results formally, let $(\xi_1(\beta), \xi_2(\beta))$ denote independent random variables from a power-law distribution with a common tail index β. The following theorem can be easily extended to any diversified portfolio of size n, in which case $(\frac{\xi_1(\alpha)+\xi_2(\alpha)}{2})$ in inequalities (4.1)–(4.2) is replaced with $\sum_{i=1}^{n} w_i \xi_i(\alpha)$.

We start by restating Theorem 2.1–2.2 from Section 2.4 for power law distributions with tail index α.

Theorem 4.1. *For sufficiently small loss probability q,*

$$VaR_q \left(\frac{\xi_1(\alpha) + \xi_2(\alpha)}{2} \right) < VaR_q(\xi_1(\alpha)), \quad \text{if} \quad \alpha > 1 \quad (4.1)$$

$$VaR_q \left(\frac{\xi_1(\alpha) + \xi_2(\alpha)}{2} \right) > VaR_q(\xi_1(\alpha)), \quad \text{if} \quad \alpha < 1. \quad (4.2)$$

As discussed in Section 2.4, an interesting boundary case corresponds to $\alpha = 1$. This is when diversification has no effect at all, i.e., it neither increases nor reduces VaR. For example, if ξ's are i.i.d. stable with $\alpha = 1$, which means they have a Cauchy distribution, it is easy to show that $\sum_{i=1}^{n} w_i \xi_i(\alpha)$ is equal in distribution to $\xi_i(\alpha)$, so a diversified and a non-diversified portfolios have identical VaRs.

Similar results are available for bounded risks concentrated on a sufficiently large interval: for such cases, VaR-based diversification is suboptimal up to a certain number of risks and then becomes optimal (Ibragimov and Walden (2007)).

It is not obvious what happens if we relax the independence assumptions. The two extreme cases, corresponding to a comonotone and countermonotone relationships between the components do not present a consistent picture. For example, if we assume that $\xi_1 = \xi_2$ (a.s.) then obviously $\mathrm{VaR}_q(w_1\xi_1(\alpha)+w_2\xi_2(\alpha)) = \mathrm{VaR}_q(\xi_1(\alpha))$ and so diversification has no effect regardless of the tails; while if we assume $\xi_1 = -\xi_2$ (a.s.) then $\mathrm{VaR}_q(w_1\xi_1(\alpha)+w_2\xi_2(\alpha)) = (w_1-w_2)\mathrm{VaR}_q(\xi(\alpha))$ and it is optimal to fully diversify regardless of the tails.

In principle, we are interested in how the aggregate loss probability for a diversified portfolio compares to that of a single risk. That is, we are interested in the behavior of

$$
\mathbb{P}\left(\frac{X_1 + X_2}{2} > x\right)
$$

$$
= \int\int_{\frac{z_1+z_2}{2}>x} f(z_1;\alpha)f(z_2;\alpha)c\left(F(z_1;\alpha),F(z_2;\alpha);\theta\right)dz_1 dz_2
$$

$$
= \mathbb{E}\left\{c\left(F(\xi_1;\alpha),F(\xi_2;\alpha);\theta\right)\mathbb{I}\left[\frac{\xi_1+\xi_2}{2}>x\right]\right\}, \qquad (4.3)
$$

where $c(u_1, u_2; \theta)$ is a copula density parameterized by θ, $f(\cdot; \alpha)$'s are power-law marginal densities of the risks X_j, $\mathbb{I}[\cdot]$ is the indicator function and ξ_j's are independent copies of X_j's.

There is no general way of expressing this in terms of $\mathbb{P}(X_1 > x)$ and whether diversification decreases or increases VaR depends on the copula family as well as on the interaction between α and θ, the parameters that characterize the properties of marginal distributions and their dependence.

However, there exist classes of copulas for which we can make explicit comparisons. In this section, we show that the results obtained for the i.i.d. setting — Theorems 2.1–2.2 of Chapter 2 — continue to hold for these classes of copulas.

4.3. Diversification and copulas

We start with reviewing the Eyraud-Farlie-Gumbel-Morgenstern
(EFGM) copula family (4.5) and with introducing its generalizations and first- or second-order approximations. This class of copulas,
which we call *power-type*, includes the power copulas (3.10) discussed
in Chapter 3, polynomial copulas of Drouet and Kotz (2009), copulas with cubic sections of Nelsen *et al.* (1997), as well as a large
number of related copulas with various dependence features such as
asymmetry, tail-dependence, comprehensiveness etc. This is done in
Section 4.3.1.

In Section 4.3.2, we show that for power law risks with these copulas whose tail exponent is below one, diversification is suboptimal
(i.e., it increases riskiness) regardless of the strength of dependence,
provided the losses are large enough.

4.3.1. *Power-type copulas*

The class of copulas of particular interest to us contains copulas that
are multiplicative or additive in powers of the margins, or can be
approximated using such copulas. We call this class *power-type*. It is
similar but more general than the power copula family discussed in
Chapter 3 and than the polynomial copula family which is considered
in this chapter.

The most common family in this class is the EFGM copula family and its generalizations dealt with in Chapter 3 that we also discuss below. The bivariate EFGM copula family can be written as
follows:

$$C(u_1, u_2) = u_1 u_2 [1 + \theta(1 - u_1)(1 - u_2)], \qquad (4.4)$$

where $\theta \in [-1, 1]$, and its density has the form $c(u_1, u_2) = 1 +
g(u_1, u_2)$ in (3.25), where $g(u_1, u_2)$ is an expansion (3.38) by linear
functions $1 - 2u_j, j = 1, 2$. This is a non-comprehensive copula
in the sense that it can accommodate a limited range of dependence. For example, Kendall's τ of the EFGM copula is restricted to
$[-\frac{2}{9}, \frac{2}{9}]$.

A multivariate version of the EFGM copula in (3.9) introduced by Cambanis (1977) has the following form:

$$C(u_1, u_2, \ldots, u_n) = u_1 u_2 \ldots u_n$$

$$\times \left[1 + \sum_{c=2}^{n} \sum_{1 \leq i_1 < i_2 < \ldots < i_c \leq n} \theta_{i_1, i_2, \ldots, i_c} (1 - u_{i_1}) \right.$$

$$\left. \times (1 - u_{i_2}) \ldots (1 - u_{i_c}) \right], \qquad (4.5)$$

where $-\infty < \theta_{i_1, i_2, \ldots, i_c} < \infty$ are such that

$$\sum_{c=2}^{n} \sum_{1 \leq i_1 < i_2 < \ldots < i_c \leq n} \theta_{i_1, i_2, \ldots, i_c} \delta_{i_1} \cdots \delta_{i_c} \geq -1$$

for all $\delta_i \in [-1, 1], i = 1, \ldots, n$. This copula family can be viewed as a special case of a wider family of n-dimensional power copulas in (3.10) introduced by Ibragimov (2009a).

The power copula family can be written as follows

$$C(u_1, \ldots, u_n) = u_1 u_2 \cdots u_n$$

$$\times \left[1 + \sum_{c=2}^{n} \sum_{1 \leq i_1 < i_2 < \ldots < i_c \leq n} \theta_{i_1, i_2, \ldots, i_c} (u_{i_1}^l - u_{i_1}^{l+1}) \right.$$

$$\left. \times (u_{i_2}^l - u_{i_2}^{l+1}) \ldots (u_{i_c}^l - u_{i_c}^{l+1}) \right], \qquad (4.6)$$

where $\theta_{i_1, i_2, \ldots, i_c} \in (-\infty, \infty)$ are such that

$$\sum_{c=2}^{n} \sum_{1 \leq i_1 < i_2 < \ldots < i_c \leq n} |\theta_{i_1, i_2, \ldots, i_c}| \leq 1.$$

This corresponds to using nonlinear rather than linear functions in the expansion of the copula density function.

Another relevant copula family, of which the EFGM copula in (4.4) is a special case, is known as a polynomial copula family (see, e.g., Drouet and Kotz (2009), p. 74). An order m $(m \geq 4)$ polynomial copula can be written as follows:

$$C(u, v) = uv \left[1 + \sum_{\substack{k \geq 1, q \geq 1}}^{k+q \leq m-2} \theta_{kq}(u^k - 1)(v^q - 1) \right], \qquad (4.7)$$

where $\theta_{kq} = \frac{\theta_{kq}}{(k+1)(q+1)}$ and $0 \leq \min(\sum_{k \geq 1, q \geq 1}^{k+q \leq m-2} q\theta_{kq}, \sum_{k \geq 1, q \geq 1}^{k+q \leq m-2} k\theta_{kq}) \leq 1$.

One example of this copula family is Nelsen *et al.*'s (1997) copula with cubic section, which is written as follows:

$$C(u, v) = uv + 2\theta uv(1 - u)(1 - v)(1 + u + v - 2uv), \qquad (4.8)$$

where $\theta \in [0, \frac{1}{4}]$.

Several other copula families can be written as approximations by EFGM copulas. For example, it is well known that EFGM copulas provide a first-order approximation to the Ali-Mikhail-Haq (AMH) copula family. The AMH copula can be written as follows:

$$C(u_1, \ldots, u_n) = (1 - \theta) \left[\prod_{i=1}^{n} \left(\frac{1 - \theta}{u_i} + \theta \right) - \theta \right]^{-1},$$

where $\theta \in [-1, 1]$.

A less known result is that the Plackett and the Frank copula families have first order Taylor approximations by EFGM copulas at independence (see, e.g., Nelsen (2006), pp. 100, 133). The n-variate Frank copula, which is comprehensive, radially symmetric and Archimedean, can be written as follows:

$$C(u_1, \ldots, u_n) = \log_\theta \left[1 + \frac{\prod_{i=1}^{n}(e^{\theta u_i} - 1)}{(e^\theta - 1)^{n-1}} \right],$$

where $\theta \geq 0$.

The n-variate Plackett copula, which is also comprehensive, is rarely discussed in the literature unless $n = 2$, in which case it has the following form:

$$C(u_1, u_2) = \frac{1}{2(\theta - 1)} \left[1 + (\theta - 1)(u_1 + u_2) \right.$$
$$\left. - \sqrt{[1 + (\theta - 1)(u_1 + u_2)]^2 - 4\theta(\theta - 1)u_1 u_2} \right],$$

where $1 \neq \theta > 0$. However, a way to generalize to $n > 2$ is presented by Molenberghs and Lesaffre (1994). It is also worth mentioning that for all the three copula families, there exist improved second-order approximations (see, e.g., Nelsen (2006), p. 83).

An interesting set of approximation results is given by Nelsen *et al.* (1997), Cuadras (2009) and Cuadras and Diaz (2012). Nelsen *et al.* (1997) provide a generalization of the bivariate EFGM copula using cubic terms as in (4.8) and show that it can be used to approximate some well-known families of copulas, both symmetric and not, such as the copulas of Kimeldorf and Sampson (1975) and Lin (1987), as well as the Sarmanov copula. They also show that copulas in (4.8) are second-degree Maclaurin approximations to members of the Frank and Plackett copula families.

Cuadras (2009) studies the power series class of copulas, obtained as weighted geometric means of the EFGM and AMH copulas, and shows that the Gumbel-Barnett and Cuadras-Auge copulas can be expressed as first-order approximations to that class. Cuadras and Diaz (2012) provide approximations of the tail-dependent Clayton-Oakes copula, which also have the form of a power-type generalization of the EFGM copula.

4.3.2. *Subadditivity of VaR*

We start with the bivariate EFGM copula in (4.4). Let (X_1, X_2) be absolutely continuous random variables with the EFGM copula and power-law marginals (1.8) with the same tail index α. Then, for any $x \geq 1$ and for $j = 1, 2$, their marginal cdf's F_j, marginal pdf's f_j and the joint cdf H and pdf h satisfy

$$F_j(x) \sim 1 - x^{-\alpha},$$
$$f_j(x) \sim \alpha x^{-\alpha - 1},$$

$$H(x_1, x_2) = F_1(x_1)F_2(x_2)\big[1 + \theta(1 - F_1(x_1))(1 - F_2(x_2))\big],$$
$$h(x_1, x_2) = f_1(x_1)f_2(x_2)\big[1 + \theta(1 - 2F_1(x_1))(1 - 2F_2(x_2))\big].$$

As before, let $(\xi_1(\alpha), \xi_2(\alpha))$ be independent random variables from power-law distributions with tail index α, often called *independent* copies of (X_1, X_2). Our key insight is that in the tail, the behavior of products and powers of power-law densities and distributions of X_j's is identical to the behavior of their independent copies. This makes it possible to provide asymptotic (with respect to the loss) comparisons between the VaR of the aggregated loss and that of a single risk. More specifically, the expression for $\mathbb{P}\left(\frac{X_1+X_2}{2} > x\right)$ in (4.3) under the EFGM copula can be written as follows:

$$\int_{\frac{s+t}{2}>x} \alpha^2 s^{-\alpha-1}t^{-\alpha-1}(2s^{-\alpha} - 1)(2t^{-\alpha} - 1)\,ds\,dt$$

$$= 4\alpha^2\mathbb{P}\left(\frac{\xi_1(2\alpha) + \xi_2(2\alpha)}{2} > x\right) - 2\alpha^2\mathbb{P}\left(\frac{\xi_1(2\alpha) + \xi_2(\alpha)}{2} > z\right)$$

$$- 2\alpha^2\mathbb{P}\left(\frac{\xi_1(\alpha) + \xi_2(2\alpha)}{2} > x\right) + \alpha^2\mathbb{P}\left(\frac{\xi_1(\alpha) + \xi_2(\alpha)}{2} > x\right),$$

where the behavior of the individual summands for large x is driven by the tail index α of the independent risks ξ_j with heavy-tailed power law distributions (1.8) (their smallest tail index in the case of not identically distributed risks X_j).

We formalize this result in the following theorem, generalized to n dependent heavy-tailed random variables $X_1, X_2, \ldots, X_n$ with multivariate EFGM copula given in (4.5) and power-law marginals with the same value of α.

Theorem 4.2. *For an asymptotically large $x > 0$, and any $n, \alpha > 0$,*

$$\mathbb{P}\left(\sum_{i=1}^{n} X_i > xn\right) \sim \mathbb{P}\left(\sum_{i=1}^{n} \xi_i(\alpha) > xn\right). \qquad (4.9)$$

The result suggests that suboptimality of diversification in the VaR framework for extremely heavy tailed losses carries over from independence to the EFGM copula. That is, diversification increases

VaR of dependent extremely heavy tailed risks within this copula family. Specifically, combining the results of Theorems 2.1–2.2 and 4.2, it is easy to see that the following corollary holds.

Corollary 4.1. *For dependent losses with the EFGM copula and sufficiently small loss probability* q,

$$VaR_q \left(\frac{X_1 + \cdots + X_n}{n} \right) < VaR_q(X_1), \quad \text{if } \alpha > 1, \quad (4.10)$$

$$VaR_q \left(\frac{X_1 + \cdots + X_n}{n} \right) > VaR_q(X_1), \quad \text{if } \alpha < 1. \quad (4.11)$$

Another interesting corollary of Theorem 4.2 can be obtained by combining this result with Theorem 1 of Sharakhmetov and Ibragimov (2002). The EFGM copula family has restrictive dependence, for example, it is not comprehensive in the sense that it cannot accommodate all possible values of Kendall's τ. Yet, as shown by Sharakhmetov and Ibragimov (2002), it can be used to represent *any* joint distribution of two-valued random variables (see also de la Peña *et al.* (2006), p. 190). Therefore, for two-valued random variables, our Theorem 4.2 applies to *all* dependence patterns.

Important generalizations of Theorem 4.2 arise if we consider the wider class of power-type copulas introduced in Section 4.3.1. Most popular members of this class such as the polynomial copula (4.7) of Drouet and Kotz (2009), the copula with cubic section (4.8) of Nelsen *et al.* (1997) and the power copula (4.6) of Ibragimov (2009a) can be written in the following general form:

$$C(u_1, \ldots, u_n) = \sum_{i_1, \ldots, i_n = 0, 1, \ldots} \theta_{i_1, i_2, \ldots, i_n} \cdot u_1^{i_1} \cdot u_2^{i_2} \cdot \ldots \cdot u_n^{i_n}, \quad (4.12)$$

for a multiple index $i = (i_1, i_2, \ldots, i_n)$ and a set of corresponding parameters θ_i with appropriate restrictions that make $C(u_1, \ldots, u_n)$ a copula. For example, Drouet and Kotz (2009, Section 4.5.2) show how to obtain the polynomial copula in (4.7) from function $f = u^k v^q$. The key feature of such copulas is that they and their densities can be expressed as powers of u_j's. This allows to apply similar arguments as for EFGM.

Theorem 4.3. *For dependent losses with a power-type copula in* (4.12) *and for an asymptotically large $z > 0$, and any $n, \alpha > 0$, the conclusions of Theorem 4.2 hold.*

One may argue that the class of copulas in (4.12) is not sufficiently general. For example, it is not clear whether it can incorporate tail dependence or comprehensive copulas. However, the power-type copulas also include copulas which can approximate or be approximated by the class in (4.12). And, as discussed in Section 4.3.1, there are comprehensive and tail-dependent copulas among these copulas. Our next corollary establishes the result for such approximations.

Corollary 4.2. *For dependent losses with copulas whose Taylor or Maclaurin expansions can be written as* (4.12)*, for an asymptotically large $z > 0$, and any $n, \alpha > 0$, the conclusions of Theorem 4.2 hold but only locally at the point of approximation.*

This corollary covers all the copula families discussed in Section 4.3.1 including the AMH, Plackett, Frank, Clayton-Oakes, Kimeldorf and Sampson, Lin, Gumbel-Barnett and many others, but only to the extent the approximations are valid. That is, the results of Theorem 4.2 hold for expansions at the point at which we expand, which often coincides with independence. Clearly, they do not have to hold when the approximation error is large. Therefore, applicability of Theorem 4.2 to a specific copula family needs to be checked on a case-by-case basis but the class of copulas to which it can be potentially applied is quite rich. It includes comprehensive copulas (Plackett, Frank), asymmetric copulas (Nelsen *et al.*'s (1997) copulas with cubic sections) and tail-dependent copulas (Clayton-Oakes).

Embrechts *et al.* (2009) consider dependent risks for which the dependence is modelled using the Archimedean copulas, and reach a similar conclusion.

We have looked at equally weighted portfolios with components having the same tail index. The restriction of equal tail indices can easily be relaxed because the tail behavior of the aggregate loss will be dominated by the component with the lowest tail index. The limits of diversification are determined by whether the lowest index in the portfolio is above or below one. A similar result can be established for unequally weighted portfolios but we leave both these extensions for another book.

4.4. Diversification and common shocks

In this section, we show that the results obtained for the i.i.d. setting — Theorems 2.1–2.2 of Chapter 2 — continue to hold for convolutions of risks with joint α-symmetric and spherical distributions, in particular for several classes of models with common shocks, multiplicative and additive.

Many important economic and financial variables are influenced by common macroeconomic, political, legal and environmental shocks. Examples of such shocks are given by (see the review and examples in Andrews (2005)) financial crises affecting individual and firm consumption, investment and production decisions; stock market shocks influencing individual wealth and firm assets; oil price shocks and business cycles affecting firm factor costs and production; inflation affecting nominal wages; and employment shocks influencing economic decisions of market participants.

Theoretical and empirical frameworks with common shocks discussed in the literature include models with output, investment and savings affected by common shocks due to, e.g., technology changes or financial crises; microeconomic models with wages influenced by prices for unmeasured skills and the worker and firm effects; factor models for financial asset returns (see the review and discussion in Section 2 of Bai (2009)); health spending and health outcomes affected by technological advances, diseases, epidemics and other health shocks, and the implementation of new health policies (see Moscone and Tosetti (2009)); co-movements of financial and insurance variables in different markets due to exogenous common shocks such as financial and economic crises (see, among others, McNeil *et al.* (2005), and references therein); and numerous others.

Several recent works provide extensions of the value at risk analysis to some dependence structures (see, e.g., Ibragimov (2009b)). However, exact closed-form solutions are usually available only in the setting with independent and uncorrelated risks. This section provides general results on portfolio value at risk comparisons for correlated heavy-tailed risks exhibiting certain types of common shocks. They include shock structures of the type

$$Y_{ij} = R_i + C_j + U_{ij}, \quad i = 1, \ldots, r, \quad j = 1, \ldots, c, \qquad (4.13)$$

where the "row effects" common shocks R_i, the "column effects" common shocks C_j and the "error" variables U_{ij} are assumed to be independent of each other and to be independent and identically distributed among themselves. We also present the VaR analysis for a particular case of (4.13) given by

$$Y_{ij} = R_i + U_{ij}, \quad i = 1, \ldots, r, \, j = 1, \ldots, c. \qquad (4.14)$$

Together with their multiplicative analogues

$$Y_{ij} = R_i U_{ij}, \quad R_i > 0, \qquad (4.15)$$

models (4.13) and (4.14) provide a natural framework for modelling risks subject to (additive) common shocks R_i and C_j. The common shocks R_i affect all risks Y_{ij}, $j = 1, \ldots, c$, in the ith row (say, in the ith country) and the common shocks C_j affect all risks Y_{ij}, $i = 1, \ldots, r$, in the jth column (say, in the jth industry). Interestingly, the dependence properties of additive common shock models (4.13) and (4.14) are more complicated than those in (4.15) and its two-shock analogues. For instance, while multiplicative common shock models (4.15) imply uncorrelatedness of the risks Y_{ij}, this, evidently, does not hold for (4.14).

This section is organized as follows. Section 4.4.1 introduces α-symmetric and spherical distributions. Section 4.4.2 discusses the results on diversification of dependent heavy-tailed risks in models with multiplicative common shocks. Section 4.4.3 contains the results on value at risk analysis and optimal portfolio choice in common shocks models (4.13).

4.4.1. α-symmetric and spherical distributions

According to the definition introduced in Cambanis *et al.* (1983), an n-dimensional distribution is called α-symmetric if its c.f. can be written as $\phi((\sum_{k=1}^{n} |t_k|^\alpha)^{1/\alpha})$, where $\phi : \mathbf{R}_+ \to \mathbf{R}$ is a continuous function (with $\phi(0) = 1$) and $\alpha > 0$. That is, a vector $(X_1, \ldots, X_n)$ has an α-symmetric distribution if, for all $t_k \in \mathbf{R}$,

$$\mathbb{E} \exp\left(\sum_{k=1}^{n} it_k X_k\right) = \phi\left(\left(\sum_{k=1}^{n} |t_k|^\alpha\right)^{1/\alpha}\right) \tag{4.16}$$

with a function ϕ that satisfies the above properties. Such distributions should not be confused with multivariate spherically symmetric stable distributions, which have characteristic functions $\exp\big[-\lambda\big(\sum_{i=1}^{n} t_i^2\big)^{\beta/2}\big]$, $0 < \beta \leq 2$. Obviously, spherically symmetric stable distributions are particular examples of α-symmetric distributions with $\alpha = 2$ (that is, of spherical distributions) and $\phi(x) = \exp(-x^\beta)$.

An important fact is that, similar to strictly stable laws, relation (1.10) holds if $(X_1, \ldots, X_n)$ has an α-symmetric distribution (see, e.g., Fang *et al.* (1990), Chapter 7). The number α is called the index and the function ϕ is called the c.f. generator of the α-symmetric distribution.

The class of α-symmetric distributions contains, as a subclass, spherical (also referred to as spherically symmetric) distributions corresponding to the case $\alpha = 2$ (see Fang *et al.* (1990), p. 184). Spherical distributions, in turn, include such examples as Kotz-type, multivariate normal, multivariate t and multivariate spherically symmetric α-stable distributions. Spherically symmetric stable distributions have characteristic functions $\exp\big[-\lambda\big(\sum_{i=1}^{n} t_i^2\big)^{\theta/2}\big]$, $0 < \theta \leq 2$, and are, thus, examples of α-symmetric distributions with $\alpha = 2$ and the c.f. generator $\phi(x) = \exp(-x^\theta)$.

Similar to the framework based on stable distributions, the stylized facts on portfolio diversification hold for convolutions of α-symmetric distributions with $\alpha > 1$. The stylized facts are reversed in the case of convolutions of α-symmetric distributions with $\alpha < 1$.

4.4.2. *Multiplicative common shocks*

For any $0 < \alpha \leq 2$, the class of α-symmetric distributions includes distributions of risks $Q_1, \ldots, Q_n$ that have the common factor representation

$$(Q_1, \ldots, Q_n) = (ZY_1, \ldots, ZY_n), \tag{4.17}$$

where $Y_i \sim S_\alpha(\sigma, 0, 0)$, $\sigma > 0$, are i.i.d. stable r.v.'s and $Z \geq 0$ is a nonnegative r.v. independent of $Y_i's$ (see Bretagnolle *et al.* (1966); Fang *et al.* (1990), p. 197). In the case $Z = 1$ (a.s.), model (4.17) represents vectors with i.i.d. symmetric stable components that have c.f.'s $\exp[-\lambda \sum_{i=1}^n |t_i|^\alpha]$ — special cases of the generator $\phi(x) = \exp(-\lambda x^\alpha)$.

According to the results of Bretagnolle *et al.* (1966) and Kuritsyn and Shestakov (1984), the function $\exp\left(-\left(|t_1|^\alpha + |t_2|^\alpha\right)^{1/\alpha}\right)$ is a c.f. of two α-symmetric r.v.'s for all $\alpha \geq 1$ (the generator of the function is $\phi(u) = \exp(-u)$). Zastavnyi (1993) demonstrates that the class of more than two α-symmetric r.v.'s with $\alpha > 2$ consists of degenerate variables (so that their c.f. generator $\phi(u) = 1$). For further review of properties and examples of α-symmetric distributions, the reader is referred to Chapter 7 of Fang *et al.* (1990) and to Gneiting (1998).

The dependence structures considered in this section include vectors $(X_1, \ldots, X_n)$ given by sums of independent random vectors $(Y_{1j}, \ldots, Y_{nj})$, $j = 1, \ldots, k$, where $(Y_{1j}, \ldots, Y_{nj})$ has an absolutely continuous α-symmetric distribution with index α_j:

$$(X_1, \ldots, X_n) = \sum_{j=1}^k (Y_{1j}, \ldots, Y_{nj}). \tag{4.18}$$

In particular, this framework includes sums of random vectors $(Z_j Y_{1j}, \ldots, Z_j Y_{nj})$, $j = 1, \ldots, k$, in (4.17) with independent r.v.'s Z_j, Y_{ij}, $j = 1, \ldots, k$, $i = 1, \ldots, n$, such that Z_j are positive and absolutely continuous and $Y_{ij} \sim S_{\alpha_j}(\sigma_j, 0, 0)$, $\alpha_j \in (0, 2]$, $\sigma_j > 0$:

$$(X_1, \ldots, X_n) = \sum_{j=1}^k (Z_j Y_{1j}, \ldots, Z_j Y_{nj}). \tag{4.19}$$

Although the dependence structure in model (4.17) alone is restrictive, convolutions (4.19) of such vectors provide a natural framework for modelling environments with multiple common shocks Z_j, such as macroeconomic or political shocks that affect all risks X_i (see Andrews (2005)).

Convolutions of α-symmetric distributions exhibit both heavy-tailedness in marginals and dependence among them. It is not difficult to show that convolutions of α-symmetric distributions with $\alpha < 1$ have extremely heavy-tailed marginals with infinite means. This is true because if one assumes that r.v.'s $X_1, \ldots, X_n$, $n \geq 2$, have an α-symmetric distribution with $\alpha < 1$ and that $\mathbb{E}|X_i| < \infty$, $i = 1, \ldots, n$, then, by the triangle inequality, $\mathbb{E}|X_1 + \cdots + X_n| \leq \mathbb{E}|X_1| + \cdots + \mathbb{E}|X_n| = n\mathbb{E}|X_1|$. This inequality, however, cannot hold since, according to (1.10), $(X_1 + \cdots + X_n) \sim n^{1/\alpha} X_1$ and thus, under the above assumptions, $\mathbb{E}|X_1 + \cdots + X_n| > n\mathbb{E}|X_1|$. Similarly, one can show that α-symmetric distributions with $\alpha < r$ have infinite marginal moments of order r.

On the other hand, convolutions of α-symmetric distributions with $1 < \alpha \leq 2$ can have marginals with power moments finite up to a certain positive order (or finite exponential moments). In particular, finiteness of moments for convolutions of models (4.17) with $1 < \alpha \leq 2$ depends on heavy-tailedness of the common shock variables Z. For instance, convolutions of models (4.17) with $1 < \alpha < 2$ and $\mathbb{E}|Z| < \infty$ have finite means but infinite variances. However, marginals of such convolutions have infinite means if the r.v.'s Z satisfy $\mathbb{E}|Z| = \infty$. Moments $\mathbb{E}|ZY_i|^p$, $p > 0$, of marginals in models (4.17) with $\alpha = 2$ (that correspond to Gaussian r.v.'s Y_i) are finite if and only if $\mathbb{E}|Z|^p < \infty$. In particular, all marginal power moments in models (4.17) with $\alpha = 2$ are finite if $\mathbb{E}|Z|^p < \infty$ for all $p > 0$.

Similarly, marginals of spherical (that is, 2-symmetric) distributions range from extremely heavy-tailed to extreme lighted-tailed ones. For example, marginal moments of spherically symmetric α-stable distributions with c.f.'s $\exp\left[-\lambda\left(\sum_{i=1}^n t_i^2\right)^{\theta/2}\right]$, $0 < \theta < 2$, are finite if and only if their order is less than θ. Marginal moments of a multivariate t-distribution with k degrees of freedom which is an example of a spherical distribution are finite if and only the order of the moments is less than k. These distributions were used in a number of works modelling heavy-tailedness with moments up to some order (see, among others, Praetz (1972); Blattberg and Gonedes (1974); Glasserman *et al.* (2002)).

Theorems 4.4 and 4.5 show that the results presented in Section 2.4 for i.i.d. risks continue to hold for convolutions (4.18) and (4.19) of α-symmetric distributions, in particular, for models with common shocks.

Let Φ denote the class of c.f. generators ϕ such that $\phi(0) = 1$, $\lim_{t \to \infty} \phi(t) = 0$, and the function $\phi'(t)$ is concave.

Theorem 4.4. *Theorem 2.1 continues to hold if any of the following is satisfied:*

— *the random vector $(X_1, \ldots, X_n)$ is given by (4.18), where, for $j = 1, \ldots, k$, $(Y_{1j}, \ldots, Y_{nj})$ has an absolutely continuous α-symmetric distribution with the c.f. generator $\phi_j \in \Phi$ and the index $\alpha_j \in (1, 2]$. In particular, Theorem 2.1 holds if $(X_1, \ldots, X_n)$ is given by (4.18), where $(Y_{1j}, \ldots, Y_{nj})$, $j = 1, \ldots, k$, have absolutely continuous spherical distributions with c.f. generators $\phi_j \in \Phi$ (the case $\alpha_j = 2$ for all j);*
— *the random vector $(X_1, \ldots, X_n)$ is given by (4.19), where $\alpha_j \in (1, 2]$, $j = 1, \ldots, k$.*

Theorem 4.5. *Theorem 2.2 continues to hold if any of the following is satisfied:*

— *the random vector $(X_1, \ldots, X_n)$ is given by (4.18), where, for $j = 1, \ldots, k$, $(Y_{1j}, \ldots, Y_{nj})$ has an absolutely continuous α-symmetric distribution with the c.f. generator $\phi_j \in \Phi$ and the index $\alpha_j \in (0, 1)$;*
— *the random vector $(X_1, \ldots, X_n)$ is given by (4.19), where $\alpha_j \in (0, 1)$, $j = 1, \ldots, k$.*

Note that similar to Theorems 4.4 and 4.5, one can obtain extensions of the sharp VaR bounds such as in Theorem 2.7 for heavy-tailed dependent risks.

4.4.3. *Additive common shocks*

As discussed in the introduction to Section 4.4, we consider portfolios of risks Y_{ij}, $i = 1, \ldots, r$, $j = 1, \ldots, c$, in model (4.13) with identically distributed "row effects" common shocks R_i, $i = 1, \ldots, r$, identically distributed "column effects" common shocks C_j, $j = 1, \ldots, c$, and identically distributed idiosyncratic components U_{ij}, $i = 1, \ldots, r$, $j = 1, \ldots, c$. The r.v.'s R_i, C_j, U_{ij}, $i = 1, \ldots, r$, $j = 1, \ldots, c$, are assumed to be independent.

For $w = (w_{11}, \ldots, w_{1c}, w_{21}, \ldots, w_{2c}, \ldots, w_{r1}, \ldots, w_{rc}) \in \mathcal{I}_{rc}$, denote $w_{0j} = \sum_{i=1}^{r} w_{ij}$, $j = 1, \ldots, c$, $w_{i0} = \sum_{j=1}^{c} w_{ij}$, $i = 1, \ldots, r$. Further, denote

$$w_0^{(row)} = (w_{10}, \ldots, w_{r0}) \in \mathcal{I}_r, \tag{4.20}$$

$$w_0^{(col)} = (w_{01}, \ldots, w_{0c}) \in \mathcal{I}_c. \tag{4.21}$$

For $w = (w_{11}, \ldots, w_{1c}, w_{21}, \ldots, w_{2c}, \ldots, w_{r1}, \ldots, w_{rc}) \in \mathcal{I}_{rc}$, the return on the portfolio of risks Y_{ij} in (4.13) with weights w is given by

$$
\begin{aligned}
Y(w) &= \sum_{i=1}^{r} \sum_{j=1}^{c} w_{ij} Y_{ij} \\
&= \sum_{i=1}^{r} w_{i0} R_i + \sum_{j=1}^{c} w_{0j} C_j + \sum_{i=1}^{r} \sum_{j=1}^{c} w_{ij} U_{ij} \\
&= R(w_0^{(row)}) + C(w_0^{(col)}) + U(w), \tag{4.22}
\end{aligned}
$$

where $R(w_0^{(row)}) = \sum_{i=1}^{r} w_{i0} R_i$, $C(w_0^{(col)}) = \sum_{j=1}^{c} w_{0j} C_j$ and $U(w) = \sum_{i=1}^{r} \sum_{j=1}^{c} w_{ij} U_{ij}$.

Consider the vector of equal weights $\underline{w}_{rc} = \underbrace{(1/(rc), 1/(rc), \ldots, 1/(rc))}_{rc} \in \mathcal{I}_{rc}$ and the vector $\overline{w}_{rc} = \underbrace{(1, 0, \ldots, 0)}_{rc} \in \mathcal{I}_{rc}$ that corresponds to the portfolio of Y_{ij}'s that consists of only one risk.

Observe that the vectors $\underline{w}_0^{(row)}$ and $\underline{w}_0^{(col)}$ that correspond to $\underline{w}_{rc}$ by (4.20) and (4.21) consist of equal weights. Namely, $\underline{w}_0^{(row)} = \underbrace{(1/r, \ldots, 1/r)}_{r} = \underline{w}_r \in \mathcal{I}_r$ and $\underline{w}_0^{(col)} = \underbrace{(1/c, \ldots, 1/c)}_{c} = \underline{w}_c \in \mathcal{I}_c$.

Similarly, for the weights $\overline{w}_0^{(row)}$ and $\overline{w}_0^{(col)}$ corresponding to $\overline{w}_{rc}$ by (4.20) and (4.21), we have $\overline{w}_0^{(row)} = \underbrace{(1, 0, \ldots, 0)}_{r} = \overline{w}_r \in \mathcal{I}_r$ and

$$\overline{w}_0^{(col)} = \underbrace{(1, 0, \ldots, 0)}_{c} = \overline{w}_c \in \mathcal{I}_c.$$

The following theorem provides value at risk comparisons for portfolio returns $Y(w)$ in (4.22) under heavy-tailedness in the risk components R_i, C_j and U_{ij}.

Theorem 4.6. *Let* $q \in (0, 1/2)$.

(i) *If* $R_i, C_j, U_{ij} \sim \overline{\mathcal{CSLC}}$, *then* $VaR_q[Y(\underline{w}_{rc})] \leq VaR_q[Y(w)] \leq VaR_q[Y(\overline{w}_{rc})]$ *for all* $w \in \mathcal{I}_{rc}$.

(ii) *If* $R_i, C_j, U_{ij} \sim \underline{\mathcal{CS}}$, *then* $VaR_q[Y(\underline{w}_{rc})] \geq VaR_q[Y(w)] \geq VaR_q[Y(\overline{w}_{rc})]$ *for all* $w \in \mathcal{I}_{rc}$.

Part (i) of Theorem 4.6 shows that, similar to the case of independence in part (ii) of Theorem 2.1, the most diversified portfolio with equal weights $\underline{w}_{rc}$ is preferred to any other portfolio of dependent risks Y_{ij} in (4.13) under moderate heavy-tailedness. In addition, the least diversified portfolio with weights $\overline{w}_{rc}$ consisting of only one risk is dominated by any other portfolio of risks Y_{ij} with additive common shocks. Part (ii) of Theorem 4.6 implies that, similar to independence in part (ii) of Theorem 2.2, the conclusions are reversed under extreme heavy-tailedness.

Extreme heavy-tailedness of common shocks R_i, C_j and idiosyncratic risks U_{ij} in (4.13) implies optimality of the least diversified portfolio with weights $\overline{w}_{rc}$ with respect to the portfolio value at risk comparisons. In contrast, the portfolio value at risk is maximal for the most diversified portfolio with equal weights $\underline{w}_{rc}$ under such assumptions.

As an immediate consequence of Theorems 2.1 and 2.2, one also obtains similar comparisons with the extremal portfolio weights $\underline{w}_{rc}$ and $\overline{w}_{rc}$ for the values at risk of the components $R(w_0^{(row)})$, $C(w_0^{(col)})$ and $U(w)$ in decomposition (4.22) with weights $w \in \mathcal{I}_{rc}$. Namely, the following conclusions hold.

Theorem 4.7. *Let* $q \in (0, 1/2)$.

(i) *If* $U_{ij} \sim \overline{\mathcal{CSLC}}$, *then* $VaR_q[U(\underline{w}_{rc})] \leq VaR_q[U(w)] \leq VaR_q[U(\overline{w}_{rc})]$ *for all* $w \in \mathcal{I}_{rc}$.

(ii) *If* $U_{ij} \sim \underline{\mathcal{CS}}$, *then* $VaR_q[U(\underline{w}_{rc})] \geq VaR_q[U(w)] \geq VaR_q[U(\overline{w}_{rc})]$ *for all* $w \in \mathcal{I}_{rc}$.

(iii) *If $R_i \sim \overline{\mathcal{CSLC}}$, then $VaR_q[R(\underline{w}_r)] \leq VaR_q[R(w_0^{(row)})] \leq VaR_q$
      $[R(\overline{w}_r)]$ for all $w \in \mathcal{I}_{rc}$.*
(iv) *If $R_i \sim \underline{\mathcal{CS}}$, then $VaR_q[R(\underline{w}_r)] \geq VaR_q[R(w_0^{(row)})] \geq VaR_q$
      $[R(\overline{w}_r)]$ for all $w \in \mathcal{I}_{rc}$.*
 (v) *If $C_j \sim \overline{\mathcal{CSLC}}$, then $VaR_q[C(\underline{w}_c)] \leq VaR_q[C(w_0^{(col)})] \leq VaR_q$
      $[C(\overline{w}_c)]$ for all $w \in \mathcal{I}_{rc}$.*
(vi) *If $C_j \sim \underline{\mathcal{CS}}$, then $VaR_q[C(\underline{w}_c)] \geq VaR_q[C(w_0^{(col)})] \geq VaR_q$
      $[C(\overline{w}_c)]$ for all $w \in \mathcal{I}_{rc}$.*

As in the case of independence in parts (i) of Theorems 2.1 and 2.2 it is of interest to also consider value at risk comparisons for general portfolio weights $v, w \in \mathcal{I}_{rc}$ satisfying $v \prec w$ (so that the portfolio with weights v is more diversified than that with weights w). However, such general comparisons cannot be obtained using majorization on $\mathcal{I}_{rc}$. This is because the values at risk

$$
VaR_q[R(w_0^{(row)})] = VaR_q\left[\sum_{i=1}^{r} w_{i0}R_i\right] = VaR_q\left[\sum_{i=1}^{r}\left(\sum_{j=1}^{c}w_{ij}\right)R_i\right],
$$

$$
VaR_q[C(w_0^{(col)})] = VaR_q\left[\sum_{j=1}^{c} w_{0j}C_j\right] = VaR_q\left[\sum_{j=1}^{c}\left(\sum_{i=1}^{r}w_{ij}\right)C_j\right]
$$

for the components $R(w_0^{(row)})$ and $C(w_0^{(col)})$ in (4.22) are not symmetric functions of w_{ij}'s. Thus, these functions are neither Schur-concave nor Schur-convex in $w \in \mathcal{I}_{rc}$. This situation is similar to the majorization-based analysis of variance decompositions for linear estimators of location in two-way classification random effects models in Section 13.B of Marshall and Olkin (1979) and Marshall *et al.* (2011) (see also Sections 4.5.1 and 4.5.3). As indicated by Marshall and Olkin (1979) Marshall *et al.* (2011), and neither Schur-convexity nor Schur-concavity (in $w \in \mathcal{I}_{rc}$) holds for the variances $\mathbb{V}[R(w_0^{(row)})] = \mathbb{V}[\sum_{i=1}^{r} w_{i0}R_i]$ and $\mathbb{V}[C(w_0^{(col)})] = \mathbb{V}[\sum_{j=1}^{c} w_{0j}C_j]$ because these functions are not symmetric functions of w'_{ij}s.

Nevertheless, the above value at risk comparisons with $v \prec w$ are possible for certain portfolio weights v and w that are different from the most diversified portfolio $\underline{w}_{rc}$ and do not correspond to the least diversification with weights $\overline{w}_{rc}$. Some of these value at risk comparisons have a natural interpretation in terms of value at risk analysis for portfolios of equally weighted indices of risks Y_{ij} in (4.13). These value at risk orderings and the settings where they arise are considered in the next two sections.

4.5. Further results for common shock models

This section covers some theoretical applications and extensions of Section 4.4. Section 4.5.1 provides extensions of the results in Theorems 4.6 and 4.7 to value at risk comparisons between portfolios that are different from the most diversified and the least diversified portfolios (Theorem 4.8).

In Section 4.5.2, we show that the majorization approach to VaR analysis for dependent risks developed in Sections 4.4.3 and 4.5.1 can be applied in the case of unbalanced models (4.13) or (4.14) that have unequal number of rows for each column or unequal number of columns for each row. Building on the interpretation in Section 4.5.1, the results in Section 4.5.2 are presented in the framework of value at risk analysis for equally weighted indices of heavy-tailed risks in (4.14).

Section 4.5.3 discusses econometric and statistical applications of the results on models with common shocks. These applications are the analogues of the value at risk results in the framework of efficiency comparisons of linear estimators of location in random effects models.

In the section on models with common shocks, we mainly work with models of the form (4.13) and (4.14) but our results can be generalized to models with varying factor loadings, such as

$$Y_{ij} = \beta_{ij}^{(r)} R_i + \beta_{ij}^{(c)} C_j + U_{ij}, \quad i = 1, \ldots, r, \ j = 1, \ldots, c, \quad (4.23)$$

(see Bai (2009)). We discuss the extensions to general factor settings (4.23) and to the case with more than two common factors in Section 4.5.4. We also note that the results can be used to obtain value at risk and diversification comparisons for the usual fixed effects models in the form $Y_{it} = X_{it}'\beta + R_i + C_t + U_{it}$, where X_{it} is a $p \times 1$ vector of (possibly heavy-tailed) regressors, β is a $p \times 1$ vector of unknown regression coefficients, and R_i and C_t are, respectively, individual and time effect variables.

4.5.1. *Further applications: Portfolio component VaR*

Let $n_{ij} \in \{0, 1\}$, $i = 1, \ldots, r$, $j = 1, \ldots, c$, be a set of indicator variables. Denote $n_{i0} = \sum_{j=1}^{c} n_{ij}$, $n_{0j} = \sum_{i=1}^{r} n_{ij}$, $n = \sum_{i=1}^{r} n_{i0} = \sum_{j=1}^{c} n_{0j} = \sum_{i=1}^{r} \sum_{j=1}^{c} n_{ij}$.

Section 13.B in Marshall and Olkin (1979) and Marshall *et al.* (2011) discusses applications of majorization theory in comparisons of variance components for linear estimators based on observations in specifications (4.13) referred to as two-way classification random effects models (see also the review in Section 4.5.3 and references therein). Marshall and Olkin (1979) and Marshall *et al.* (2011) consider the equal weights $\underline{w}_{ij} = 1/(rc)$ dealt with in the previous section and also the portfolio weights $\tilde{v}_{ij} = n_{i0}/(nc)$, $\tilde{\tilde{v}}_{ij} = n_{0j}/(nr)$ and $\tilde{w}_{ij} = \frac{(n - n_{i0} - n_{0j} + n_{ij})n_{ij}}{n^2 - \sum_{i=1}^{r} n_{i0}^2 - \sum_{j=1}^{c} n_{0j}^2 + n}$, $i = 1, \ldots, r$, $j = 1, \ldots, c$, discussed in Koch (1967a,b). Denote the weight vectors corresponding to the last three choices by $\tilde{v}$, $\tilde{\tilde{v}}$ and $\tilde{w}$.

In the context of portfolio choice, some of the properties of the portfolios with weights $\underline{w}$, $\tilde{v}$, $\tilde{\tilde{v}}$ and $\tilde{w}$ may be summarized as follows. Consider r equally weighted indices comprised, for $i = 1, \ldots, r$, of risks Y_{ij}, $j = 1, \ldots, c$, in the ith row of (4.13). The return on index i is thus given by $Z_i^{(col)} = R_i + \frac{1}{c}\sum_{j=1}^{c} C_j + \frac{1}{c}\sum_{j=1}^{c} U_{ij}$, $i = 1, \ldots, r$. For $w = (w_1, \ldots, w_r) \in \mathcal{I}_r$, the return on the portfolio of the indices $i = 1, \ldots, r$ with returns $Z_i^{(col)}$ and weights w is given by

$$Z^{(col)}(w) = \sum_{i=1}^{r} w_i Z_i^{(col)} = \sum_{i=1}^{r} w_i R_i + \frac{1}{c}\sum_{j=1}^{c} C_j$$

$$+ \sum_{i=1}^{r} \frac{(U_{i1} + \cdots + U_{ic})}{c} w_i. \qquad (4.24)$$

Similarly, consider, for $j = 1, \ldots, c$, the equally weighted indices comprised of risks Y_{ij}, $i = 1, \ldots, r$, in the jth column of (4.13). The return on index j is thus given by $Z_j^{(row)} = C_j + \frac{1}{r}\sum_{i=1}^{c} R_i +$

$\frac{1}{r}\sum_{i=1}^{r} U_{ij}$, $j = 1, \ldots, c$. For $w = (w_1, \ldots, w_c) \in \mathcal{I}_c$, the return on the portfolio of the indices $j = 1, \ldots, c$ with returns $Z_j^{(row)}$ and weights w is given by

$$Z^{(row)}(w) = \sum_{j=1}^{c} w_j Z_j^{(row)}$$

$$= \frac{1}{r}\sum_{i=1}^{r} R_i + \sum_{j=1}^{c} w_j C_j + \sum_{j=1}^{c} \frac{(U_{1j} + \cdots + U_{rj})}{r} w_j.$$

$$(4.25)$$

The risk $Y(\tilde{v})$ obtained using (4.22) with weights $\tilde{v}$ is the same as the return on the portfolio of the indices $Z_i^{(col)}$, $i = 1, \ldots, r$, with weights $w = \underbrace{(n_{10}/n, \ldots, n_{r0}/n)}_{r} \in \mathcal{I}_r$ in (4.24). The return $Y(\tilde{\tilde{v}})$ in (4.22) with weights $\tilde{\tilde{v}}$ is the same as the return on the portfolio of the risks $Z_j^{(row)}$, $j = 1, \ldots, c$, with weights $w = \underbrace{(n_{01}/n, \ldots, n_{0c}/n)}_{c} \in \mathcal{I}_c$ in (4.25).

As discussed in the previous section, the equal weights $\underline{w}_{rc} = \underbrace{\left(1/(rc), 1/(rc), \ldots, 1/(rc)\right)}_{rc} \in \mathcal{I}_{rc}$ correspond to the most diversified portfolio of the risks Y_{ij}. The return $Y(\underline{w}_{rc})$ with these weights in (4.22) is the same as the returns $Z^{(col)}(\underline{w}_r)$ and $Z^{(row)}(\underline{w}_c)$ on the portfolios of indices in (4.24) and (4.25) with equal weights $\underline{w}_r = \underbrace{(1/r, \ldots, 1/r)}_{r} \in \mathcal{I}_r$ and $\underline{w}_c = \underbrace{(1/c, \ldots, 1/c)}_{c} \in \mathcal{I}_c$.

The weights $\tilde{w}$ correspond to a portfolio of Y_{ij} where, in contrast to $\underline{w}_{rc}$, $\tilde{v}$ and $\tilde{\tilde{v}}$, some of the risks Y_{ij} are taken with zero weights that may be due to the risks' unavailability.

Lemma 13.B.2.a in Marshall and Olkin (1979) (provided by Lemma 13.B.3 in Marshall *et al.* (2011)) and relation (2.1) with $N = rc$ show that the following majorization comparisons hold for the vectors $\underline{w}_{rc}$, $\tilde{v}$, $\tilde{\tilde{v}}$, $\tilde{w}$ and $\overline{w}_{rc}$:

$$\underline{w}_{rc} \prec \tilde{v} \prec \tilde{w} \prec \overline{w}_{rc}, \qquad (4.26)$$

$$\underline{w}_{rc} \prec \tilde{\tilde{v}} \prec \tilde{w} \prec \overline{w}_{rc}. \qquad (4.27)$$

Theorems 4.6 and 4.7 imply value at risk comparisons for the portfolio returns $Y(w)$ and its components $R(w_0^{(row)})$, $C(w_0^{(col)})$ and $U(w)$ in (4.22) between an arbitrary $w \in \mathcal{I}_{rc}$ (for instance, $w = \tilde{v}, \tilde{\tilde{v}}, \tilde{w}$) and the extremal portfolio weights $\underline{w}_{rc}$ and $\overline{w}_{rc}$. In particular, from parts (iii)–(vi) of Theorem 4.7, it follows that the following comparisons hold for all $q \in (0, 1/2)$ and all $w \in \mathcal{I}_{rc}$ (e.g., for $w = \tilde{v}, \tilde{\tilde{v}}, \tilde{w}$). These comparisons are direct consequences of the results for the independent case in Theorems 2.1 and 2.2.

$$\mathrm{VaR}_q[R(\tilde{v}_0^{(row)})] = \mathrm{VaR}_q[R(\underline{w}_r)] \leq \mathrm{VaR}_q[R(w_0^{(row)})] \text{ if } R_i \sim \overline{\mathcal{CSLC}},$$
(4.28)

$$\mathrm{VaR}_q[C(\tilde{v}_0^{(col)})] = \mathrm{VaR}_q[C(\underline{w}_c)] \leq \mathrm{VaR}_q[C(w_0^{(col)})] \text{ if } C_j \sim \overline{\mathcal{CSLC}},$$
(4.29)

$$\mathrm{VaR}_q[R(\tilde{v}_0^{(row)})] = \mathrm{VaR}_q[R(\underline{w}_r)] \geq \mathrm{VaR}_q[R(w_0^{(row)})] \text{ if } R_i \sim \underline{\mathcal{CS}},$$
(4.30)

$$\mathrm{VaR}_q[C(\tilde{v}_0^{(col)})] = \mathrm{VaR}_q[C(\underline{w}_c)] \geq \mathrm{VaR}_q[C(w_0^{(col)})] \text{ if } C_j \sim \underline{\mathcal{CS}}.$$
(4.31)

Similar value at risk comparisons for $\mathrm{VaR}_q[U(w)]$ are also obtained between the portfolio with weights $\tilde{w}$ and those with weights $w = \tilde{v}, \tilde{\tilde{v}}$. Namely, the following result holds.

Theorem 4.8. *Let $q \in (0, 1/2)$. The following value at risk comparisons hold for the component $U(w)$ of decomposition (4.22):*

(i) *If $U_{ij} \sim \overline{\mathcal{CSLC}}$, then $\mathrm{VaR}_q[U(\tilde{w})] \geq \mathrm{VaR}_q[U(\tilde{v})]$ and $\mathrm{VaR}_q[U(\tilde{w})] \geq \mathrm{VaR}_q[U(\tilde{\tilde{v}})]$.*
(ii) *If $U_{ij} \sim \underline{\mathcal{CS}}$, then $\mathrm{VaR}_q[U(\tilde{w})] \leq \mathrm{VaR}_q[U(\tilde{v})]$ and $\mathrm{VaR}_q[U(\tilde{w})] \leq \mathrm{VaR}_q[U(\tilde{\tilde{v}})]$.*

The main conclusions from the above value at risk comparisons for the portfolio components $U(w)$, $R(w_0^{(row)})$ and $C(w_0^{(row)})$ in (4.22) with weights $\underline{w}_{rc}$, $\tilde{v}$, $\tilde{\tilde{v}}$, $\tilde{w}$ and $\overline{w}_{rc}$ are summarized as follows. In the case of moderately heavy-tailed common shocks R_i, C_j and idiosyncratic risks U_{ij}, we conclude that full diversification on the level of underlying Y_{ij} with weights $\underline{w}_{rc}$ is preferred to any other portfolio choice, provided all the rc risks are available (left inequality in part (i) of Theorem 4.6). In particular, the equal weights $\underline{w}_{rc}$ are preferred to less diversification with weights $\tilde{w}$ where some of the risks may be taken with zero weights. Full diversification with weights $\underline{w}_{rc}$ at Y_{ij} is also preferred to the portfolio of equally weighted indices $Z_i^{(col)}$, $i = 1, \ldots, r$, with weights $w = \underbrace{(n_{10}/n, \ldots, n_{r0}/n)}_{r} \in \mathcal{I}_r$ and the

portfolio of indices with returns $Z_j^{(row)}$, $j = 1, \ldots, c$, and the weight vector $w = \underbrace{(n_{01}/n, \ldots, n_{0c}/n)}_{c} \in \mathcal{I}_c$. In turn, comparisons (4.28) and

(4.29) for the common shock parts $R(w_0^{(row)})$ and $C(w_0^{(row)})$ in (4.22) and part (i) of Theorem 4.8 for $U(w)$ suggest that the weight vectors $\tilde{v}$ and $\tilde{\tilde{v}}$ may be preferred to the vector $\tilde{w}$ where some of the risks are included with zero weights. These conclusions are similar to those in Section 13.B in Marshall and Olkin (1979) and Marshall *et al.* (2011) for variance comparisons of linear estimators based on observations (4.13) with $ER_i = \mu$, $EC_j = 0$, $EU_{ij} = 0$, $var(R_i) = \sigma_R^2$, $var(C_j) = \sigma_C^2$, $var(U_{ij}) = \sigma_U^2$, $i = 1, \ldots, r$, $j = 1, \ldots, c$.

These results are reversed for extremely heavy-tailed risks R_i, C_j and U_{ij}. In such settings, the equal weights $\underline{w}_{rc}$ are dominated by any other portfolio weights (left inequality in part (ii) of Theorem 4.6). In contrast, the smallest VaR is achieved at the weights $\overline{w}_{rc}$ and the portfolio consisting of only one risk. In particular, the portfolio of indices with returns $Z_i^{(col)}$, $i = 1, \ldots, r$, and weights $w = \underbrace{(n_{10}/n, \ldots, n_{r0}/n)}_{r} \in \mathcal{I}_r$ and the portfolio of indices with returns

$Z_j^{(row)}$, $j = 1, \ldots, c$, and $w = \underbrace{(n_{01}/n, \ldots, n_{0c}/n)}_{c} \in \mathcal{I}_c$ are preferred to

the fully diversified portfolio of Y'_{ij}s with equal weights $\underline{w}_{rc}$. Inequalities (4.30) and (4.31) for $R(w_0^{(row)})$ and $C(w_0^{(row)})$ and part (ii) of Theorem 4.8 for $U(w)$ suggest that the weights $\tilde{w}$ may dominate the weights $\tilde{v}$ and $\tilde{\tilde{v}}$.

The results in (4.28)–(4.31) and Theorem 4.8 also indicate that, especially in the cases where heavy-tailedness of the common shocks

R_i and C_j and that of the idiosyncratic risks U_{ij} is of different degree (say, in the case of the assumptions $U_{ij} \sim \overline{CSLC}$ combined with $R_i \sim \underline{CS}$ and $C_j \sim \underline{CS}$), the VaR comparisons for the components $R(w_0^{(row)})$, $C(w_0^{(col)})$ and $U(w)$ in (4.22) do not point out to an optimal portfolio choice among $\tilde{v}$, $\tilde{\tilde{v}}$ and $\tilde{w}$. Thus, the optimal portfolio selection depends crucially on the distributional properties of common shocks R_i and C_j and idiosyncratic risks U_{ij}.

4.5.2. *When heavy-tailedness helps: VaR for financial indices*

In many real world situations, the sets of available risks in (4.13) or (4.14) are unbalanced and include unequal number of rows for each column or unequal number of columns for each row. In this section, we show how the approach presented in Section 4.4 can be applied in the analysis of such settings. We obtain the results for unbalanced analogues of models (4.14) with one set of "row effects" common shocks R_i and focus on the framework of value at risk comparisons for indices based on risks Y_{ij} in such models. The results can be extended to more general models, including those corresponding to settings with two sets of common shocks in (4.13), risks in form (4.23) with varying factor loadings and multiple common shock models (see the discussion and the results in Section 4.5.4).

Let $n_1 \geq n_2 \geq \cdots \geq n_r \geq 1$, $\sum_{i=1}^{r} n_i = n$. Similar to the interpretation of weights $\tilde{v}_{ij}$ and $\tilde{\tilde{v}}_{ij}$ in Section 4.5.1, consider r equally weighted indices comprised, for $i = 1, \ldots, r$, of risks

$$Y_{ij} = R_i + U_{ij}, \quad j = 1, \ldots, n_i, \qquad (4.32)$$

with identically distributed common shocks R_i, $i = 1, \ldots, r$, and identically distributed idiosyncratic risks U_{ij}, $j = 1, \ldots, n_i$, $i = 1, \ldots, r$.

Equally weighted indices are quite common in financial markets, and their examples include the majority of hedge fund indices (see Lhabitant (2000)), the Value Line Arithmetic Index, Global Dow, the Thomson Reuters Equal Weight Continuous Commodity Index, the Thomson Reuters/Jefferies CRB (Commodity Research Bureau)

Index, and others. We also note that since the results in this section can be generalized to models with varying factor loadings (see Section 4.5.4), non-equal weighted indices can also be covered by the analysis if certain restrictions are met.

As before, the r.v.'s R_i, U_{ij} $j = 1, \ldots, n_i$, $i = 1, \ldots, r$, are assumed to be independent. The return on the index i is given by

$$Z_i = \frac{1}{n_i} \sum_{j=1}^{n_i} Y_{ij} = R_i + \frac{1}{n_i} \sum_{j=1}^{n_i} U_{ij}. \tag{4.33}$$

As in Sections 4.4 and 4.5.1, for $w = (w_1, \ldots, w_r) \in \mathcal{I}_r$, denote by $Z(w)$ the return on the portfolio of the indices $i = 1, \ldots, r$ with returns Z_i in (4.33) and weights w:

$$Z(w) = \sum_{i=1}^{r} w_i Z_i = \sum_{i=1}^{r} \frac{(Y_{i1} + \cdots + Y_{in_i})}{n_i} w_i$$

$$= \sum_{i=1}^{r} w_i R_i + \sum_{i=1}^{r} \frac{(U_{i1} + \cdots + U_{in_i})}{n_i} w_i. \tag{4.34}$$

In what follows, for $N \geq 1$, $e_N = \underbrace{(1, \ldots, 1)}_{N} \in \mathbf{R}^N$ will denote the N-vector of ones. In addition, for m row vectors $x^{(k)} \in \mathbf{R}^{1 \times N_k}$, $N_k \geq 1$, $k = 1, \ldots, m$, we will denote by $x = (x^{(1)}, \ldots, x^{(m)}) \in \mathbf{R}^{1 \times N}$, $N = N_1 + \cdots + N_m$, the vector with the first N_1 components equal to those of $x^{(1)}$, the next N_2 components equal to those of $x^{(2)}$, and so on: $x_i = x_i^{(1)}$, $i = 1, \ldots, N_1$; $x_{N_1+i} = x_i^{(2)}$, $i = 1, \ldots, N_2$; $\ldots$, $x_{N_1+N_2+\cdots+N_{m-1}+i} = x_i^{(m)}$, $i = 1, \ldots, N_m$.

Decomposition (4.34) can be written as

$$Z(w) = R(w) + U(\tilde{w}), \tag{4.35}$$

where $R(w)$ is the return on the portfolio of common shocks R_i with weights $w = (w_1, \ldots, w_r) \in \mathcal{I}_r$, and $U(\tilde{w}) = \sum_{i=1}^{r} \sum_{j=1}^{n_i} \tilde{w}_{ij} U_{ij}$ is the return on the portfolio of idiosyncratic risks U_{ij} with weights $\tilde{w}_{ij} = w_i/n_i, j = 1, \ldots, n_i, i = 1, \ldots, r$, and the corresponding weight vector

$$\tilde{w} = (\tilde{w}_{11}, \ldots, \tilde{w}_{1n_1}, \ldots, \tilde{w}_{r1}, \ldots, \tilde{w}_{rn_r}) = \left(\frac{w_1}{n_1} e_{n_1}, \ldots, \frac{w_r}{n_r} e_{n_r} \right) \in \mathcal{I}_n.$$
(4.36)

The return on the portfolio of the indices with equal weights

$$w^{(1)} = \underline{w}_r = \underbrace{(1/r, \ldots, 1/r)}_{r} \in \mathcal{I}_r$$
(4.37)

is given by the sample mean of the risks Z_i, $i = 1, \ldots, r$:

$$Z(w^{(1)}) = \frac{1}{r} \sum_{i=1}^{r} Z_i = \frac{1}{r} \sum_{i=1}^{r} \frac{(Y_{i1} + \cdots + Y_{in_i})}{n_i}$$

$$= \frac{1}{r} \sum_{i=1}^{r} R_i + \frac{1}{r} \sum_{i=1}^{r} \frac{(U_{i1} + \cdots + U_{in_i})}{n_i}.$$
(4.38)

Similarly, the choice of portfolio weights

$$w^{(2)} = \left(\frac{n_1}{n}, \frac{n_2}{n}, \ldots, \frac{n_r}{n} \right)$$
(4.39)

produces the return $Z(w^{(2)})$ equal to the sample mean of the underlying risks Y_{ij} in (4.32):

$$Z(w^{(2)}) = \sum_{i=1}^{r} \frac{n_i}{n} Z_i = \frac{1}{n} \sum_{i=1}^{r} \sum_{j=1}^{n_i} Y_{ij},$$
(4.40)

that is, in the notations of Sections 4.4.2 and 4.5.1, $Z(w^{(2)}) = Y(\underline{w}_n)$, where $\underline{w}_n$ is the portfolio with equal weights $\underline{w}_n = \underbrace{(1/n, \ldots, 1/n)}_{n} \in$

$\mathcal{I}_n$. In other words, while the weights $w^{(1)}$ reflect diversification on the level of indices with returns Z_i, the weights $w^{(2)}$ correspond to (full) diversification on the level of the underlying risks Y_{ij} that comprise these indices (see the discussion in Sections 4.4.2 and 4.5.1 for formalization of the notions of diversification in terms of majorization relations for portfolio weights at Y_{ij} and Z_i).

Following the works in the statistics literature reviewed in Section 4.5.3, we also consider the portfolio weights $w^{(3)} =$

$(w_1^{(3)}, w_2^{(3)}, \ldots, w_r^{(3)}) \in \mathcal{I}_r$ and $w(c) = (w_1(c), \ldots, w_r(c)) \in \mathcal{I}_r$, $c \in [0, 1]$, with

$$w_i^{(3)} = \frac{n_i(n - n_i)}{n^2 - \sum_{s=1}^m n_s^2}, \quad i = 1, \ldots, r, \tag{4.41}$$

and

$$w_i(c) = \frac{n_i[(n_i - 1)c + 1]^{-1}}{\sum_{i=1}^r n_i[(n_i - 1)c + 1]^{-1}}. \tag{4.42}$$

For the weights $w(c)$, one has $w(0) = w^{(2)}$ and $w(1) = w^{(1)} = \underline{w}_r$. We also note that the weight vectors $w^{(1)}$, $w^{(2)}$, $w^{(3)}$ and $w(c)$ in (4.37), (4.39), (4.41) and (4.42) become the same in the special case of balanced models with $n_1 = n_2 = \ldots = n_r : w^{(l)} = w(c) = \underline{w}_r$ for all $l = 1, 2, 3$, and $c \in [0, 1]$.

Theorems 4.9 and 4.10 provide value at risk comparisons for equally weighted indices comprised of risks Y_{ij} spanned by heavy-tailed common shocks R_i and idiosyncratic risks U_{ij}. In both of them, the degree of heavy-tailedness of common shocks R_i is different from that for idiosyncratic risks U_{ij}. Theorem 4.9 concerns the case of extremely heavy-tailed R_i and moderately heavy-tailed U_{ij}.

Theorem 4.9. *Let $q \in (0, 1/2)$. Suppose that, in (4.32), $R_i \sim \underline{CS}$, $i = 1, \ldots, r$, and $U_{ij} \sim \overline{CSLC}$, $j = 1, \ldots, n_i$, $i = 1, \ldots, r$. Then the function $VaR_q[Z(w(c))]$ is non-decreasing in $c \in [0, 1]$. In particular,*

$$VaR_q[Z(w^{(1)})] \geq VaR_q[Z(w(c))] \geq VaR_q[Z(w^{(2)})]$$

for all $c \in [0, 1]$. In addition,

$$VaR_q[Z(w^{(1)})] \geq VaR_q[Z(w^{(3)})] \geq VaR_q[Z(w^{(2)})].$$

Theorem 4.10 shows that the conclusions of Theorem 4.9 are reversed in the case where the common shocks R_i are moderately heavy-tailed and the idiosyncratic risks U_{ij} are extremely heavy-tailed.

Theorem 4.10. *Let $q \in (0, 1/2)$. Suppose that, in (4.32), $R_i \sim \overline{CSLC}$, $i = 1, \ldots, r$, and $U_{ij} \sim \underline{CS}$, $j = 1, \ldots, n_i$, $i = 1, \ldots, r$.*

Then the function $VaR_q[Z(w(c))]$ is non-increasing in $c \in [0,1]$. In particular,

$$VaR_q[Z(w^{(2)})] \geq VaR_q[Z(w(c))] \geq VaR_q[Z(w^{(1)})]$$

for all $c \in [0,1]$. In addition,

$$VaR_q[Z(w^{(2)})] \geq VaR_q[Z(w^{(3)})] \geq VaR_q[Z(w^{(1)})].$$

Similar to Theorems 4.6 and 4.7, the results provided by Theorems 4.9 and 4.10 hold for the value at risk comparisons for the components $R(w)$ and $U(\tilde{w})$ in decomposition (4.35) for the portfolio returns $Z(w)$ (here and below, for weights $w = (w_1, \ldots, w_r) \in \mathcal{I}_r$ at Z_i's, $\tilde{w} \in \mathcal{I}_n$ is the vector of weights at U_{ij} in (4.35) that corresponds to w by (4.36)). These comparisons are provided in Theorem 4.11.

By the left majorization comparisons in (2.1), the portfolio weight vector $w^{(1)} = \underline{w}_r$ provides the most diversified portfolio of common shocks R_i and the vector $\tilde{w}^{(2)} = \underline{w}_n$ provides the most diversified portfolio of idiosyncratic risks U_{ij}. The portfolio weights w_i at R_i in decomposition (4.35) that correspond to the weights $\tilde{w}_{ij}^{(2)} = 1/n$, $j = 1, \ldots, n_i$, $i = 1, \ldots, r$, at the risks U_{ij} (the risks Y_{ij}) are given by $w_i = n_i/n$, $i = 1, \ldots, r$, and are thus different from the equal components $w_i^{(1)} = 1/r$ of $w^{(1)}$. This provides the intuition for the opposite results for the common shocks and idiosyncratic risks in Theorem 4.11 and for different degrees of heavy-tailedness in R_i and U_{ij} needed for the VaR comparisons in Theorems 4.9 and 4.10 to hold.

Theorem 4.11. *Let $q \in (0, 1/2)$. The following comparisons hold for the components $R(w)$ and $U(\tilde{w})$ in decomposition (4.35).*

(i) *Suppose $R_i \sim \overline{\mathcal{CSLC}}$. Then the function $VaR_q[R(w(c))]$ is non-increasing in $c \in [0,1]$. In particular,*

$$VaR_q[R(w^{(2)})] \geq VaR_q[R(w(c))] \geq VaR_q[R(w^{(1)})]$$

for all $c \in [0,1]$. In addition, $VaR_q[R(w^{(3)})] \leq VaR_q[R(w^{(2)})]$ and $VaR_q[R(w^{(1)})] \leq VaR_q[R(w)]$ for all $w \in \mathcal{I}_r$.

(ii) *Suppose* $R_i \sim \underline{CS}$. *Then the function* $VaR_q[R(w(c))]$ *is non-decreasing in* $c \in [0, 1]$. *In particular,*

$$VaR_q[R(w^{(2)})] \leq VaR_q[R(w(c))] \leq VaR_q[R(w^{(1)})]$$

for all $c \in [0, 1]$. *In addition,* $VaR_q[R(w^{(3)})] \geq VaR_q[R(w^{(2)})]$ *and* $VaR_q[R(w^{(1)})] \geq VaR_q[R(w)]$ *for all* $w \in \mathcal{I}_r$.

(iii) *Suppose* $U_{ij} \sim \overline{CSLC}$. *Then the function* $VaR_q[U(\tilde{w}(c))]$ *is non-decreasing in* $c \in [0, 1]$. *In particular,*

$$VaR_q[U(\tilde{w}^{(2)})] \leq VaR_q[U(\tilde{w}(c))] \leq VaR_q[U(\tilde{w}^{(1)})].$$

In addition, $VaR_q[U(\tilde{w}^{(2)})] \leq VaR_q[U(\tilde{w}^{(3)})]$ *and* $VaR_q[U(\tilde{w}^{(2)})] = VaR_q[U(\underline{w}_n)] \leq VaR_q[U(\tilde{w})]$ *for all* $w \in \mathcal{I}_r$.

(iv) *Suppose* $U_{ij} \sim \underline{CS}$. *Then the function* $VaR_q[U(\tilde{w}(c))]$ *is non-increasing in* $c \in [0, 1]$. *In particular,*

$$VaR_q[U(\tilde{w}^{(2)})] \geq VaR_q[U(\tilde{w}(c))] \geq VaR_q[U(\tilde{w}^{(1)})].$$

In addition, $VaR_q[U(\tilde{w}^{(2)})] \geq VaR_q[U(\tilde{w}^{(3)})]$ *and* $VaR_q[U(\tilde{w}^{(2)})] = VaR_q[U(\underline{w}_n)] \geq VaR_q[U(\tilde{w})]$ *for all* $w \in \mathcal{I}_r$.

It is interesting to compare the results provided by Theorems 4.9–4.11 with the inequalities for the variances $var[Z(w)]$ in relations (4.44)–(4.46) in Section 4.5.3. If both the common shock variables R_i and the idiosyncratic risks U_{ij} have finite second moments and are thus thin-tailed, then solving the optimal portfolio choice problem with minimization of the variance $var[Z(w)]$ is problematic in the following sense.

First, the optimal solution is given by the portfolio weights $w_i(\theta)$ that depend on the value of the intra-class correlation θ which is typically unknown. Second, in (4.45) and (4.46), the contributions to the variances of the risks $Z(w)$ from the common shock and the idiosyncratic risk parts V_R and V_U are ordered in the opposite way for the diversified portfolio weights $w^{(1)} = \underline{w}_r$ and $w^{(2)}$. This holds regardless of the values of the variances σ_R^2 and σ_U^2 of the r.v.'s R_i and U_{ij}. Thus, minimization of the variance $\mathbb{V}[Z(w)]$ does not point out, even in the case of identically distributed R_i's and U_{ij}'s, to diversification either on the level of indices $i = 1, \ldots, r$ with returns Z_i

(the case of weights $w^{(1)} = \underline{w}_r$) or on the level of underlying risks Y_{ij} (the case of weights $w^{(2)}$).

These conclusions are in sharp contrast with the results for the value at risk portfolio choice under independence discussed in the Introduction and reviewed in Chapter 2 (see Theorems 2.1 and 2.2). They are also in contrast with the results for balanced models (4.13) presented in Sections 4.4 and 4.5.1. Namely, portfolio value at risk under independence or in balanced models (4.13) is minimized at equal weights for all moderately heavy-tailed risks with finite first moments (part (ii) of Theorem 2.1 and part (i) of Theorem 4.6).

Similarly, the solution to the value at risk minimization in such settings is given by the portfolio consisting of one risk within the whole class of extremely heavy-tailed risks with infinite first moments (part (ii) of Theorem 2.2 and part (ii) of Theorem 4.6).

In the settings of Theorem 4.9, the value at risk $\mathrm{VaR}_q[Z(w(c))]$ of the portfolios of indices $i = 1, \ldots, r$, with weights $w(c)$ defined in (4.42) is non-decreasing in $c \in [0, 1]$. Thus, the choice of portfolio weights $w(0) = w^{(2)}$ in (4.39) and diversification on the level of underlying risks Y_{ij} is preferred, in terms of value at risk comparisons, to $w(c)$ with any $c \in (0, 1]$. In particular, $w(0) = w^{(2)}$ is preferred to the portfolio of equal weights $w(1) = w^{(1)} = \underline{w}_r$ and the implied diversification on the level of indices. In addition, as shown by Theorem 4.9, the weight vector $w^{(2)}$ is preferred to $w^{(3)}$ that, in turn, dominates $w^{(1)} = \underline{w}_r$ in terms of the value at risk comparisons for $Z(w)$.

Under the assumptions of Theorem 4.10, the value at risk $\mathrm{VaR}_q[Z(w(c))]$ is non-increasing in $c \in [0, 1]$ for the weights $w(c)$ in (4.42). Thus, in terms of VaR comparisons, the choice of equal weights $w(1) = w^{(1)} = \underline{w}_r$ and diversification on the level of indices $i = 1, \ldots, r$, is preferred to $w(c)$ with any $c \in [0, 1)$. The equal weights $w(1) = w^{(1)} = \underline{w}_r$ for the portfolio of indices are preferred, in particular, to the weights $w(0) = w^{(2)}$ and the implied diversification on the level of individual risks Y_{ij}. Theorem 4.10 shows that the weight vector $w^{(1)} = \underline{w}_r$ is also preferred to $w^{(3)}$ and $w^{(3)}$ is preferred to $w^{(2)}$.

Parts (i) and (iii) of Theorem 4.11 show that, if all the variables R_i, U_{ij} (and, thus, the risks Y_{ij} in (4.32)) are moderately heavy-tailed, then the orderings of the value at risks for the portfolio components $R(w)$ and $U(\tilde{w})$ in (4.35) with weights $w = w^{(1)}, w^{(2)}, w^{(3)}$ are the same as in the case of variance comparisons in relations (4.45) and (4.46) in Section 4.5.3.

In addition, these comparisons do not point out to optimality of $w^{(1)}$ or $w^{(2)}$ within these three weight vectors or among the weights $w(c)$, $c \in [0,1]$, in (4.42). In other words, similar to variance minimization used as the portfolio choice criterion, the value at risk comparisons do not point out to diversification either on the level of indices $i = 1, \ldots, r$ with returns Z_i (the case of weights $w^{(1)}$) or on the level of underlying risks Y_{ij} (the case of weights $w^{(2)}$).

Parts (ii) and (iv) of Theorem 4.11 show that the above conclusions are reversed in the case where all the variables R_i, U_{ij} (and, thus, the risks Y_{ij} in (4.32)) are extremely heavy-tailed. In such setting, the orderings of the value at risks for the portfolio components $R(w)$ and $U(\tilde{w})$ in (4.35) with weights $w = w^{(1)}, w^{(2)}, w^{(3)}$ are the opposite to those in the case of variance comparisons in (4.45) and (4.46) and the case of VaR under moderate heavy-tailedness given by parts (i) and (iii) of Theorem 4.11. However, again, these orderings do not imply optimality of $w^{(1)}$ or $w^{(2)}$ within these three weight vectors or within the weight vectors $w(c)$, $c \in [0,1]$. Thus, the orderings do not imply optimality of diversification either on the level of indices $i = 1, \ldots, r$ with returns Z_i or the underlying risks Y_{ij}.

4.5.3. *From risk management to econometrics: Efficiency of random effects estimators*

The value at risk results presented in Section 4.4 can be reformulated in the framework of the analysis of efficiency of linear estimators in random effects models. As an example, in this section we discuss the implications of the results in Section 4.5.2 for efficiency of linear estimators in unbalanced two-stage nested design random effects models like (4.32). Using the results in Sections 4.4.2 and 4.5.1 and in the next section, similar extensions can be obtained for linear estimators of location in two-way classification random effects models (4.13) and their analogues with more than two common shocks and varying factor loadings.

Similar to (4.32), we consider observations from the model

$$Y_{ij} = \mu + R_i + U_{ij}, \quad j = 1, \ldots, n_i, \quad i = 1, \ldots, r, \qquad (4.43)$$

where $\mu \in \mathbf{R}$ and R_i and U_{ij} are symmetric and unimodal r.v.'s. Similar to Sections 4.4.3–4.5.2, in this section it is assumed that the variables R_i and U_{ij} are independent of each other and are independent and identically distributed among themselves.

A number of works in statistics and its applications have focused on the estimation of location in models (4.43) that are referred to in the fields as two-stage nested design, random effects location models. Several authors have considered variance decompositions and efficiency comparisons for location estimators in such models (see the discussion and reviews by Weiler and Culpin (1970), Section 13.B in Marshall and Olkin (1979) and Marshall *et al.* (2011), Birkes *et al.* (1981), El-Bassiouni and Abdelhafez (2000), and references therein).

Suppose that $\mathbb{E}R_i = 0$, $\mathbb{V}(R_i) = \sigma_R^2$, $\mathbb{E}U_{ij} = 0$, $\mathbb{V}(U_{ij}) = \sigma_U^2$, $j = 1, \ldots, n_i, i = 1, \ldots, r$. Evidently, for the variables $Z(w)$ in (4.47), one has

$$\mathbb{V}[Z(w)] = \sigma_R^2 \mathbb{V}_R(w) + \sigma_U^2 \mathbb{V}_U(w), \qquad (4.44)$$

where $\mathbb{V}_R(w) = \sum_{i=1}^{r} w_i^2$ and $\mathbb{V}_U(w) = \sum_{i=1}^{r} \frac{w_i^2}{n_i}$.

In the framework of inference on the location μ using linear unbiased estimators, Cochran (1954) recommends using the unweighted $(Z(w^{(1)})$ in (4.47) with $w^{(1)}$ in (4.37)) and weighted $(Z(w^{(2)})$ with the weights $w^{(2)}$ in (4.39)) averages of group means, for large and small values of the intra-class correlation $\theta = \sigma_R^2/(\sigma_R^2 + \sigma_U^2)$, respectively. Birkes *et al.* (1981) show that the minimal complete class of linear unbiased estimators of μ is given by $Z(w(c))$ in (4.47) with the weights $w(c)$ in (4.42).

It is straightforward to show that if the intra-class correlation $\theta = \sigma_R^2/(\sigma_R^2 + \sigma_U^2)$ is known, then the variance $\mathbb{V}[Z(w)]$, $w \in \mathcal{I}_r$, in (4.44) is minimized under the choice of weights $w(\theta)$.

Birkes *et al.* (1981) further focus on the analysis of efficiency $eff(c, \theta)$ for estimators $Z(w(c))$ defined as the ratio $eff(c, \theta) = \mathbb{V}[Z(w(c))]/\mathbb{V}[Z(w(\theta))]$ of the variance of $Z(w(c))$ to the least possible variance $\mathbb{V}[Z(w(\theta))]$ of linear unbiased estimators of μ in (4.43). The authors identify the maximin efficiency estimator $Z(w(c^*))$ that maximizes (over $c \in [0, 1]$) the minimum possible efficiency $\min_{\theta \in [0,1]} eff(c, \theta)$. The value c^* is found from the equation $n\mathbb{V}_U(w(c^*)) = r\mathbb{V}_R(w(c^*))$, that is, $n\sum_{i=1}^r w_i^2(c^*)/n_i = r\sum_{i=1}^r w_i^2(c^*)$.[1]

Koch (1967a) discusses variance decompositions (4.44) for the averages $Z(w^{(1)})$ and $Z(w^{(2)})$ in (4.47) with $w^{(1)}$ and $w^{(2)}$ in (4.37) and (4.39). He shows that the statistics $Z(w^{(1)})$ and $Z(w^{(2)})$ have the opposite orderings of the contributions to their variances in (4.44) from the row effects and the idiosyncratic error parts $\mathbb{V}_R$ and $\mathbb{V}_U$. More precisely, as shown in Koch (1967a), $\mathbb{V}_R(w^{(1)}) \leq \mathbb{V}_R(w^{(2)})$ and $\mathbb{V}_U(w^{(1)}) \geq \mathbb{V}_U(w^{(2)})$.[2]

Koch (1967a) further conjectures that for the weights $w^{(3)}$ in (4.41) one has

$$\mathbb{V}_R[Z(w^{(1)})] \leq \mathbb{V}_R[Z(w^{(3)})] \leq \mathbb{V}_R[Z(w^{(2)})] \qquad (4.45)$$

and

$$\mathbb{V}_U[Z(w^{(1)})] \geq \mathbb{V}_U[Z(w^{(3)})] \geq \mathbb{V}_U[Z(w^{(2)})]. \qquad (4.46)$$

This conjecture was proven by Low (1970) using some inequalities implied by majorization theory. An alternative more direct proof of the conjecture is provided in Section 13.B in Marshall and Olkin (1979) and Marshall and Olkin (2011). As discussed by Birkes *et al.*

---

[1]If one compares the estimators $Z(w)$ by variances instead of efficiencies, then it is easy to show that, as discussed in Birkes *et al.* (1981), the unweighted average $Z(\underline{w}_r)$ of group means in (4.47) with $\underline{w}_r$ in (4.37) has the optimal property of being the "minimax variance" linear unbiased estimator of μ in models (4.43) with fixed $\sigma_R^2 + \sigma_U^2$. More precisely, $\mathbb{V}[Z(\underline{w}_r)] = \min_{w \in \mathcal{I}_r} \max_{\theta \in [0,1]} \mathbb{V}[Z(w)]$, where, from (4.44), $\mathbb{V}[Z(w)] = (\sigma_R^2 + \sigma_U^2)[\theta\mathbb{V}_R(w) + (1-\theta)\mathbb{V}_U(w)]$, and $\max_{\theta \in [0,1]} \mathbb{V}[Z(w)] = (\sigma_R^2 + \sigma_U^2)\mathbb{V}_R(w)$ since $\mathbb{V}_R(w) \geq \mathbb{V}_U(w)$ for all $w \in \mathcal{I}_r$.
[2]Due to a typo, the inequality sign in the second of these two relations is reversed in the review on p. 393 in Marshall and Olkin (1979) and p. 537 in Marshall and Olkin (2011).

(1981), the maximin efficiency of $Z(w(c^*))$ compares favorably with efficiency of $Z(w^{(k)})$, $k = 1, 2, 3$.

A natural approach to comparisons of estimators of a population parameter under heavy-tailedness is that based on the likelihood of observing large deviations of these estimators from the true value of the parameter.

Let $\hat{\theta}^{(1)}$ and $\hat{\theta}^{(2)}$ be two estimators of the location parameter μ in model (4.43). Following the above approach, we say, similar to Ibragimov (2007), that the estimator $\hat{\theta}^{(1)}$ is (weakly) more efficient than $\hat{\theta}^{(2)}$ in the sense of peakedness (P-more efficient than $\hat{\theta}^{(2)}$ for short) if $\mathbb{P}(|\hat{\theta}^{(1)} - \mu| > \epsilon) \leq \mathbb{P}(|\hat{\theta}^{(2)} - \mu| > \epsilon)$ for all $\epsilon > 0$.

For $w = (w_1, \ldots, w_r) \in \mathcal{I}_r$, consider the linear estimators $Z(w)$ of the location parameter μ in form (4.34), with Z_i, $i = 1, \ldots, r$, defined in (4.33):

$$Z(w) = \sum_{i=1}^{r} w_i Z_i, \quad Z_i = \frac{1}{n_i} \sum_{j=1}^{n_i} Y_{ij}. \tag{4.47}$$

As in the Section 4.5.2, we deal with the weight vectors $w = w^{(1)}, w^{(2)}, w^{(3)}, w(c)$ defined in (4.37), (4.39), (4.41) and (4.42). Theorems 4.12 and 4.13 below provide P-efficiency comparisons for linear estimators $Z(w)$ with the above weights considered, in the context of value at risk analysis, in Section 4.5.2.

Theorem 4.12. *Let* $\epsilon > 0$. *Suppose that, in (4.14),* $R_i \sim \underline{\mathcal{CS}}$, $i = 1, \ldots r$, *and* $U_{ij} \sim \overline{\mathcal{CSLC}}$, $j = 1, \ldots, n_i$, $i = 1, \ldots, r$. *Then the function* $\tau(c) = \mathbb{P}[|Z(w(c)) - \mu| > \epsilon]$ *is non-decreasing in* $c \in [0, 1]$. *In particular,*

$$\mathbb{P}\left[|Z(w^{(1)}) - \mu| > \epsilon\right] \geq \mathbb{P}\left[|Z(w(c)) - \mu| > \epsilon\right] \geq \mathbb{P}\left[|Z(w^{(2)}) - \mu| > \epsilon\right].$$

In addition,

$$\mathbb{P}\left[|Z(w^{(1)}) - \mu| > \epsilon\right] \geq \mathbb{P}\left[|Z(w^{(3)}) - \mu| > \epsilon\right] \geq \mathbb{P}\left[|Z(w^{(2)}) - \mu| > \epsilon\right].$$

Theorem 4.13. *Let* $\epsilon > 0$. *Suppose that, in (4.14),* $R_i \sim \overline{\mathcal{CSLC}}$, $i = 1, \ldots, r$, *and* $U_{ij} \sim \underline{\mathcal{CS}}$, $j = 1, \ldots, n_i$, $i = 1, \ldots, r$. *Then the* *function* $\tau(c) = \mathbb{P}[|Z(w(c)) - \mu| > \epsilon]$ *is non-increasing in* $c \in [0,1]$. *In particular,*

$$\mathbb{P}\Big[\big|Z(w^{(2)}) - \mu\big| > \epsilon\Big] \geq \mathbb{P}\Big[\big|Z(w(c)) - \mu\big| > \epsilon\Big] \geq \mathbb{P}\Big[\big|Z(w^{(1)}) - \mu\big| > \epsilon\Big].$$

In addition,

$$\mathbb{P}\Big[\big|Z(w^{(2)}) - \mu\big| > \epsilon\Big] \geq \mathbb{P}\Big[\big|Z(w^{(3)}) - \mu\big| > \epsilon\Big] \geq \mathbb{P}\Big[\big|Z(w^{(1)}) - \mu\big| > \epsilon\Big].$$

Theorems 4.12 and 4.13 show that $\mathbb{P}$-efficiency comparisons in models (4.43) under heavy-tailedness are similar to VaR results in risk models (4.32) dealt with in the previous section. In particular, the results in Theorems 4.12 and 4.13 are in contrast to the case of variance comparisons for linear estimators $Z(w)$ in the literature discussed in Sections 4.5.2 and 4.5.3. Namely, in contrast to the results for variances, under extreme heavy-tailedness in the common shocks R_i and moderate heavy-tailedness in the idiosyncratic risks U_{ij}, $\mathbb{P}$-efficiency comparisons point out to optimality of $Z(w^{(2)}) = Z(w(0))$ among the estimators $Z(w(c))$, $c \in [0,1]$ (Theorem 4.12). The estimator $Z(w^{(2)})$ is also more $\mathbb{P}$-efficient than $Z(w^{(1)})$ and $Z(w^{(3)})$.

Similarly, in the case of moderate heavy-tailedness in R_i and extreme heavy-tailedness in U_{ij}, $\mathbb{P}$-efficiency of the estimators $Z(w(c))$, $c = [0,1]$, is maximal under equal weights $\underline{w}_r = w(1)$ (Theorem 4.13). The estimator $Z(w^{(1)})$ is also more $\mathbb{P}$-efficient than $Z(w^{(2)})$ and $Z(w^{(3)})$.

4.5.4. *Extensions: Multiple additive and multiplicative common shocks*

The analysis presented in the preceding sections can be extended to the case where the underlying risks Y in the portfolios exhibit dependence with more than two common shocks. For instance, let $m \geq 1$, and let $N_1, N_2, \ldots, N_m \in \mathbf{N}$. Denote $L = \prod_{s=1}^{m} N_s$. One can show that the analogues of the results in Section 4.4.2 also hold for the multiple shock extensions of (4.13) given by

$$Y_{i_1,i_2,\ldots,i_m} = \sum_{s=1}^{m} \sum_{1 \leq j_1 < \cdots < j_s \leq m} U_{i_{j_1},\ldots,i_{j_s}}^{(j_1,\ldots,j_s)}, \qquad (4.48)$$

$1 \leq i_k \leq N_k$, $k = 1, \ldots, m$, where the variables $U_{i_{j_1}, \ldots, i_{j_s}}^{(j_1, \ldots, j_s)}$ are independent over all the indices $1 \leq j_1 < \cdots < j_s \leq m$, $s = 1, \ldots, m$, $1 \leq i_k \leq N_k$, $k = 1, \ldots, m$. The underlying risks Y in (4.48) have dependence structures exhibited by sums of U-statistics dealt with in Chapter 3. Such dependence structures and a number of probabilistic and statistical results for them are discussed, among others, in de la Peña *et al.* (2002, 2003) and references therein.

A particular case of models (4.48) with $m = 3$ is given by the risks

$$Y_{i,j,k} = U_i^{(1)} + U_j^{(2)} + U_k^{(3)} + U_{ij}^{(4)} + U_{ik}^{(5)} + U_{jk}^{(6)} + U_{ijk}, \quad (4.49)$$

$1 \leq i \leq N_1$, $1 \leq j \leq N_2$, $1 \leq k \leq N_3$, where the summands are independent over all the indices. Specifications (4.48) also include, for instance, multi-stage nested design random effects models with $U_{i_{j_1}, \ldots, i_{j_s}}^{(j_1, \ldots, j_s)} = 0$ for $(j_1, \ldots, j_s) \neq (1, \ldots, s) : Y_{i_1, i_2, \ldots, i_m} = U_{i_1}^{(1)} + U_{i_1, i_2}^{(2)} + \cdots + U_{i_1, i_2, \ldots, i_m}^{(m)}$, and multi-way classification random effects models and their analogues (see Koch, 1967a,b).

Consider the portfolios of risks $Y_{i_1, i_2, \ldots, i_m}$ in (4.48) with weights $w_{i_1, i_2, \ldots, i_m} \in \mathbf{R}_+$ such that $\sum_{i_1=1}^{N_1} \cdots \sum_{i_m=1}^{N_m} w_{i_1, i_2, \ldots, i_m} = 1$. Let $w \in \mathcal{I}_L$ denote the corresponding weight vector with components $w_{i_1, i_2, \ldots, i_m}$. The return on the portfolio with weights $w_{i_1, i_2, \ldots, i_m}$ is given by $Y(w) = \sum_{i_1=1}^{N_1} \cdots \sum_{i_m=1}^{N_m} w_{i_1, i_2, \ldots, i_m} Y_{i_1, i_2, \ldots, i_m}$. As in Section 4.4.2, one concludes that, in the case of moderately heavy-tailed $U_{i_{j_1}, \ldots, i_{j_s}}^{(j_1, \ldots, j_s)} \sim \overline{\mathcal{CSLC}}$, the value at risk $\mathrm{VaR}_q[Y(w)]$ of $Y(w)$, $w \in \mathcal{I}_L$ is minimized in the case of the most diversified portfolio with equal weights $\underline{w}_{i_1, i_2, \ldots, i_m} = \underbrace{(1/L, \ldots, 1/L)}_{L} \in \mathcal{I}_L$. In such settings, the value at risk $\mathrm{VaR}_q[Y(w)]$, $w \in \mathcal{I}_L$, is maximized in the case of the least diversified portfolio with weights $\overline{w}_{i_1, i_2, \ldots, i_m} = \underbrace{(1, 0, \ldots, 0)}_{L} \in \mathcal{I}_L$ that consists of only one risk.

These comparisons are reversed for extremely heavy-tailed $U_{i_{j_1}, \ldots, i_{j_s}}^{(j_1, \ldots, j_s)} \sim \underline{\mathcal{CS}}$. Under extreme heavy-tailedness, the equal weights $\underline{w}_{i_1, i_2, \ldots, i_m} = (1/L, \ldots, 1/L) \in \mathcal{I}_L$ maximize the portfolio value at risk

$\mathrm{VaR}_q[Y(w)]$ over $w \in \mathcal{I}_L$. In contrast, the minimal portfolio value at risk over $w \in \mathcal{I}_L$ is achieved for the least diversified portfolio with weights $\overline{w}_{i_1,i_2,\dots,i_m} = (1,0,\dots,0) \in \mathcal{I}_L$.

The analysis in the sections can also be generalized to the settings where the summands R_i, C_j and U_{ij} in model (4.13) and its analogues (including the case of multiple additive common shocks $U_{i_{j_1},\dots,i_{j_s}}^{(j_1,\dots,j_s)}$ in (4.48)) exhibit dependence. For instance, using the extensions of the results in Theorems 2.1 and 2.2 to the case of dependence discussed by Ibragimov (2009b) and Ibragimov and Walden (2007), one obtains that all the results in the section also hold in settings where the risks R_i, C_j and U_{ij} in (4.13) and (4.14) are dependent among themselves or are bounded. These generalizations include models (4.13) and (4.14) in which the vectors of common shocks $(R_1,\dots,R_r)$ and $(C_1,\dots,C_c)$ and the vector of idiosyncratic errors $(U_{11},\dots,U_{1c},\dots,U_{r1},\dots,U_{rc})$ have distributions which are convolutions of α-symmetric distributions (see Fang *et al.* (1990) and the review by Ibragimov (2009b); Ibragimov and Walden (2007)).

As discussed earlier, for any $0 < \alpha \le 2$, the class of α-symmetric distributions includes distributions of vectors of risks $(X_1,\dots,X_N)$ that have the common factor representation

$$(X_1,\dots,X_N) = (ZY_1,\dots,ZY_N), \tag{4.50}$$

where $Y_i \sim S_\alpha(\sigma,0,0)$ are i.i.d. symmetric stable r.v.'s with $\sigma > 0$ and the index of stability α and $Z \ge 0$ is a nonnegative r.v. independent of $Y_i's$ (see, e.g., Fang *et al.* (1990), p. 197).

Multiplicative common shock specifications (4.50) provide extensions of models (4.13) with $R_i = \sum_{s=1}^{m_1} F_s \tilde{R}_{is}$, $C_j = \sum_{s=1}^{m_2} G_s \tilde{C}_{js}$, $U_{ij} = \sum_{s=1}^{m_3} H_s \tilde{U}_{ijs}$, where the risks $F_s, G_s, H_s > 0$ and $\tilde{R}_{is}, \tilde{C}_{js}, \tilde{U}_{ijs}$ are independent of each other and among themselves. In these extensions, in addition to the two common shocks $\tilde{R}$ and $\tilde{C}$ as in (4.50), the risks Y_{ij} are also affected by $m_1+m_2+m_3$ common multiplicative shocks F, G and H.

Let Φ stand for the class of c.f. generators ϕ such that $\phi(0) = 1$, $\lim_{t\to\infty} \phi(t) = 0$, and the function $\phi'(t)$ is concave. For $0 < \alpha \le 2$, denote by $\mathcal{G}_N(\alpha)$ the class of random vectors $(X_1, \ldots, X_N)$ with dependent components that satisfy one of the following conditions:

(C1) $(X_1, \ldots, X_N)$ is a sum of k independent random vectors $(Y_{1j}, \ldots, Y_{Nj})$, $j = 1, \ldots, k$, where $(Y_{1j}, \ldots, Y_{Nj})$ has an absolutely continuous α-symmetric distribution with $\phi_j \in \Phi$ and $\alpha_j \in (0, 2]$;

(C2) $(X_1, \ldots, X_N)$ is a sum of k random vectors $(Y_{1j}, \ldots, Y_{Nj}) = (Z_j V_{1j}, \ldots, Z_j V_{Nj})$, $j = 1, \ldots, k$, in (4.50) with independent r.v.'s Z_j, V_{ij}, $j = 1, \ldots, k$, $i = 1, \ldots, N$, such that Z_j are positive and absolutely continuous and $V_{ij} \sim S_{\alpha_j}(\sigma_j, 0, 0)$, $\sigma_j > 0$, $\alpha_j \in (0, 2]$. That is, $X_i = \sum_{j=1}^{k} Z_j V_{ij}$ for $i = 1, \ldots, N$.

Theorems 5.1 and 5.2 in Ibragimov (2009b) provide the following extensions of the VaR and diversification comparisons in Theorems 2.1 and 2.2 to the case of dependence.

Proposition 4.1. *Let $(X_1, X_2, \ldots, X_N) \in \mathcal{G}_N(\alpha)$, $\alpha \in (0, 2]$. Then Theorem 2.1 holds if $\alpha \in [1, 2]$, and Theorem 2.2 holds if $\alpha \in (0, 1]$.*

Using the VaR comparisons under dependence given by Proposition 4.1, one can show that the results in the paper continue to hold for the case of dependent common shocks R_i and C_j and idiosyncratic risks U_{ij}. In particular, the following results hold.

Theorem 4.14. *Let $(R_1, R_2, \ldots, R_r) \in \mathcal{G}_r(\alpha_r)$, $(C_1, C_2, \ldots, C_c) \in \mathcal{G}_c(\alpha_c)$ and $(U_{11}, U_{12}, \ldots, U_{rc}) \in \mathcal{G}_{rc}(\alpha_u)$ with $\alpha_r, \alpha_c, \alpha_u \in (0, 2]$. Part (i) of Theorem 4.6; parts (i), (iii) and (v) of Theorem 4.7; relations (4.28) and (4.29); part (i) of Theorem 4.8 and parts (i) and (iii) of Theorem 4.11 hold if $\alpha_r, \alpha_c, \alpha_u \in [1, 2]$. Part (ii) of Theorem 4.6; parts (ii), (iv) and (vi) of Theorem 4.7; relations (4.30) and (4.31); part (ii) of Theorem 4.8 and parts (ii) and (iv) of Theorem 4.11 hold if $\alpha_r, \alpha_c, \alpha_u \in (0, 1]$. Theorems 4.9 and 4.12 hold if $\alpha_r \in (0, 1]$ and $\alpha_u \in [1, 2]$. Theorems 4.10 and 4.13 hold if $\alpha_r \in [1, 2]$ and $\alpha_u \in (0, 1]$.*

The results on common shocks can also be extended to portfolio choice problems for non-identically distributed risks. In addition, the results continue to hold for risk settings more general than common shock structures (4.13) and (4.14) considered so far for simplicity of presentation and the arguments. In particular, the results continue to hold for models (4.23) with varying factor loadings. As an example of the above generalizations, Theorem 4.15 provides the analogues of the results in Theorems 4.9 and 4.10 for the unbalanced one-factor case of (4.23) given by

$$Y_{ij} = \beta_{ij} R_i + U_{ij}, \quad j = 1, \ldots, n_i, i = 1, \ldots, r, \qquad (4.51)$$

$n_1 \geq n_2 \geq \cdots \geq n_r$, $\sum_{i=1}^{r} n_i = n$, where $R_i \sim S_\alpha(\sigma_i, 0, 0)$, $U_{ij} \sim S_{\alpha'}(\sigma'_{ij}, 0, 0)$, $\alpha, \alpha' \in (0, 2]$, $\sigma_i, \sigma'_{ij} > 0$, $j = 1, \ldots, n_i$, $i = 1, \ldots, r$, are independent not necessarily identically distributed heavy-tailed stable r.v.'s. Denote $\tilde{\sigma}_i = \frac{1}{n_i} \left(\sum_{j=1}^{n_i} \beta_{ij} \right) \sigma_i$. Observe that the within-group orderings $\sigma'_{i1} \leq \cdots \leq \sigma'_{i,n_i}$ and $\sigma'_{i1} \geq \cdots \geq \sigma'_{i,n_i}$, $i = 1, \ldots, r$, for the scale parameters σ'_{ij} in Theorem 4.15 do not restrict generality.

Theorem 4.15. *Theorem 4.9 holds for* (4.51) *if* $\alpha \leq 1$, $\alpha' \geq 1$, $\tilde{\sigma}_1 \leq \cdots \leq \tilde{\sigma}_r$ *and* $\sigma'_{11} \leq \cdots \leq \sigma'_{1,n_1} \leq \sigma'_{21} \leq \cdots \leq \sigma'_{2,n_2} \leq \cdots \leq \sigma'_{r1} \leq \cdots \leq \sigma'_{r,n_r}$. *Theorem 4.10 holds for* (4.51) *if* $\alpha \geq 1$, $\alpha' \leq 1$, $\tilde{\sigma}_1 \geq \cdots \geq \tilde{\sigma}_r$ *and* $\sigma'_{11} \geq \cdots \geq \sigma'_{1,n_1} \geq \sigma'_{21} \geq \cdots \geq \sigma'_{2,n_2} \geq \cdots \geq \sigma'_{r1} \geq \cdots \geq \sigma'_{r,n_r}$.

As discussed in Section 4.5.2, the degree of heavy-tailedness of the common shocks R_i in Theorems 4.9, 4.10 and 4.15 is different from that for the idiosyncratic risks U_{ij}. It is interesting to compare the theorems with the results in Theorem 4.16 below for models (4.51) with varying factor loadings and heavy-tailed risks R_i and U_{ij} with non-identical distributions. In contrast to Theorems 4.9, 4.10 and 4.15, the degrees of heavy-tailedness of the common factors R_i and the errors U_{ij} in Theorem 4.16 are the same. Denote $\tilde{\sigma}'_i = \frac{1}{n_i} \left(\sum_{j=1}^{n_i} (\sigma'_{ij})^{\alpha'} \right)^{1/\alpha'}$, $i = 1, \ldots, r$.

Theorem 4.16. *Theorem 4.9 holds for* (4.51) *if* $\alpha, \alpha' \leq 1$, $\tilde{\sigma}_1 \leq \cdots \leq \tilde{\sigma}_r$ *and* $\tilde{\sigma}'_1 \leq \cdots \leq \tilde{\sigma}'_r$. *Theorem 4.10 holds for* (4.51) *if* $\alpha, \alpha' \geq 1$, $\tilde{\sigma}_1 \geq \cdots \geq \tilde{\sigma}_r$ *and* $\tilde{\sigma}'_1 \geq \cdots \geq \tilde{\sigma}'_r$.

Theorems 4.15 and 4.16 illustrate that portfolio diversification decisions are affected by the risk structures dealt with and the interplay between dependence and heterogeneity properties of the risks in consideration. Note, in particular, that the assumptions in Theorems 4.15 include the homogeneous case $\beta_{ij} = \beta$, $\sigma_i = \sigma$, $\sigma'_{ij} = \sigma'$, $j = 1, \ldots, n_i$, $i = 1, \ldots, r$. However, the assumptions in Theorem 4.16 require heterogeneity in the idiosyncratic risks U_{ij}.

As is easy to see, since $n_1 \geq n_2 \geq \cdots \geq n_r$, the orderings $\tilde{\sigma}'_1 \leq \cdots \leq \tilde{\sigma}'_r$ for $\alpha' \leq 1$ and $\tilde{\sigma}'_1 \geq \cdots \geq \tilde{\sigma}'_r$ for $\alpha' \geq 1$ in the latter theorem cannot hold under homogeneity $\sigma'_{ij} = \sigma'$ in the r.v.'s U_{ij} unless $n_1 = n_2 = \ldots = n_r$ or $\alpha' = 1$. As indicated in Section 4.5.2, in the balanced case $n_1 = n_2 = \ldots = n_r$, the vectors $w^{(1)}$, $w^{(2)}$, $w^{(3)}$ and $w(c)$ in Theorems 4.9 and 4.10 become the same: $w^{(l)} = w(c) = \underline{w}_r$ for all $l = 1, 2, 3$, and $c \in [0, 1]$. In the special case $\alpha' = 1$ with the stable risks $U_{ij} \sim S_1(\sigma', 0, 0)$, the conditions in Theorem 4.16 involve only the assumptions on the scale parameters σ_i and the tail index α for the common shocks R_i. The latter assumptions on the degrees of heavy-tailedness of R'_is ($\alpha \leq 1$ or $\alpha \geq 1$) are similar to those in Theorems 4.9 and 4.10. We further note that the VaR comparisons in Theorems 4.9 and 4.10 hold as equalities in the case $\alpha = \alpha' = 1$.

Similar to the arguments for the results in Section 4.5.3, from Theorems 4.15 and 4.16 it follows that Theorems 4.12 and 4.13 also hold (with the same assumptions on the common shocks R_i and the idiosyncratic risks U_{ij} as in Theorems 4.15 and 4.16) for the factor models $Y_{ij} = \mu + \beta_{ij}R_i + U_{ij}$, $j = 1, \ldots, n_i$, $i = 1, \ldots, r$. Furthermore, similar to the proof of Theorems 4.1–4.16, one can also obtain their analogues for risk models with more than two additive shocks, like those in (4.48). In addition, similar to Proposition 4.1 and Theorem 4.16, one can also obtain extensions of the results to dependent and possibly non-identically distributed risks, including convolutions of scaled α-symmetric random vectors.

4.6. Conclusion

We have revisited the limits of diversification for dependent risks. The revisit focused on a wide class of copulas that are additive in powers of margins and on dependence structures induced by convolutions of α-symmetric distributions.

This class of copulas we considered covers some well-known families such as EFGM, power and polynomial families but also contains a number of other copula classes which do not have this form but can be approximated using Taylor-type expansions. So the resulting class we consider is very wide — comprehensive, tail-dependent and asymmetric copula families can be considered within this class (see also Mo (2013) and Burns (2014) for an extensive numerical analysis of diversification (sub-)optimality for a wide range of copula models and the ratios of the value at risk of a portfolio and a sum of VaR's of its components).

The class of dependence encompassed by convolutions of α-symmetric distributions is also quite wide. Convolutions of α-symmetric distributions contain, as subclasses, convolutions of several models with common shocks affecting all heavy-tailed risks as well as spherical distributions which are α-symmetric with $\alpha = 2$. Spherical distributions, in turn, include such examples as Kotz type, multinormal, logistic and multivariate α-stable distributions. In addition, they include a subclass of mixtures of normal distributions as well as multivariate t-distributions that were used in the literature to model heavy-tailedness phenomena with dependence and with finite moments up to a certain order.

The main result of the chapter is that within the classes, diversification typically increases riskiness in a VaR framework if the power index of the individual risks falls below one. This makes dependent risks within this class no different from independent in the sense that the same threshold value of the tail index delineates the benefits of diversification.

As by-products, we could formulate a number of counterintuitive results that are extensions or applications of the limits of diversification. For example, these results in Sections 4.5.2 and 4.5.3 are in contrast to variance comparisons for portfolio returns implied by the results in the literature on linear location estimation in random effects models. The results in these sections further emphasize the importance of both heavy-tailedness and dependence structure of risks in consideration in the analysis of (non)robustness of diversification analysis and other models in economics, finance, econometrics and related fields.

Our analysis illustrates the generality of the majorization-based approach to the study of portfolio diversification and value at risk. In particular, this chapter shows that the approach can be used in a wide range of dependent models, including those with multiple additive common shocks.

Similar to the case of independence, typically, the tail index threshold $\alpha = 1$ and finiteness of first moments of some of the risk components is the boundary between the robustness and reversals of the standard results in the variance minimization framework. Usually, these reversals under extreme heavy-tailedness point away from diversification.

Surprisingly, however, for some important problems — including the optimal portfolio choice for indices of dependent heavy-tailed risks — the implications are opposite and diversification is optimal when risks are extremely heavy-tailed — see Section 4.5.2. The value of diversification and (non)robustness of key models in economics and finance thus depend crucially on the interplay among heavy-tailedness, dependence and heterogeneity properties of the risks involved.

4.7. Appendix: Proofs

Proof of Theorem 4.2. We start with the case $n = 2$. Due to independence between ξ_1 and ξ_2, we have that

$$\mathbb{P}\left(\frac{\xi_1(\beta_1) + \xi_2(\beta_2)}{2} > z\right) = \beta_1\beta_2 \int_{\frac{s+t}{2}>z} s^{-\beta_1-1}t^{-\beta_2-1}dsdt. \quad (4.52)$$

Now for non-independent (X_1, X_2) under the EFGM copula, we can write using (4.52):

$$\mathbb{P}\left(\frac{X_1 + X_2}{2} > z\right) = \int_{\frac{s+t}{2}>z} f_1(s)f_2(t)$$
$$\times [1 + \theta(1 - 2F_1(s))(1 - 2F_2(t))]dsdt$$

$$= \mathbb{P}\left(\frac{\xi_1(\alpha) + \xi_2(\alpha)}{2} > z\right)$$

$$+ \theta \int_{\frac{s+t}{2} > z} f_1(s) f_2(t)$$

$$\times (1 - 2F_1(s))(1 - 2F_2(t)) ds dt$$

$$= \mathbb{P}\left(\frac{\xi_1(\alpha) + \xi_2(\alpha)}{2} > z\right)$$

$$+ \theta \mathbb{E}(1 - 2F_1(\xi))(1 - 2F_2(\eta)) I\left(\frac{\xi_1 + \xi_2}{2} > z\right),$$

where $I(\cdot)$ denotes the indicator function.

Now consider the last term:

$$\int_{\frac{s+t}{2} > z} f_1(s) f_2(t)(1 - 2F_1(s))(1 - 2F_2(t)) ds dt$$

$$= \int_{\frac{s+t}{2} > z} \alpha^2 s^{-\alpha-1} t^{-\alpha-1} (2s^{-\alpha} - 1)(2t^{-\alpha} - 1) ds dt$$

$$= 4\alpha^2 \int_{\frac{s+t}{2} > z} s^{-2\alpha-1} t^{-2\alpha-1} ds dt$$

$$- 2\alpha^2 \int_{\frac{s+t}{2} > z} s^{-2\alpha-1} t^{-\alpha-1} ds dt$$

$$- 2\alpha^2 \int_{\frac{s+t}{2} > z} s^{-\alpha-1} t^{-2\alpha-1} ds dt$$

$$+ \alpha^2 \int_{\frac{s+t}{2} > z} s^{-\alpha-1} t^{-\alpha-1} ds dt$$

$$= 4\alpha^2 \mathcal{I}_1 - 2\alpha^2 \mathcal{I}_2 - 2\alpha^2 \mathcal{I}_3 + \alpha^2 \mathcal{I}_4,$$

where $\mathcal{I}_1 = \mathbb{P}(\frac{\xi_1(2\alpha) + \xi_2(2\alpha)}{2} > z)$, $\mathcal{I}_2 = \mathbb{P}(\frac{\xi_1(2\alpha) + \xi_2(\alpha)}{2} > z)$, $\mathcal{I}_3 = \mathbb{P}(\frac{\xi_1(\alpha) + \xi_2(2\alpha)}{2} > z)$ and $\mathcal{I}_4 = \mathbb{P}(\frac{\xi_1(\alpha) + \xi_2(\alpha)}{2} > z)$.

Thus we obtain

$$\mathbb{P}\left(\frac{X+Y}{2} > z\right) = (1 + \theta\alpha^2)\mathbb{P}\left(\frac{\xi_1(\alpha) + \xi_2(\alpha)}{2} > z\right)$$

$$-2\theta\alpha^2\mathbb{P}\left(\frac{\xi_1(\alpha) + \xi_2(2\alpha)}{2} > z\right)$$

$$-2\theta\alpha^2\mathbb{P}\left(\frac{\xi_1(2\alpha) + \xi_2(\alpha)}{2} > z\right)$$

$$+4\theta\alpha^2\mathbb{P}\left(\frac{\xi_1(2\alpha) + \xi_2(2\alpha)}{2} > z\right).$$

It is a well-known result in the power law literature (see, among others, Corollary 1.3.2 in Embrechts *et al.* (1997)) that, asymptotically as $z \to \infty$,

$$\mathbb{P}\left(\frac{\xi_1(\beta) + \xi_2(\beta)}{2} > z\right) \sim 2\mathbb{P}(\xi_1(\beta) > 2z) \sim 2^{1-\beta}z^{-\beta}$$

$$(4.53)$$

for all $\beta > 0$. In addition, if $\beta_1 < \beta_2$, then

$$\mathbb{P}\left(\frac{\xi_1(\beta_1) + \xi_2(\beta_2)}{2} > z\right) \sim \mathbb{P}(\xi_1(\beta_1) > 2z) \sim 2^{-\beta_1}z^{-\beta_1}.$$

$$(4.54)$$

It follows from (4.53)–(4.54) that, as $z \to \infty$,

$$\mathbb{P}\left(\frac{X+Y}{2} > z\right)$$

$$\sim (1 + \theta\alpha^2)2^{1-\alpha}z^{-\alpha} - 2\theta\alpha^2 2^{1-\alpha}z^{-\alpha} + 4\theta\alpha^2 2^{1-2\alpha}z^{-2\alpha}$$

$$\sim (1 - \theta\alpha^2)2^{1-\alpha}z^{-\alpha}$$

$$\sim \mathbb{P}\left(\frac{\xi_1(\alpha) + \xi_2(\alpha)}{2} > z\right).$$

$$(4.55)$$

We now provide a generalization for any n. Let $X_1, X_2, \ldots, X_n$ have a multidimensional EFGM copula

$$C(u_1, u_2, \ldots, u_n)$$

$$= u_1 u_2 \ldots u_n \left[1 + \sum_{c=2}^{n} \sum_{1 \le i_1 < i_2 < \ldots < i_c \le n} \theta_{i_1, i_2, \ldots, i_c} \right.$$

$$\left. \times (1 - u_{i_1})(1 - u_{i_2}) \ldots (1 - u_{i_c}) \right], \tag{4.56}$$

where $\theta_{i_1, i_2, \ldots, i_c}$ are real constants satisfying certain inequalities that guarantee that (4.56) represents a proper copula.

Let $X_1, X_2, \ldots, X_n$ have power law distributions with the same parameter $\alpha > 0$. It follows from (4.56) that the joint cdf of $X_1, X_2, \ldots, X_n$ has the form

$$F(x_1, x_2, \ldots, x_n)$$

$$= F_1(x_1) F_2(x_2) \ldots F_n(x_n)$$

$$\times \left[1 + \sum_{c=2}^{n} \sum_{1 \le i_1 < i_2 < \ldots < i_c \le n} \theta_{i_1, i_2, \ldots, i_c} (1 - F_{i_1}(x_{i_1})) \right.$$

$$\left. \times (1 - F_{i_2}(x_{i_2})) \ldots (1 - F_{i_c}(x_{i_c})) \right]. \tag{4.57}$$

Let, $\xi_1(\beta_1), \xi_2(\beta_2), \ldots, \xi_n(\beta_n)$ denote the independent random variables with power law distributions with tail indices $\beta_1, \beta_2, \ldots, \beta_n$, respectively. That is,

$$\mathbb{P}(\xi_i(\beta_i) > x) = x^{-\beta_i}, \tag{4.58}$$

$x \ge 1$, $i = 1, 2, \ldots, n$. In particular, $\xi_1(\alpha), \xi_2(\alpha), \ldots, \xi_n(\alpha)$ are independent copies of $X_1, X_2, \ldots, X_n$.

Then, it follows that

$$\mathbb{P} \left(\sum_{i=1}^{n} X_i > zn \right)$$

$$= \mathbb{P}\left(\sum_{i=1}^{n} \xi_i(\alpha) > zn\right)$$

$$+ \sum_{c=2}^{n} \sum_{1 \le i_1 < i_2 < \ldots < i_c \le n} \theta_{i_1, i_2, \ldots, i_c}$$

$$\times \mathbb{E}\left[(1 - 2F_{i_1}(\xi_{i_1}(\alpha)))(1 - 2F_{i_2}(\xi_{i_2}(\alpha))) \ldots\right.$$

$$\left. \times (1 - 2F_{i_c}(\xi_{i_c}(\alpha))) \mathbb{I}\left(\sum_{i=1}^{n} \xi_i(\alpha) > zn\right)\right]. \qquad (4.59)$$

Thus, since the random variables $\xi_1(\alpha), \xi(\alpha), \ldots, \xi_n(\alpha)$ are i.i.d.,

$$\mathbb{P}\left(\sum_{i=1}^{n} X_i > zn\right)$$

$$= \mathbb{P}\left(\sum_{i=1}^{n} \xi_i(\alpha) > zn\right)$$

$$+ \left(\sum_{c=2}^{n} \sum_{1 \le i_1 < i_2 < \ldots < i_c \le n} \theta_{i_1, i_2, \ldots, i_c}\right)$$

$$\times \mathbb{E}\left[(1 - 2F_1(\xi_1(\alpha)))(1 - 2F_2(\xi_2(\alpha))) \ldots (1 - 2F_c(\xi_c(\alpha)))\right.$$

$$\left. \times \mathbb{I}\left(\sum_{i=1}^{n} \xi_i(\alpha) > zn\right)\right]. \qquad (4.60)$$

Now consider the last term

$$\mathbb{E}\left[(1 - 2F_1(\xi_1(\alpha)))(1 - 2F_2(\xi_2(\alpha))) \ldots (1 - 2F_c(\xi_c(\alpha)))\right.$$

$$\left. \times \mathbb{I}\left(\sum_{i=1}^{n} \xi_i(\alpha) > zn\right)\right]$$

$$= \sum_{s=0}^{c} \sum_{1 \le j_1 < j_2 < \ldots < j_s \le c} (-1)^{c-s} \int_{\sum_{i=1}^{n} x_i > zn} \prod_{k \in \{j_1, j_2, \ldots, j_s\}} (2\alpha) x_k^{-2\alpha-1}$$

$$\times \prod_{k \in \{1,2,\ldots,n\} \backslash \{j_1, j_2, \ldots, j_s\}} \alpha x_k^{-\alpha-1} dx_1 dx_2 \ldots dx_n$$

$$= \sum_{s=0}^{c} \sum_{1 \le j_1 < j_2 < \ldots < j_s \le c} (-1)^{c-s} \mathbb{P}$$

$$\times \left(\sum_{k \in \{j_1, j_2, \ldots, j_s\}} \xi_k(2\alpha) + \sum_{k \in \{1,2,\ldots,n\} \backslash \{j_1, j_2, \ldots, j_s\}} \xi_k(\alpha) > z \right),$$
$$(4.61)$$

where $1 \le j_1 < j_2 < \ldots < j_s \le c$, $s = 0, 1, \ldots, c$, $c = 2, \ldots, n$, $(s, c) \ne (n, n)$ (and, thus, $(j_1, j_2, \ldots, j_c)$ is different from $(1, 2, \ldots, n)$).
Consequently, for large z, we obtain

$$\mathbb{P} \left(\sum_{k \in \{j_1, j_2, \ldots, j_s\}} \xi_k(2\alpha) + \sum_{k \in \{1,2,\ldots,n\} \backslash \{j_1, j_2, \ldots, j_s\}} \xi_k(\alpha) > z \right)$$

$$\sim \mathbb{P} \left(\sum_{k \in \{1,2,\ldots,n\} \backslash \{j_1, j_2, \ldots, j_s\}} \xi_k(\alpha) > zn \right). \qquad (4.62)$$

In addition, by Corollary 1.3.2 of Embrechts *et al.* (1997), we have, for large $z > 0$,

$$\mathbb{P} \left(\sum_{k \in \{1,2,\ldots,n\} \backslash \{j_1, j_2, \ldots, j_s\}} \xi_k(\alpha) > z \right) \sim (n - s) \mathbb{P}(\xi_1(\alpha) > zn)$$

$$\sim \frac{n - s}{z^\alpha n^\alpha}. \qquad (4.63)$$

So, for $s = c = n$, $(j_1, j_2, \ldots, j_n) = (1, 2, \ldots, n)$,

$$\mathbb{P} \left(\sum_{k=1}^{n} \xi_k(2\alpha) > zn \right) \sim n\mathbb{P}(\xi_1(2\alpha) > zn) \sim \frac{n}{z^{2\alpha} n^{2\alpha}}. \qquad (4.64)$$

From (4.61)–(4.64), it follows that, with $1 \leq j_1 < j_2 < \ldots < j_s \leq c$, $s = 0, 1, \ldots, c$, $c = 2, \ldots, n$, $(s, c) \neq (n, n)$,

$$
\mathbb{E}\left[(1 - 2F_1(\xi_1(\alpha)))(1 - 2F_2(\xi_2(\alpha))) \ldots (1 - 2F_c(\xi_c(\alpha)))\mathbb{I} \right.
$$

$$
\left. \times \left(\sum_{i=1}^{n} \xi_i(\alpha) > zn \right) \right]
$$

$$
\sim \sum_{s=0}^{c} \sum_{1 \leq j_1 < j_2 < \ldots < j_s \leq c} (-1)^{c-s} \frac{n-s}{z^\alpha n^\alpha}
$$

$$
= \left(\sum_{s=0}^{c} (-1)^{c-s} C_c^s \right) z^{-\alpha} n^{1-\alpha} - \left(\sum_{s=0}^{c} (-1)^{c-s} s C_c^s \right) z^{-\alpha} n^{-\alpha},
$$

$$
\tag{4.65}
$$

where $C_c^s = c!/(s!(c-s)!)$ denotes binomial coefficients.

Now, by the well-known identity for binomial coefficients,

$$
\sum_{s=0}^{c} (-1)^{c-s} C_c^s = \sum_{s=0}^{c} (-1)^s C_c^s = 0, \tag{4.66}
$$

$$
\sum_{s=0}^{c} (-1)^{c-s} s C_c^s = c \sum_{s=1}^{c} (-1)^{c-s} C_{c-1}^{s-1}
$$

$$
= -c \sum_{s=0}^{c-1} (-1)^{c-1-s} C_{c-1}^s = 0. \tag{4.67}
$$

It thus follows that $\mathbb{P}(\sum_{i=1}^{n} X_i > zn) \sim \mathbb{P}(\sum_{i=1}^{n} \xi_i(\alpha) > zn)$.

Proof of Theorem 4.3. The density corresponding to (4.12) is a polynomial of a lower order, which we write in the following generic form:

$$
c(u_1, \ldots, u_n) = \sum_{k_1, \ldots, k_n = 0, 1, \ldots} \phi_{k_1, k_2, \ldots, k_n} \cdot u_1^{k_1} \cdot u_2^{k_2} \cdot \ldots \cdot u_n^{k_n}, \tag{4.68}
$$

Then, using arguments similar to Theorem 4.2,

$$\mathbb{P}\left(\sum_{i=1}^n X_i > zn\right)$$

$$= \mathbb{E}\left[\sum_{k_i \in \{0,1,\ldots\}} \phi_{k_1,k_2,\ldots,k_n} F_1^{k_1}(\xi_1(\alpha)) F_2^{k_2}(\xi_2(\alpha)) \ldots F_n^{k_n}(\xi_n(\alpha))\right.$$

$$\left. \times \mathbb{I}\left(\sum_{i=1}^n \xi_i(\alpha) > zn\right)\right]$$

$$= \mathbb{P}\left(\sum_{i=1}^n \xi_i(\alpha) > zn\right) + \mathbb{E}\left[\sum_{k_i \in \{0,1,\ldots\}\setminus\{k_i=0 \forall i\}} \phi_{k_1,k_2,\ldots,k_n}\right.$$

$$\left. \times F_1^{k_1}(\xi_1(\alpha)) F_2^{k_2}(\xi_2(\alpha)) \ldots F_n^{k_n}(\xi_n(\alpha)) \mathbb{I}\left(\sum_{i=1}^n \xi_i(\alpha) > zn\right)\right].$$

$$(4.69)$$

Now consider the last term.

$$\mathbb{E}\left[\sum_{k_i \in \{0,1,\ldots\}\setminus\{k_i=0 \forall i\}} \phi_{k_1,k_2,\ldots,k_n} F_1^{k_1}(\xi_1(\alpha)) F_2^{k_2}(\xi_2(\alpha)) \ldots F_n^{k_n}(\xi_n(\alpha))\right.$$

$$\left. \times \mathbb{I}\left(\sum_{i=1}^n \xi_i(\alpha) > zn\right)\right]$$

$$= \int_{\sum_{i=1}^n s_i > nz} \sum_{k_i \in \{0,1,\ldots\}} \psi_{k_1,k_2,\ldots,k_n} s_1^{-\alpha(k_1+1)}$$

$$\times s_2^{-\alpha(k_2+1)} \ldots s_n^{-\alpha(k_n+1)} ds_1 \ldots ds_n$$

$$= \sum_{k_i \in \{0,1,\ldots\}} \psi_{k_1,k_2,\ldots,k_n} \int_{\sum_{i=1}^n s_i > nz} s_1^{-\alpha(k_1+1)}$$

$$\times s_2^{-\alpha(k_2+1)} \ldots s_n^{-\alpha(k_n+1)} ds_1 \ldots ds_n$$

$$= \sum_{k_i \in \{0,1,\dots\}} \psi_{k_1, k_2, \dots, k_n}$$

$$\mathbb{P}\left(\frac{\xi_1(\alpha(k_1+1)) + \cdots + \xi_n(\alpha(k_n+1))}{n} > z\right), \qquad (4.70)$$

where the new coefficients ψ's are different from ϕ's because we have expressed $(1 - s_i^{-\alpha})^{k_i}$ in terms of powers of s_i^{α}. Now, using the same arguments as for (4.53)–(4.54),

$$\mathbb{P}\left(\frac{\xi_1(\alpha) + \cdots + \xi_n(\alpha)}{n} > z\right) \sim n\mathbb{P}(\xi_1(\alpha) > nz)$$

$$\sim n^{1-\alpha} z^{-\alpha}$$

$$\mathbb{P}\left(\frac{\xi_1(\alpha(k_1+1)) + \cdots + \xi_n(\alpha(k_n+1))}{n} > z\right) \sim \mathbb{P}(\xi_1(\alpha) > nz)$$

$$\sim n^{-\alpha} z^{-\alpha},$$

for all $k_i \geq 0$. It thus follows that $\mathbb{P}(\sum_{i=1}^n X_i > zn) \sim \mathbb{P}(\sum_{i=1}^n \xi_i(\alpha) > zn)$.

Proof of Theorems 4.4 and 4.5. The proof of the extensions of Theorems 2.1 and 2.2 to the dependent case follows the same lines as the proof of the above theorems since the following properties hold:

Relation (1.10) holds if $(X_1, \dots, X_n)$ has an α-symmetric distribution (see Fang *et al.* (1990), Chapter 7, and Section 4.4 in this chapter).

The densities of the r.v.'s $\sum_{i=1}^n v_i Y_{ij}$, $j = 1, \dots, k$, and $\sum_{i=1}^n w_i Y_{ij}$, $j = 1, \dots, k$, are symmetric and unimodal by Proposition 2.3.

The densities of the r.v.'s $Z_j \sum_{i=1}^n v_i Y_{ij}$, $j = 1, \dots, k$, and $Z_j \sum_{i=1}^n w_i Y_{ij}$, $j = 1, \dots, k$, are symmetric and unimodal if $Y_{ij} \sim S_{\alpha_j}(\sigma_j, 0, 0)$, $i = 1, \dots, n$, $j = 1, \dots, k$, and Z_j are absolutely continuous positive r.v.'s independent of Y_{ij} (this follows by symmetry and unimodality of $\sum_{i=1}^n v_i Y_{ij}$ and $\sum_{i=1}^n w_i Y_{ij}$ implied by Propositions 2.1 and 2.4, Definition 2.1 and conditioning arguments).

Proof of Theorem 4.7. Parts (i) and (ii) of Theorem 4.7 follow from Theorems 2.1 and 2.2 and majorization comparisons $\underline{w}_{rc} \prec w \prec \overline{w}_{rc}$ for all $w \in \mathcal{I}_{rc}$ implied by (2.1) with $N = rc$.

Using (2.1) with $N = r$ and $N = c$, we conclude that, for the vectors $w_0^{(row)}$ and $w_0^{(col)}$ in (4.20) and (4.21) one has

$$\underline{w}_0^{(row)} \prec w_0^{(row)} \prec \overline{w}_0^{(row)}, \tag{4.71}$$

$$\underline{w}_0^{(col)} \prec w_0^{(col)} \prec \overline{w}_0^{(col)}, \tag{4.72}$$

where, as in Section 4.4.2, $\underline{w}_0^{(row)} = \underline{w}_r = \underbrace{(1/r, 1/r, \ldots, 1/r)}_{r} \in \mathcal{I}_r$,

$\underline{w}_0^{(col)} = \underline{w}_c = \underbrace{(1/c, 1/c, \ldots, 1/c)}_{c} \in \mathcal{I}_c$, $\overline{w}_0^{(row)} = \underbrace{(1, 0, \ldots, 0)}_{r} = \overline{w}_r \in$

$\mathcal{I}_r$ and $\overline{w}_0^{(col)} = \underbrace{(1, 0, \ldots, 0)}_{c} = \overline{w}_c \in \mathcal{I}_c$ are the vectors that corre-

spond to $\underline{w}_{rc}$ and $\overline{w}_{rc}$ by (4.20) and (4.21). Majorization comparisons (4.71) and (4.72), together with Theorems 2.1 and 2.2, imply parts (iii)–(vi) of Theorem 4.7.

Proof of Theorem 4.6. Let $R_i, C_j, U_{ij} \sim \mathcal{CS}$, $i = 1, \ldots, r$, $j = 1, \ldots, c$, and let $w \in \mathcal{I}_{rc}$. From part (ii) of Theorem 4.7, it follows that the risks $U(w)$ in decomposition (4.22) satisfy

$$\text{VaR}_q[U(\underline{w}_{rc})] \geq \text{VaR}_q[U(w)] \geq \text{VaR}_q[U(\overline{w}_{rc})], \quad q \in (0, 1/2). \tag{4.73}$$

In addition, from parts (iv) and (vi) of Theorem 4.7, we conclude that the following value at risk comparisons hold for the components $R(w_0^{(row)})$ and $C(w_0^{(col)})$ in decomposition (4.22):

$$\text{VaR}_q[R(\underline{w}_0^{(row)})] \geq \text{VaR}_q[R(w_0^{(row)})]$$

$$\geq \text{VaR}_q[R(\overline{w}_0^{(row)})], \quad q \in (0, 1/2), \tag{4.74}$$

$$\text{VaR}_q[C(\underline{w}_0^{(col)})] \geq \text{VaR}_q[C(w_0^{(col)})]$$

$$\geq \text{VaR}_q[C(\overline{w}_0^{(col)})], \quad q \in (0, 1/2), \tag{4.75}$$

where $\underline{w}_0^{(row)} = \underline{w}_r \in \mathcal{I}_r$, $\overline{w}_0^{(row)} = \overline{w}_r \in \mathcal{I}_r$, $\underline{w}_0^{(col)} = \underline{w}_c \in \mathcal{I}_c$ and $\overline{w}_0^{(col)} = \overline{w}_c \in \mathcal{I}_c$ are the vectors that correspond to $\underline{w}_{rc}$ and $\overline{w}_{rc}$ by (4.20) and (4.21).

From Theorem 2.7.6 in Zolotarev (1986), p. 134, and Theorems 1.6 and 1.10 in Dharmadhikari and Joag-Dev (1988), pp. 13 and 20, by induction it follows that the densities of the r.v.'s $R(w_0^{(row)})$, $C(w_0^{(col)})$ and $U(w)$ are symmetric and unimodal if the assumptions of Theorem 4.6 hold. From Lemma in Birnbaum (1948) (see also Theorem 3.D.4 on p. 173 in Shaked and Shanthikumar (2007)) it follows that if $X_1, \ldots, X_n$ and $Y_1, \ldots, Y_n$ are independent absolutely continuous symmetric unimodal r.v.'s such that, for $i = 1, 2, \ldots, n$, and all $q \in (0, 1/2)$, $\mathrm{VaR}_q(X_i) \le \mathrm{VaR}_q(Y_i)$, then $\mathrm{VaR}_q(\sum_{i=1}^n X_i) \le \mathrm{VaR}_q(\sum_{i=1}^n Y_i)$ for all $q \in (0, 1/2)$.

This, together with inequalities (4.73)–(4.75), implies that, for all $q \in (0, 1/2)$,

$$\mathrm{VaR}_q[R(\underline{w}_0^{(row)}) + C(\underline{w}_0^{(col)}) + U(\underline{w}_{rc})]$$

$$\ge \mathrm{VaR}_q[R(w_0^{(row)}) + C(w_0^{(col)}) + U(w)]$$

$$\ge \mathrm{VaR}_q[R(\overline{w}_0^{(row)}) + C(\overline{w}_0^{(col)}) + U(\overline{w}_{rc})]. \qquad (4.76)$$

Consequently,

$$\mathrm{VaR}_q[Y(\underline{w}_{rc})] \ge \mathrm{VaR}_q[Y(w)] \ge \mathrm{VaR}_q[Y(\overline{w}_{rc})] \qquad (4.77)$$

for all $q \in (0, 1/2)$. Thus, part (ii) of Theorem 4.6 holds. Part (i) of Theorem 4.6 may be proven in a similar way, with the use of parts (i), (iii) and (v) of Theorem 4.7 instead of parts (ii), (iv) and (vi) of the theorem.

Proof of Theorem 4.8. The theorem follows from parts (i) of Propositions 2.1 and 2.2 and the majorization comparisons between $\tilde{v}$ and $\tilde{w}$ and between $\tilde{\tilde{v}}$ and $\tilde{\tilde{w}}$ given by (4.26) and (4.27).

For the proof of Theorems 4.9–4.11, we need a lemma that follows from Proposition 5.B.1 in Section 5.B in Marshall and Olkin (1979) and Marshall et al. (2011) applied with condition (a') in that section.

Lemma 4.1. *If $a_1 \geq \cdots \geq a_r > 0$, $b_1 \geq \cdots \geq b_r > 0$ and b_i/a_i is non-increasing in $i = 1, \ldots, r$, then*

$$\left(\frac{a_1}{\sum_{i=1}^r a_i}, \ldots, \frac{a_r}{\sum_{i=1}^r a_i} \right) \prec \left(\frac{b_1}{\sum_{i=1}^r b_i}, \ldots, \frac{b_r}{\sum_{i=1}^r b_i} \right). \quad (4.78)$$

If $0 < a_1 \leq \cdots \leq a_r$, $0 < b_1 \leq \cdots \leq b_r$ and b_i/a_i is non-decreasing in $i = 1, \ldots, r$, then

$$\left(\frac{a_1}{\sum_{i=1}^r n_i a_i} e_{n_1}, \ldots, \frac{a_r}{\sum_{i=1}^r n_i a_i} e_{n_r} \right)$$

$$\prec \left(\frac{b_1}{\sum_{i=1}^r n_i b_i} e_{n_1}, \ldots, \frac{b_r}{\sum_{i=1}^r n_i b_i} e_{n_r} \right), \quad (4.79)$$

where, as in Section 4.5.2, for $N \geq 1$, $e_N = \underbrace{(1, \ldots, 1)}_{N} \in \mathbf{R}^N$ denotes the N-vector of ones.

Proof of Theorem 4.11. Consider the vectors $w^{(1)} = \underline{w}_r = \underbrace{(1/r, \ldots, 1/r)}_{r} \in \mathcal{I}_r$, $w^{(2)}$, $w^{(3)}$ and $w(c)$, $0 \leq c \leq 1$, defined in (4.37), (4.39), (4.41) and (4.42). From the left majorization comparison in (2.1), it follows that

$$w^{(1)} \prec w \quad (4.80)$$

for all $w \in \mathcal{I}_r$ and, since $(\frac{w_1^{(2)}}{n_1} e_{n_1}, \ldots, \frac{w_r^{(2)}}{n_r} e_{n_r}) = \underline{w}_n = \underbrace{(1/n, \ldots, 1/n)}_{n} \in \mathcal{I}_n$,

$$\left(\frac{w_1^{(2)}}{n_1} e_{n_1}, \ldots, \frac{w_r^{(2)}}{n_r} e_{n_r} \right) \prec w \quad (4.81)$$

for all $w \in \mathcal{I}_n$. Let us show, using Lemma 4.1, that the following majorization relations hold:

$$w^{(3)} \prec w^{(2)} \quad (4.82)$$

(relation (4.82) is a part of Lemma 13.B.1.a in Marshall and Olkin (1979) and Marshall *et al.* (2011));

$$\left(\frac{w_1^{(3)}}{n_1} e_{n_1}, \ldots, \frac{w_r^{(3)}}{n_r} e_{n_r} \right) \prec \left(\frac{w_1^{(1)}}{n_1} e_{n_1}, \ldots, \frac{w_r^{(1)}}{n_r} e_{n_r} \right), \quad (4.83)$$

and

$$w(c') \prec w(c), \tag{4.84}$$

$$\left(\frac{w_1(c)}{n_1} e_{n_1}, \ldots, \frac{w_r(c)}{n_r} e_{n_r} \right) \prec \left(\frac{w_1(c')}{n_1} e_{n_1}, \ldots, \frac{w_r(c')}{n_r} e_{n_r} \right), \tag{4.85}$$

if $0 \le c < c' \le 1$.

To obtain (4.82), take $a_i = n_i(n - n_i)$ and $b_i = n_i$, $i = 1, \ldots, r$, in Lemma 4.1. Under the assumptions of the theorem, $b_1 \ge \cdots \ge b_r$. As indicated in the proof of Lemma 13.B.1.a in Marshall and Olkin (1979) and Marshall *et al.* (2011), because $z_1 \ge z_2$ and $z_1 + z_2 \le 1$ together imply $z_1(1 - z_1) \ge z_2(1 - z_2)$, one also has $a_1 \ge \cdots \ge a_r$. In addition, evidently, $b_i/a_i = 1/(n - n_i)$ is non-increasing in $i = 1, \ldots, r$. Consequently, by (4.78), (4.82) indeed holds.

To establish (4.83), take $a_i = n - n_i$ and $b_i = 1/(rn_i)$. Then $a_1 \le \cdots \le a_r$, $b_1 \le \cdots \le b_r$ and $b_i/a_i = 1/(rn_i(n - n_i))$ is non-decreasing in $i = 1, \ldots, r$. Consequently, (4.83) holds by (4.79).

Relation (4.84) is a consequence of (4.78) applied to $a_i = n_i/((n_i - 1)c' + 1)$ and $b_i = n_i/((n_i - 1)c + 1)$.

Majorization (4.85) follows from (4.79) applied to $a_i = 1/((n_i - 1)c + 1)$ and $b_i = 1/((n_i - 1)c' + 1)$.

Theorem 4.11 now follows from parts (i) of Propositions 2.1 and 2.2 and majorization comparisons (4.80)–(4.85).

Proof of Theorems 4.9 and 4.10. Suppose that, in (4.32), $R_i \sim \overline{\mathcal{CSLC}}$, $i = 1, \ldots, r$, and $U_{ij} \sim \underline{\mathcal{CS}}$, $j = 1, \ldots, n_i$, $i = 1, \ldots, r$. Let $0 \le c < c' \le 1$. Using parts (i) and (ii) of Theorem 4.11, we obtain

$$\mathrm{VaR}_q[R(w(c'))] \le \mathrm{VaR}_q[R(w(c))], \tag{4.86}$$

$$\mathrm{VaR}_q[U(\tilde{w}(c'))] \le \mathrm{VaR}_q[U(\tilde{w}(c))], \tag{4.87}$$

$$\mathrm{VaR}_q[R(w^{(1)})] \le \mathrm{VaR}_q[R(w^{(3)})] \le \mathrm{VaR}_q[R(w^{(2)})], \tag{4.88}$$

$$\mathrm{VaR}_q[U(\tilde{w}^{(1)})] \le \mathrm{VaR}_q[U(\tilde{w}^{(3)})] \le \mathrm{VaR}_q[U(\tilde{w}^{(2)})] \tag{4.89}$$

for all $q \in (0, 1/2)$.

Similar to the proof of Theorem 4.6, from Theorem 2.7.6 in Zolotarev (1986), p. 134, and Theorems 1.6 and 1.10 in Dharmadhikari and Joag-Dev (1988), pp. 13 and 20, we conclude that the densities of the r.v.'s $R(w)$ and $U(\tilde{w})$ are symmetric and unimodal under the assumptions of Theorems 4.9–4.11. As in the proof of Theorem 4.6, inequalities (4.86)–(4.89), together with Lemma in Birnbaum (1948) and Theorem 3.D.4 on p. 173 in Shaked and Shanthikumar (2007), imply

$$\mathrm{VaR}_q[R(w(c')) + U(w(c'))] \leq \mathrm{VaR}_q[R(w(c)) + U(w(c))], \qquad (4.90)$$

$$\mathrm{VaR}_q[R(w^{(1)}) + U(\tilde{w}^{(1)})] \leq \mathrm{VaR}_q[R(w^{(3)}) + U(\tilde{w}^{(3)})]$$

$$\leq \mathrm{VaR}_q[R(w^{(2)}) + U(\tilde{w}^{(2)})] \qquad (4.91)$$

for all $q \in (0, 1/2)$. That is, $\mathrm{VaR}_q[Z(w(c'))] \leq \mathrm{VaR}_q[Z(w(c))]$ and $\mathrm{VaR}_q[Z(w^{(1)})] \leq \mathrm{VaR}_q[Z(w^{(3)})] \leq \mathrm{VaR}_q[Z(w^{(2)})]$ for all $q \in (0, 1/2)$. This proves Theorem 4.10.

Theorem 4.9 for $R_i \sim \underline{\mathcal{CS}}$ and $U_{ij} \sim \overline{\mathcal{CSLC}}$ may be proven in a similar way, with the reversals of the inequality signs in (4.86)–(4.91) implied by parts (ii) and (iv) of Theorem 4.11.

Proof of Theorems 4.12 and 4.13. As is easy to see, for symmetric r.v.'s X_1 and X_2, $\mathbb{P}(|X_1| > \epsilon) \leq \mathbb{P}(|X_2| > \epsilon)$ for all $\epsilon > 0$ if and only if $\mathrm{VaR}_q(X_1) \leq \mathrm{VaR}_q(X_2)$ for all $q \in (0, 1/2)$. Therefore, Theorems 4.12 and 4.13 follow from the value at risk comparisons in Theorems 4.9 and 4.10.

Proof of Theorem 4.14. Denote $\mathcal{R} = (R_1, R_2, \ldots, R_r)$, $\mathcal{C} = (C_1, C_2, \ldots, C_c)$ and $\mathcal{U} = (U_{11}, U_{12}, \ldots, U_{rc})$. The arguments for extensions of relations (4.28)-(4.31) and the results in Theorems 4.7, 4.8 and 4.11 for $\mathcal{R} \in \mathcal{G}_r(\alpha_r)$, $\mathcal{C} \in \mathcal{G}_c(\alpha_c)$ and $\mathcal{U} \in \mathcal{G}_{rc}(\alpha_u)$ are completely similar to their proof in the case of classes $\overline{\mathcal{CSLC}}$, $\overline{\mathcal{CS}}$ and $\underline{\mathcal{CS}}$, with the use of Proposition 4.1 instead of Propositions 2.1 and 2.2.

Chapter 5

Robustness of Econometric Methods to Copula Misspecification and Heavy Tails

This chapter provides a survey of estimation methods for copula models with a focus on robustness of copula-based models to copula misspecification. We also consider selected aspects of robustness to heavy tails.

5.1. Introduction

Robustness of a copula-based econometric methodology to copula misspecification means that a method remains valid, i.e., produces consistent estimators and reliable inference, even if the copula is misspecified.

As an example, consider a situation where we have panel data, with individuals $i = 1, \ldots, N$ and time periods $j = 1, \ldots, d$. Presume that N is large and d is small, and consider the standard asymptotic arguments as $N \to \infty$ with d fixed. In this setting, suppose that we have a correctly specified likelihood-based model (e.g., a logit model, or a parametric duration model, or a stochastic frontier model, etc.) which could be consistently estimated from any of the d cross sections. However, we would like to be able to use the data from the whole panel as using more information usually leads to a more precise estimation. For example, we would be comfortable estimating our logit model from a single cross section, but we wish to combine information over time to get a more efficient estimate.

In this setting, robustness to copula misspecification means that a copula-based Maximum Likelihood Estimation (MLE) remains consistent even if the copula is wrong. Clearly, for a single cross section at time t, we would have a log-likelihood of the generic form

$$\ln L = \sum_{i=1}^{N} \ln f(x_{ij}, \theta),$$

where x_{ij} represents the data for person i at time j, and this is what we would use if we ignored dependence over time. The obvious extension is to specify a joint log-likelihood of the generic form

$$\ln L = \sum_{i=1}^{N} \ln f(x_{i1}, \ldots, x_{id}, \theta).$$

The problem, of course, is that there are many joint (all d time periods) distributions consistent with the marginal (individual time period) distributions. Therefore, there are issues of whether we wish to specify a joint distribution and what it may cost us if we do so.

The chapter is organized as follows. Section 5.2 discusses various estimation approaches based on i.i.d. observations. Section 5.3 considers the case of copula-based time series. Section 5.4 provides improved and robust parametric-likelihood based estimators. Section 5.5 discusses robustness and efficiency in models with nonparametric copulas.

5.2. Copula estimation

Depending on the assumptions made on the data generating process, estimation procedures lead to parametric, semi-parametric and non-parametric inference methods. Clearly, stronger parametric assumptions limit robustness of a method as its properties are affected by correctness of those assumptions.

Section 5.2.1 reviews parametric methods; Sections 5.2.2 and 5.2.3 review nonparametric and semiparametric approaches to inference on copulas using random samples with dependent marginals.

5.2.1. *Parametric models: MLE and IFM*

Consider the key relation in copula theory given by formula (1.1) in Proposition 1.1 under absolute continuity assumptions. In the notations of Sklar's theorem in Proposition 1.1, the density f of the d-dimensional d.f. F with univariate margins $F_1, F_2, \ldots, F_d$ and the corresponding univariate densities $f_1, f_2, \ldots, f_d$ can be represented as

$$f(x_1, x_2, \ldots, x_d) = c(F_1(x_1), F_2(x_2), \ldots, F_d(x_d)) \prod_{j=1}^{d} f_j(x_j), \quad (5.1)$$

where $c(u_1, u_2, \ldots, u_d) = \frac{\partial C(u_1, u_2, \ldots, u_d)}{\partial u_1 \partial u_2 \ldots \partial u_d}$ is the density of the d-dimensional copula $C(u_1, u_2, \ldots, u_d; \theta)$ in (1.1) in Proposition 1.1.

Representation (5.1) implies the following decomposition for the log-likelihood function $L = \sum_{i=1}^{N} \log f(x_1^{(i)}, x_2^{(i)}, \ldots, x_d^{(i)})$ of a random sample of (i.i.d.) vectors $x^{(i)} = (x_1^{(i)}, x_2^{(i)}, \ldots, x_d^{(i)})$, $i = 1, 2, \ldots, N$, with density f:

$$L = \underbrace{L_C}_{\text{dependence}} + \underbrace{\sum_{j=1}^{d} L_j}_{\text{marginals}}, \quad (5.2)$$

where $L_C = \sum_{i=1}^{N} \log \, c(F_1(x_1^{(i)}), F_2(x_2^{(i)}), \ldots, F_d(x_d^{(i)}))$ is the log-likelihood contribution from dependence represented by the copula C and $L_j = \sum_{i=1}^{N} \log \, f_j(x_j^{(i)})$, $j = 1, 2, \ldots, d$, are the log-likelihood contributions from each margin. Observe that $\sum_{j=1}^{d} L_j$ in (5.2) is exactly the log-likelihood of the sample under the independence assumption.

Suppose that the copula C belongs to a family of copulas indexed by a (vector) parameter θ so that $C = C(u_1, u_2, \ldots, u_d; \theta)$ and the univariate margins F_j along with the corresponding densities f_j are indexed by (vector) parameters $\alpha_j : F_j = F_j(x_j; \alpha_j)$, $f_j = f_j(x_j; \alpha_j)$.

For, example, the α's can represent the tail index of a heavy tailed marginal distribution.

Let $(\hat{\alpha}_1^{MLE}, \hat{\alpha}_2^{MLE}, \ldots, \hat{\alpha}_d^{MLE}, \hat{\theta}^{MLE})$ denote the maximum likelihood estimator (MLE) of the model parameters $(\alpha_1, \alpha_2, \ldots, \alpha_d, \theta)$. The MLE is obtained by simultaneous maximization of the log-likelihood L in (5.2):

$$(\hat{\alpha}_1^{MLE}, \ldots, \hat{\alpha}_d^{MLE}, \hat{\theta}^{MLE}) = \arg\max_{\alpha_1, \ldots, \alpha_d, \theta} L(\alpha_1, \ldots, \alpha_d, \theta),$$

where

$$L(\alpha_1, \ldots, \alpha_d, \theta) = L_C(\alpha_1, \ldots, \alpha_d, \theta) + \sum_{j=1}^{d} L_j(\alpha_j)$$

$$= \sum_{i=1}^{N} \log\ c(F_1(x_1^{(i)}; \alpha_1), \ldots, F_d(x_d^{(i)}; \alpha_d); \theta)$$

$$+ \sum_{j=1}^{d} \sum_{i=1}^{N} \log\ f_j(x_j^{(i)}; \alpha_j). \tag{5.3}$$

Joe (1997, Section 10.1) provides an excellent review of MLE in multivariate copula models (see, also, Joe (2014)). He also discusses a computationally attractive alternative to MLE, which avoids the simultaneous maximization over the dependence (θ) and marginal $(\alpha_1, \alpha_2, \ldots, \alpha_d)$ parameters.

The alternative estimation approach is motivated by decomposition (5.2) and is often referred to as the method of inference functions for margins (IFM). In the first stage of the IFM procedure, α_j's are estimated from the log-likelihood L_j of each margin in (5.2)–(5.3). Let $\hat{\alpha}_j^{IFM}$ denote these estimators. Then, they can be written as follows:

$$\hat{\alpha}_j^{IFM} = \arg\max_{\alpha_j} L_j(\alpha_j).$$

That is, $(\hat{\alpha}_1^{IFM}, \ldots, \hat{\alpha}_d^{IFM})$ is defined as MLE of the model parameters under independence.

In the second stage of the procedure, the copula parameter θ is estimated by maximizing the copula likelihood contribution L_C in

(5.2)–(5.3) with the marginal parameters α_j replaced by their first-stage estimates $\hat{\alpha}_j^{IFM}$. Let $\hat{\theta}^{IFM}$ denote that estimator. Then it can be obtained as follows:

$$\hat{\theta}^{IFM} = \arg\max_{\theta} L_C(\hat{\alpha}_1^{IFM}, \ldots, \hat{\alpha}_d^{IFM}, \theta).$$

While, under regularity conditions, the MLE estimator $(\hat{\alpha}_1^{MLE}, \hat{\alpha}_2^{MLE}, \ldots, \hat{\alpha}_d^{MLE}, \hat{\theta}^{MLE})$ solves

$$(\partial L/\partial \alpha_1, \partial L/\partial \alpha_2, \ldots, \partial L/\partial \alpha_d, \partial L/\partial \theta)' = 0,$$

the two-stage IFM estimator $(\hat{\alpha}_1^{IFM}, \hat{\alpha}_2^{IFM}, \ldots, \hat{\alpha}_d^{IFM}, \hat{\theta}^{IFM})$ solves

$$(\partial L_1/\partial \alpha_1, \partial L_2/\partial \alpha_2, \ldots, \partial L_d/\partial \alpha_d, \partial L/\partial \theta) = 0.$$

It is well known that MLE can be viewed as Generalized Method of Moments (GMM) estimator based on the score function (see Godambe (1960, 1976)) — we will return to this point in Section 5.4.2.

As discussed in Joe (1997), the MLE and IFM estimation procedures are equivalent in the case of multivariate normal d.f.'s that have multivariate Gaussian copulas and univariate normal margins. Naturally, however, this equivalence does not hold in general. We return to this point in Section 5.4.

Similar to the MLE, the IFM estimator $(\hat{\alpha}_1^{IFM}, \ldots, \hat{\alpha}^{IFM}, \hat{\theta}^{IFM})$ is consistent and asymptotically normal under the usual regularity conditions (see, e.g., Serfling (1980)) for the multivariate model and for each of its margins, provided the parametric form of the marginals and the copula are correctly specified. Estimation of the corresponding covariance matrices is difficult both analytically and numerically due to the need to compute many derivatives, and jackknife and related methods may be used in inference. Efficiency comparisons based on estimation of the asymptotic covariance matrices and Monte-Carlo simulation for different dependence models suggest that the IFM approach to inference provide a highly efficient alternative. However, full MLE remains the efficient estimator. Both are biased and inconsistent if the marginals or the copula are misspecified.

5.2.2. *Nonparametric models: Empirical and Bernstein copulas*

Many nonparametric estimation procedures for copulas are based on the inversion formula (1.1) in Proposition 1.1. An estimator $\hat{C}(u_1, u_2, \ldots, u_d)$ of a d-copula $C(u_1, u_2, \ldots, u_d)$ is typically given by an empirical analogue of the inversion formula (see, e.g., Durante and Sempi (2010), Equation (12)). That is,

$$\hat{C}(u_1, u_2, \ldots, u_d) = \hat{F}(\hat{F}_1^{-1}(u_1), \hat{F}_2^{-1}(u_2), \ldots, \hat{F}_d^{-1}(u_d)), \quad (5.4)$$

where $\hat{F}$ is a nonparametric estimator of the d-dimensional d.f. F and $\hat{F}_1^{-1}, \hat{F}_2^{-1}, \ldots, \hat{F}_d^{-1}$ are nonparametric estimators of the pseudo-inverses $F_j^{-1}(s) = \{t|F_j(t) \geq s\}$ of the univariate margins $F_1, F_2, \ldots, F_d$. Often $\hat{F}$ is taken to be the empirical d-dimensional d.f.

$$\hat{F}(x_1, x_2, \ldots, x_d) = \frac{1}{N} \sum_{i=1}^{N} \mathbb{I}(X_{1i} \leq x_1, X_{2i} \leq x_2, \ldots, X_{di} \leq x_d),$$

where $\mathbb{I}(\cdot)$ denotes the indicator functions. Similarly, F_j^{-1} are often estimated using the pseudo-inverses

$$\hat{F}_j^{-1}(s) = \{t|\hat{F}_j(t) \geq s\}$$

of the empirical univariate d.f.'s

$$\hat{F}_j = \frac{1}{N} \sum_{i=1}^{N} \mathbb{I}(X_{ji} \leq x_j)$$

or their rescaled versions.

Deheuvels (1979, 1981) established consistency and asymptotic normality of the empirical copula process for i.i.d. observations of random vectors with independent margins, i.e., for the case of the product copula $C = \Pi$ (see Durante and Sempi (2010), Section 3). Fermanian *et al.* (2004) provides a proof of consistency and asymptotic normality of the empirical copula process for general copulas C with continuous partial derivatives (also see Fermanian and Scaillet (2003)).

Fermanian *et al.* (2004) further show that, under regularity conditions, asymptotic normality also holds for smoothed copula processes $\hat{C}(u_1, u_2) = \hat{F}(\hat{F}_1^{-1}(u_1), \hat{F}_2^{-1}(u_2))$ that are constructed using the nonparametric kernel estimators

$$\hat{F}(x_1, x_2) = \frac{1}{N} \sum_{i=1}^{N} K\left(\frac{x_1 - X_{2i}}{a_N}, \frac{x_2 - X_{2i}}{a_N}\right)$$

of the joint d.f. F. Here $K(x_1, x_2) = \int_{-\infty}^{x_1} \int_{-\infty}^{x_2} k(u, v) du dv$ for some bivariate kernel function $k : \mathbf{R}^2 \to \mathbf{R}$ with $\int k(x, y) dx dy = 1$, and the sequence of bandwidths $a_N > 0$ satisfies $a_N \to 0$ as $N \to \infty$ (see Carrasco and Chen (2002), for the normal asymptotics of smoothed copula processes based on local linear versions of K rather than kernels known to have boundary biases).

Popular alternatives to nonparametric estimators based on empirical distributions are copulas based on series approximations known as sieves (see, e.g., Chen (2007), for a review of sieve theory). One such estimator uses the Bernstein polynomial (see, e.g., Sancetta and Satchell (2004)).

Let $[0, 1]^2$ denote the unit cube in $\mathbb{R}^2$. For a distribution function $P_c : [0, 1]^2 \to \mathbb{R}$, a bivariate Bernstein polynomial of order $\mathbf{k} = (k_1, k_2)$ associated with P_c is defined as

$$B_{\mathbf{k},P_c}(\mathbf{u}) = \sum_{i_1=0}^{k_1} \sum_{i_2=0}^{k_2} P_c\left(\frac{i_1}{k_1}, \frac{i_2}{k_2}\right) q_{i_1 k_1}(u_1) q_{i_2 k_2}(u_2), \qquad (5.5)$$

where $\mathbf{u} = (u_1, u_2) \in [0, 1]^2$, $q_{i_s k_s}(u_s) = \binom{k_s}{i_s} u_s^{i_s} (1 - u_s)^{k_s - i_s}$. The polynomial is dense in the space of distribution functions on $[0, 1]^2$ and its order $\mathbf{k}$ controls the smoothness of $B_{\mathbf{k},P_c}$, with a smaller k_s associated with a smoother function along dimension s. Moreover, with the conditions $P_c(0, 1) = P_c(1, 0) = 0$ and $P_c(1, 1) = 1$, $B_{\mathbf{k},P_c}(\mathbf{u})$ is a copula function and is referred to as the Bernstein copula associated with P_c. As $\min\{\mathbf{k}\} \to \infty$, $B_{\mathbf{k},P_c}(\mathbf{u})$ converges to P_c at each continuity point of P_c and if P_c is continuous then the convergence is uniform on the unit cube $[0, 1]^2$ (see, e.g., Zheng (2011)).

The derivative of (5.5) is the bivariate Bernstein density function

$$b_{\mathbf{k},P_c}(\mathbf{u}) = \frac{\partial^2}{\partial u_1 \partial u_2} B_{\mathbf{k},P_c}(\mathbf{u})$$

$$= \sum_{i_1=1}^{k_1} \sum_{i_2=1}^{k_2} w_{\mathbf{k}}(\mathbf{i}) \prod_{s=1}^{2} \beta(u_s; i_s, k_s - i_s + 1) \qquad (5.6)$$

where, for $\mathbf{i} = (i_1, i_2)$, $w_{\mathbf{k}}(\mathbf{i}) = \mathbf{\Delta} P_c(\frac{i_1-1}{k_1}, \frac{i_2-1}{k_2})$ are weights derived using the forward difference operator $\mathbf{\Delta}$, and $\beta(\cdot; \gamma, \delta)$ denotes the probability density function of the β-distribution with parameters γ and δ.

In order to give a mixing interpretation to $w_{\mathbf{k}}$, let Cube$(\mathbf{i}, \mathbf{k})$ denote a cube given by $((i_1 - 1)/k_1, i_1/k_1] \times ((i_2 - 1)/k_2, i_2/k_2]$ with the convention that if $i_s = 0$ then the interval $((i_s - 1)/k_s, i_s/k_s]$ is replaced by the point $\{0\}$. Then, the mixing weights $w_{\mathbf{k}}(\mathbf{i})$ are the probabilities of Cube$(\mathbf{i}, \mathbf{k})$ under P_c. The Bernstein density function $b_{\mathbf{k},P_c}(\mathbf{u})$ can thus be viewed as a mixture of beta densities, and if P_c is a copula, $b_{\mathbf{k},P_c}(\mathbf{u})$ is itself a copula density.

Alternatively, if we interpret P_c as an empirical copula on $[\frac{1}{k_1}, \frac{2}{k_1}, \ldots, \frac{k_1}{k_1}] \times [\frac{1}{k_2}, \frac{2}{k_2}, \ldots, \frac{k_2}{k_2}]$ then $b_{\mathbf{k},P_c}(\mathbf{u})$ can be viewed as a smoothed copula histogram using β-densities as smoothing functions.

The Bernstein copula density has several attractive properties as a sieve for the space of copula densities, which makes it preferable to other types of sieve such as trigonometric or power series. Being a mixture of (a product of) β-densities, it assigns no weights outside $[0, 1]^2$ and it easily extends to dimensions higher than two. Other sieves known to approximate well smooth functions and densities on $\mathbb{R}$ are often subject to the boundary problem and do not extend easily to multivariate settings (see, e.g., Chen (2007); Bouezmarni and Rombouts (2010)). The Bernstein sieve is a copula density by construction; at the same time, it does not impose symmetry, contrary to other conventional kernels used in mixture models such as multivariate Gaussian (see, e.g., Burda and Prokhorov (2014)).

Most importantly, as a density corresponding to $B_{\mathbf{k},P_c}(\mathbf{u})$, $b_{\mathbf{k},P_c}(\mathbf{u})$ converges, as $\min\{\mathbf{k}\} \to \infty$, to $p_c(\mathbf{u}) \equiv \frac{\partial^2}{\partial u_1 \partial u_2} P_c(\mathbf{u})$ at every point on $[0, 1]^2$ where $p_c(\mathbf{u})$ exists, and if p_c is continuous and bounded then the convergence is uniform (Lorentz (1986)). Uniform approximation results for the univariate and bivariate Bernstein density estimator can be found in Vitale (1975) and Tenbusch (1994).

Nonparametric copula estimators are attractive because they do not suffer from misspecification biases typical for parametric estimators. In this sense, they can be viewed as robust to copula misspecification.

5.2.3. *Semiparametric estimation: Copulas vs marginals*

Similar to the parametric inference procedures, semiparametric estimation methods are usually motivated by the density representations and decompositions of the log-likelihood as in (5.1) and (5.2). As a rule, semiparametric estimation in copula settings refers to one of the following two scenarios.

Scenario one is when the marginal distributions are parametrically specified while the copula part of the likelihood is left unspecified, i.e., is nonparametric. In this case, marginal distributions F_j are estimated using the parametric methods discussed in Section 5.2.1. Primarily this is done by IFM, which is MLE under the assumption of independence. Given estimates of the marginals $\hat{F}_j$, the copula is then estimated using nonparametric methods discussed in Section 5.2.2.

Scenario two is when univariate margins F_j are estimated non-parametrically, e.g., by the empirical d.f.'s $\hat{F}_j$ discussed in Section 5.2.2, while the copula is specified as a parametric family and its parameters are estimated using maximization of the copula contribution to the log-likelihood as described in Section 5.2.1.

The difference between the two scenarios is in where the researcher prefers to remain agnostic, in the dependence structure (scenario one) or in the functional form of marginals (scenario two). The latter setting is arguably more common in finance where estimation of the copula dependence parameter and copula-based dependence measures is of interest while marginal parameters carry no interpretation. The former may be more common in economics, where interest is in efficient estimation of parameters in marginal distributions, which provide such important quantities as partial effects and elasticities.

Let us start with scenario two and consider, as in Genest *et al.* (1995a), the problem of estimation of the (vector) parameter θ of a family of d-dimensional copulas $C(u_1, u_2, \ldots, u_d, \theta)$ with the density $c(u_1, u_2, \ldots, u_d, \theta)$. Given non-parametric estimators $\hat{F}_j$ of the univariate margins F_j, it is natural to estimate the copula parameter θ as

$$\hat{\theta} = \arg\max_{\theta} L_C(\theta)$$

$$= \arg\max_{\theta} \sum_{i=1}^{N} \log \, c\left(\hat{F}_1(x_1^{(i)}), \hat{F}_2(x_2^{(i)}), \ldots, \hat{F}_d(x_d^{(i)}); \theta\right).$$

The resulting semiparametric estimator $\hat{\theta}$ of the dependence parameter θ is consistent and asymptotically normal under suitable regularity conditions. Genest *et al.* (1995a) propose a consistent estimator of the limiting variance-covariance matrix of $\hat{\theta}$. They further show that, under additional copula regularity assumptions that are satisfied for a large class of bivariate copulas, including bivariate Gaussian, Eyraud-Farlie-Gumbel-Morgenstern (EFGM), Clayton and Frank families (see Section 1.2.1), the estimator $\hat{\theta}$ is fully efficient at independence.

Numerical results presented by Genest *et al.* (1995a) demonstrate that efficiency of $\hat{\theta}$ compares favorably to the alternative estimators. In addition, according to the numerical results, the coverage probability of confidence intervals for θ based on asymptotic normality of $\hat{\theta}$ is close to the nominal level for random samples with some copulas exhibiting small to medium range of dependence.

In scenario one — when copula is nonparametric — the benefit of copula modelling also depends on the strength of dependence between marginals. Panchenko and Prokhorov (2016) study semiparametric estimation of the parameters in marginals using the Bernstein polynomial to model the copula contribution to the log-likelihood. They demonstrate that under regularity conditions on the copula and marginal distributions, the resulting estimator reaches the semiparametric efficiency bound, and provided the dependence is strong, the efficiency gains from using the dependence information are substantial. We will return to this point in Section 5.5.

5.3. Copula-based estimation of time series models

This section provides an overview of results available for estimation of copula based time series models. In particular, Sections 5.3.1 and 5.3.2 discuss parametric, semiparametric and nonparametric estimation methods for copulas in the time series context. Section 5.3.3 reviews weak dependence properties of copula-based time series.

5.3.1. *Parametric and semiparametric estimation of Markov processes*

The results in Section 3.3 provide a copula-based framework for the analysis of (possibly higher-order) Markov processes. The results show that any Markov process of order $k \geq 1$ is completely characterized by its $(k+1)$-dimensional copulas and marginal distributions of its components. In particular, under stationarity, these processes are characterized by a $(k+1)$-copula C and its marginal one-dimensional cdf.

As discussed by Joe (1997, Section 10.4), under suitable regularity conditions, asymptotic results for ML estimation under i.i.d. observations (see Section 5.2.1) also hold for stationary Markov processes. Chen and Fan (2006b) propose semiparametric estimation procedures for copula-based Markov processes. These procedures generalize the two-stage semiparametric inference approaches discussed in Section 5.2.3 to the time series framework.

Consider a stationary Markov process based on a bivariate copula $C = C(u_1, u_2; \theta)$. Similar to Section 5.2.3, the semiparametric estimator $\hat{\theta}$ proposed by Chen and Fan (2006b) can be written as follows:

$$\hat{\theta} = \arg\max_{\theta} L_C(\theta) = \arg\max_{\theta} \sum_{i=1}^{N} \log \, c(\hat{F}(x_1^{(i)}), \hat{F}(x_2^{(i)}); \theta),$$

where $\hat{F}(x)$ is a nonparametric estimator of the univariate margin of X_j : e.g., the rescaled empirical d.f. $\hat{F}(x) = \frac{1}{N+1} \sum_{j=1}^{N} \mathbb{I}\{X_j \leq x\}$.

Chen and Fan (2006b) show that the semiparametric estimator $\hat{\theta}$ is consistent and asymptotically normal under suitable regularity assumptions. These assumptions include a condition that the process $\{X_j\}$ is β-mixing with polynomial decay rate. Chen and Fan (2006b) provide copula-based sufficient conditions for the latter weak dependence assumption and further verify the assumptions implying consistency and asymptotic normality of $\hat{\theta}$ for Markov processes based on Gaussian, Clayton and Frank copulas. They show that the asymptotic variance of $\hat{\theta}$ can be estimated using heteroskedasticity autocorrelation consistent (HAC) estimators (see, for instance, Andrews (1991); Newey and West (1987b); Hamilton (1994), Chapter 10). Alternatively, the asymptotic distributions of the estimator can be approximated using bootstrap (see Chen and Fan (2006b), Section 4.3).

Chen *et al.* (2009) consider efficient sieve MLE methods for copula-based stationary Markov processes. The authors show that sieve MLE's of any smooth functionals of the copula parameter and marginal d.f. are $\sqrt{N}$-consistent, asymptotically normal and efficient; and that their sieve likelihood ratio statistics are asymptotically chi-square distributed. The numerical results in Chen *et al.* (2009) further indicate that the sieve MLE of copula parameters, the margins and the conditional quantiles all perform very well in finite samples even for Markov models based on tail dependent copulas, such as Clayton and Gumbel-Hougaard (see Durante and Sempi (2010), Section 6.1 and Equations (18)–(19)). In addition, in the case of Markov models generated via tail dependent copulas, the sieve MLE has much smaller biases and smaller variances than the two-step semiparametric estimation procedures of Chen and Fan (2006b) discussed above.

5.3.2. *Nonparametric copula inference for time series*

Doukhan *et al.* (2009) discuss extensions of the results on copula processes discussed in Section 5.2.2 to time series models. Their results imply that empirical copula process (5.4) is asymptotically Gaussian for weakly dependent vector-valued processes $\{X_i\}$, including the case of strongly mixing and β-mixing sequences with polynomial decay rates.

More generally, asymptotic normality of the copula processes holds for random sequences that satisfy multivariate functional central limit theorem for empirical processes (see Doukhan (1994)). Applied to the vector-valued process (X_j, X_{j+1}) of a C-based stationary Markov sequence $\{X_j\}$ satisfying mixing conditions (see Section 3.3), these results imply asymptotic normality of the empirical copula process $\hat{C}(u_1, u_2) = \hat{G}(\hat{F}^{-1}(u_1), \hat{F}^{-1}(u_2))$, where $\hat{F}(x)$ is a nonparametric estimator of the univariate margin of X_j and $\hat{G}$ is a nonparametric estimator of the bivariate d.f. of (X_j, X_{j+1}). As an example, one can consider empirical d.f.'s $\hat{F}(x) = \frac{1}{N} \sum_{j=1}^{N} \mathbb{I}(X_j \leq x)$, and $\hat{G}(x) = \frac{1}{N} \sum_{j=1}^{N} \mathbb{I}(X_j \leq x) \mathbb{I}(X_{j+1} \leq x)$.

Doukhan *et al.* (2009) further established asymptotic normality of the smoothed copula process with kernel estimates of mutivariate d.f. and univariate margins in (5.4) for weakly dependent vector-valued sequences. Similar asymptotic results are also shown to hold for smoothed copula densities (see also Gijbels and Mielniczuk (1990), for asymptotics of kernel estimators of the copula density for i.i.d. vector observations with dependent components).

5.3.3. *Dependence properties of copula-based time series*

As discussed in Sections 5.3.1 and 5.3.2, consistency and asymptotic normality of copula estimators are obtained under the assumptions of weak dependence. Among other results, Beare (2010), Chen *et al.* (2009) and Lentzas and Ibragimov (2008) provide studies of persistence properties of stationary copula-based Markov processes.

Lentzas and Ibragimov (2008) show via simulations that stationary Markov processes based on tail-dependent Clayton copulas (see Durante and Sempi (2010), equation (19)) can behave as long memory time series with copulas exhibiting high persistence, which is common in financial and economic applications. This long memory-like behavior is captured by an extremely slow decay of copula-based dependence measures between lagged values of the processes for commonly used lags. Lentzas and Ibragimov (2008) further show that, in contrast, the Gaussian and EFGM copulas (see Durante and

Sempi (2010), Section 6.1 and Equation (22)) always produce short memory stationary Markov processes.

Beare (2010) shows that a C-based stationary Markov process exhibits weak dependence properties, including α- and β-mixing with exponential decay rates, if C is a symmetric absolutely continuous copula with a square integrable density and the maximal correlation coefficient of C is less than one. These results imply that stationary Markov processes based on the Gaussian and EFGM copulas are weakly dependent and mixing. Beare (2010) also provides numerical results that suggest exponential decay in β-mixing coefficients and, thus, also in α-mixing coefficients of Clayton copula-based stationary Markov processes.

Chen *et al.* (2009) obtain theoretical results that show that tail-dependent Clayton, survival Clayton, Gumbel and t-copulas always generate Markov processes that are geometric ergodic and hence geometric β-mixing. The conclusions in Chen *et al.* (2009) imply that, although, according to the numerical results in Lentzas and Ibragimov (2008), Clayton copula-based Markov processes can behave like long memory time series on copula levels exhibiting high persistence for commonly used lag numbers, they are in fact weakly dependent and short memory in terms of mixing properties.

5.4. Improved and robust parametric estimators

This section considers likelihood based estimation of copula models, where for each marginal we have a correctly specified model that could be estimated by MLE. The section shows how to improve on the QMLE. It then considers MLE based on joint distributions constructed using copulas. It discusses the efficiency gain from using the true copula, and shows that knowledge of the true copula is redundant only if the variance matrix of the relevant set of moment conditions is singular. It also discusses the question of robustness against misspecification of the copula. The Generalized Method of Moments (GMM) is argued to be useful analytically, and also for reasons of efficiency if the copula is robust but not correct.

In Section 5.4.1, we ask what can be done without specifying a joint distribution. It is well known that the QMLE based on the likelihood that would be correct under independence is robust to non-independence. Using GMM methods, we suggest an improved QMLE estimator (IQMLE) that dominates the QMLE, and we derive the condition under which the efficiency gain is positive.

Section 5.4.2 considers construction of the joint distribution using copulas. We provide a GMM interpretation of the MLE estimator based on the joint distribution: it adds another set of moment conditions, which we call the "copula score," to the moment conditions that the IQMLE uses.

In Section 5.4.3, under the assumption that the copula is correct, we ask under what circumstances the MLE based on the joint distribution is more efficient than the IQMLE. This is so when the copula score is not redundant. We show that in this setting, redundancy can only occur when the covariance matrix of the full set of moment conditions is singular.

In Section 5.4.4, we discuss the question of robustness. In general, if we misspecify the copula we have misspecified the joint distribution and the MLE will be inconsistent. However, we show that it is possible that the copula score has mean zero even if the copula is incorrect, so that the MLE is robust to the misspecification of the copula. In this case, the GMM estimator based on the full set of moment conditions dominates the (pseudo) MLE based on the incorrectly specified joint likelihood. We show this in Section 5.4.5.

5.4.1. *QMLE and improved QMLE*

In this section, we ask what can be done using only the marginal distributions. In non-copula setting, the MLE estimator of the marginals assuming independence between them is often call the quasi-MLE (QMLE). It is just the first stage of the IFM from Section 5.2.1. It is known that the QMLE (which maximizes the likelihood that would arise if there were independence between marginals) is consistent even if there is dependence. We discuss this result briefly, and show how to define an improved QMLE (IQMLE).

Assume that $(x_{i1}, \ldots, x_{id})$ are i.i.d. over i, but are not independent over j, where j indexes the marginals, $j = 1, \ldots, d$. Then, the following discussion is textbook material (e.g., Hayashi (2000), Section 8.7). The QMLE is the value of α that maximizes the quasi-likelihood

$$\ln L^Q = \sum_i \sum_j \ln f(x_{ij}, \alpha). \tag{5.7}$$

The expression in (5.7) is the likelihood if we have independence over i and j. However, the QMLE remains consistent if we have dependence over j.

Define the score functions

$$s_{ij}(\alpha) = \nabla_\alpha \ln f(x_{ij}, \alpha), \tag{5.8}$$

$$s_i(\alpha) = \sum_j s_{ij}(\alpha), \tag{5.9}$$

where "∇_α" means partial derivative with respect to α. Then the QMLE $\hat{\alpha}$ solves the first-order condition

$$\sum_i s_i(\hat{\alpha}) = 0. \tag{5.10}$$

As mentioned in Section 5.2.1, it can be viewed as a GMM estimator based on the moment condition

$$\mathbb{E} s_i(\alpha_o) = 0. \tag{5.11}$$

But this condition holds so long as the marginal density $f(x_{ij}, \alpha)$ is correctly specified, because (subject to the usual regularity conditions for MLE) correct specification of the marginal density $f(x_{ij}, \alpha)$ implies that $\mathbb{E} s_{ij}(\alpha_o) = 0$ for all j, and therefore $\mathbb{E} s_i(\alpha_o) = \sum_j \mathbb{E} s_{ij}(\alpha_o) = 0$.

We further define the Hessian

$$H_{ij}(\alpha) = \nabla_\alpha s_{ij}(\alpha) = \nabla_\alpha^2 \ln f(x_{ij}, \alpha) \tag{5.12}$$

and correspondingly $H_i(\alpha) = \sum_j H_{ij}(\alpha)$ and $\mathbb{H} = \mathbb{E} H_i(\alpha_o)$. Also we define the variance matrix of the score:

$$\mathbb{V} = \mathbb{E} s_i(\alpha_o) s_i(\alpha_o)'. \tag{5.13}$$

Then under suitable regularity conditions the asymptotic variance of $\hat{\alpha}$ is $(\mathbb{H}\mathbb{V}^{-1}\mathbb{H})^{-1} = \mathbb{H}^{-1}\mathbb{V}\mathbb{H}^{-1}$. (We use the standard terminology that "the asymptotic variance of $\hat{\alpha}$ is Σ" means that $\sqrt{N}(\hat{\alpha} - \alpha_o)$ converges in distribution to $N(0, \Sigma)$.) The "sandwich form" is necessary because $\mathbb{H} = -\mathbb{V}$ under independence over j, but not (in general) otherwise.

Prokhorov and Schmidt (2009) make the observation that, except under independence, summation is not generally the optimal way to combine the observations. That is, instead of summing over j, we can stack all d values and let GMM perform the optimal weighting. Therefore instead of $\mathbb{E}s_i(\alpha_o) = 0$ as in (5.11), we use the stacked moment conditions

$$\mathbb{E}s_i^*(\alpha_o) = 0, \tag{5.14}$$

where

$$s_i^*(\alpha) = \begin{bmatrix} s_{i1}(\alpha) \\ s_{i2}(\alpha) \\ \vdots \\ s_{id}(\alpha) \end{bmatrix}. \tag{5.15}$$

We will call the optimal GMM estimator based on the moment conditions in (5.14) the improved QMLE (IQMLE). Because we deal only in asymptotics here, we will not be explicit about issues like how to estimate the optimal GMM weighting matrix. Nor would it matter if instead of GMM we considered other asymptotically equivalent estimators that fall under the umbrella of Generalized Empirical Likelihood (GEL) such as empirical likelihood, exponential tilting, etc.

Like the QMLE, the IQMLE should be consistent so long as the marginal distributions are correctly specified, since $\mathbb{E}s_i^*(\alpha_o) = 0$ if $\mathbb{E}s_{ij}(\alpha_o) = 0$ for all j. Define $\mathbb{H}_* = \mathbb{E}\nabla_\alpha s_i^*(\alpha_o)$ and $\mathbb{V}_* = \mathbb{E}s_i^*(\alpha_o)s_i^*(\alpha_o)'$. Then, if $\tilde{\alpha}$ is the IQMLE estimator, standard results would indicate that the asymptotic variance of $\tilde{\alpha}$ is $(\mathbb{H}_*'\mathbb{V}_*^{-1}\mathbb{H}_*)^{-1}$.

It is obvious from basic principles that the IQMLE estimator is efficient relative to the QMLE estimator (optimal weighting is optimal), and the only remaining question is when they are equally efficient. The following result states the first result formally, and answers the remaining question.

To state the result, we define some additional notation. Let $A = 1'_d \otimes I_p$, where 1_d is a $d \times 1$ vector of ones, and where $p = \dim(\alpha)$. The matrix A arises naturally because $s_i = As_i^*$, and correspondingly $\mathbb{H} = A\mathbb{H}_*$, and $\mathbb{V} = A\mathbb{V}_*A'$.

Theorem 5.1. (a) *The IQMLE estimator is efficient relative to the QMLE estimator.* (b) *The two estimators are equally efficient if and only if* $\mathbb{H}_*$ *is in the space spanned by* $(\mathbb{V}_*A')$.

The condition in part (b) of Theorem 5.1 is not very intuitive. However, we can identify two cases when it holds.

Theorem 5.2. *The QMLE and IQMLE estimators are equally efficient if either of the two following conditions holds.*

(a) x_{ij} *is i.i.d. over both i and j.*
(b) x_{ij} *is identically distributed over j and the scores are "equicorrelated," in the sense that* $\mathbb{V}_* = E \otimes V_o$, *where*

$$E = \begin{bmatrix} 1 & \theta & \cdots & \theta \\ \theta & 1 & \cdots & \theta \\ \vdots & \vdots & \ddots & \vdots \\ \theta & \theta & \cdots & 1 \end{bmatrix}, \quad d \times d$$

and V_o is a positive definite $p \times p$ matrix and $\theta > -\frac{1}{d-1}$.

Condition 5.2(a) is obvious. Condition 5.2(b) is less so. However, in panel data models, it could arise in a variety of random effects models (where the random effects error structure generates E) with regressors that are the same for all j (so that V_o would be the variance matrix of the marginal score for all j).

5.4.2. *Full MLE as GMM*

Section 5.2.1 showed that the difference between the MLE based on the full joint distribution (FMLE) and the IQMLE estimator based on the marginal scores (the first stage of the IFM) lies in a term we call the copula score.

Let H be an d-dimensional distribution function with marginals $F_1, \ldots, F_d$. Then, by Proposition 1.1 introduced in Section 1.2 (Sklar's theorem), there exists a d-dimensional copula C such that for all $(x_1, \ldots, x_d)$,

$$H(x_1, \ldots, x_d) = C(F_1(x_1), \ldots, F_d(x_d)). \qquad (5.16)$$

Conversely, if C is an d-dimensional copula and $F_1, \ldots, F_d$ are distribution functions, then the function H in (5.16) is an d-dimensional distribution function with marginals $F_1, \ldots, F_d$.

Our interest in copulas in this section stems from the first part of Sklar's theorem. This says that any continuous joint distribution uniquely implies the marginal distributions and the copula (the case of non-unique copulas corresponding to discrete joint distributions is not considered here). Therefore as in Section 5.2.1, we can examine the properties of the MLE based on the joint distribution of $(x_{i1}, \ldots, x_{id})$ in terms of its components, the marginal distributions of the x_{ij}, and the copula cdf C.

In what follows we will restrict our exposition to the bivariate case, $d = 2$. (Most of the copula literature follows this expositional convention, just to keep the notation under control.) Also, for notational simplicity, we will henceforth dispense with the index i except where needed. Thus (x_1, x_2) will be the data for which we seek a joint distribution.

Suppose that the joint cdf of (x_1, x_2) is $H(x_1, x_2; \alpha, \theta)$, the marginal cdf's are $F_1(x_1; \alpha)$ and $F_2(x_2; \alpha)$, and the copula cdf is $C(\cdot, \cdot; \theta)$. Note that the parameters of interest are the α's. For example, they may contain the tail indexes of power-law marginals, as before. The nuisance parameter "θ" is present only in the copula. Being a copula dependence parameter, "θ" need not be a correlation, except for the Gaussian copula (see Section 1.2). Then Sklar's theorem says:

$$H(x_1, x_2; \alpha, \theta) = C(F_1(x_1; \alpha), F_2(x_2; \alpha); \theta). \qquad (5.17)$$

Differentiating with respect to (x_1, x_2), we obtain Equation (5.1) specialized to $d = 2$:

$$h(x_1, x_2; \alpha, \theta) = c(F_1(x_1; \alpha), F_2(x_2; \alpha); \theta) \cdot f_1(x_1; \alpha) \cdot f_2(x_2; \alpha). \qquad (5.18)$$

Finally, taking logs, we obtain

$$\ln h(x_1, x_2; \alpha, \theta) = \ln c(F_1(x_1; \alpha), F_2(x_2; \alpha); \theta)$$

$$+ \ln f_1(x_1; \alpha) + \ln f_2(x_2; \alpha). \qquad (5.19)$$

Now consider summing this expression over the suppressed index i. Similar to Equation (5.2), the left-hand side would be the log of the joint likelihood. The first two terms on the right-hand side would be the quasi-log-likelihood. As in Equation (5.2), the difference between the joint log-likelihood and the quasi-log-likelihood is the sum of the log copula density terms.

Finally, we consider the score functions (with respect to α and to θ) corresponding to (5.19). As noted in Section 5.2.1, MLE can be viewed as GMM based on the score function (see Godambe (1960, 1976)). The expected value of the score function for the correctly specified joint log-likelihood is zero at the true value of parameters. Furthermore, if the marginal densities are correctly specified, the same is true for the marginal log-likelihoods. Hence, under classical regularity conditions, the following four sets of moment conditions hold at the true values of the parameters (α_o, θ_o):

$$\begin{array}{ll} \mathbb{E}\nabla_\alpha \ln f_1(x_1; \alpha_o) = 0, & (\text{A}) \\ \mathbb{E}\nabla_\alpha \ln f_2(x_2; \alpha_o) = 0, & (\text{B}) \\ \mathbb{E}\nabla_\alpha \ln c(F_1(x_1; \alpha_o), F_2(x_2; \alpha_o); \theta_o) = 0, & (\text{C}) \\ \mathbb{E}\nabla_\theta \ln c(F_1(x_1; \alpha_o), F_2(x_2; \alpha_o); \theta_o) = 0. & (\text{D}) \end{array} \qquad (5.20)$$

We call (A) and (B) the "marginal scores" and (C) and (D) the "copula scores." Note that the GMM problem as stated in (5.20) is overidentified. If the dimension of α is $p \times 1$, and the dimension of θ is $q \times 1$, we have $p + q$ parameters and $3p + q$ moment conditions.

We can now consider how the various estimators that we consider relate to each other.

The QMLE is based on a moment condition that equals (A)+(B), the sum of the two marginal scores.

The IQMLE is based on the moment conditions (A)&(B). Here "&" is used to indicate that the set of moment conditions includes (A) *and* (B), which will be weighted as appropriate by the GMM machinery.

If there are no parameters in the copula, (D) does not exist. In this case the MLE is based on (A)+(B)+(C). It is asymptotically equivalent to the GMM estimator based on (A)&(B)&(C), since (assuming the copula is correct) the optimal weighting is indeed summation. The efficiency gain from MLE, as opposed to IQMLE, is due to the additional information in the copula score (C).

If there are parameters (θ) in the copula, (D) does exist. Now the MLE is based on [(A)+(B)+(C)]&(D), whereas the asymptotically equivalent GMM estimator is based on (A)&(B)&(C)&(D). Once again the difference between these estimators and the IQMLE lies in the copula scores, (C) and (D).

5.4.3. *Efficiency and redundancy of copulas*

In this section, we assume that the joint distribution is correctly specified, so that both the marginal distributions and the copula are correct. Therefore the (full) MLE is efficient. Prokhorov and Schmidt (2009) ask the question under what circumstances the IQMLE is efficient; that is, under what circumstances the MLE is no more efficient than the IQMLE. This is the question of when the copula scores are redundant in the sense of Breusch *et al.* (1999). Perhaps surprisingly, this turns out to be the case only when the full set of scores (A)–(D) is linearly dependent.

The following lemma reveals the structure of the variance and derivative matrices of the moment functions in (5.20). We repeat that correct specification of the copula is assumed.

Lemma 5.1. *Denote the covariance matrix of the moment functions in (5.20) by* **V**, *their expected derivative matrix with respect to* (α, θ) *by* **D**. *Then,*

$$
\mathbf{V} = \left[\begin{array}{cc|cc} \mathbf{A} & \mathbf{G} & -\mathbf{G} & \mathbf{0} \\ \mathbf{G}' & \mathbf{B} & -\mathbf{G}' & \mathbf{0} \\ \hline -\mathbf{G}' & -\mathbf{G} & \mathbf{J} & \mathbf{E} \\ \mathbf{0} & \mathbf{0} & \mathbf{E}' & \mathbf{F} \end{array} \right] \tag{5.21}
$$

and

$$D = \left[\begin{array}{c|c} -A & 0 \\ -B & 0 \\ \hline G + G' - J & -E \\ -E' & -F \end{array} \right], \tag{5.22}$$

where A, B, E, F, G, J *are matrix-functions of* (α, θ) *defined in the Appendix.*

Several interesting observations follow from the Lemma. First, the covariance of the first marginal score with the copula score with respect to α equals minus the covariance of the first marginal score with the second marginal score. Thus the marginal scores are uncorrelated with the copula score with respect to α if and only if they are uncorrelated with each other. Second, both marginal scores are uncorrelated with the copula score with respect to θ. Third, it is easy to see that, if V is nonsingular, the optimal GMM estimate based on (5.20) is asymptotically equivalent to the MLE. To see this, note that the optimally weighted GMM estimate based on a set of moment conditions "g" is asymptotically equivalent to the GMM estimate based on the exactly identified set of moment conditions $D'V^{-1}g$. Let $\mathbb{I}$ denote the identity matrix. By Lemma 5.1,

$$D' = - \begin{bmatrix} \mathbb{I} & \mathbb{I} & \mathbb{I} & 0 \\ 0 & 0 & 0 & \mathbb{I} \end{bmatrix} V$$

and so

$$D'V^{-1} = - \begin{bmatrix} \mathbb{I} & \mathbb{I} & \mathbb{I} & 0 \\ 0 & 0 & 0 & \mathbb{I} \end{bmatrix} VV^{-1} = - \begin{bmatrix} \mathbb{I} & \mathbb{I} & \mathbb{I} & 0 \\ 0 & 0 & 0 & \mathbb{I} \end{bmatrix}. \tag{5.23}$$

Then $D'V^{-1}g = 0$, with "g" as given in (5.20), gives the first-order conditions for the MLE.

For nonsingular V, the asymptotic variance matrix of the optimal GMM estimator based on (5.20) is of the familiar form

$\mathbb{V}_{\text{GMM}} = (\mathbf{D}'\mathbf{V}^{-1}\mathbf{D})^{-1}$. By Lemma 5.1, this is identical to the asymptotic variance matrix of the MLE estimator of (α, θ)

$$\mathbb{V}_{\text{MLE}} = -\left(\begin{bmatrix} \mathbb{I} & \mathbb{I} & \mathbb{I} & \mathbf{0} \\ \mathbf{0} & \mathbf{0} & \mathbf{0} & \mathbb{I} \end{bmatrix} \mathbf{D}\right)^{-1} = \left(\begin{bmatrix} \mathbb{I} & \mathbb{I} & \mathbb{I} & \mathbf{0} \\ \mathbf{0} & \mathbf{0} & \mathbf{0} & \mathbb{I} \end{bmatrix} \mathbf{V} \begin{bmatrix} \mathbb{I} & \mathbf{0} \\ \mathbb{I} & \mathbf{0} \\ \mathbb{I} & \mathbf{0} \\ \mathbf{0} & \mathbb{I} \end{bmatrix}\right)^{-1}. \quad (5.24)$$

In contrast to $\mathbb{V}_{\text{GMM}}$, $\mathbb{V}_{\text{MLE}}$ is defined even if $\mathbf{V}$ is singular. In fact the last representation in (5.24) involves the outer-product-of-the-score form of the information matrix, while the one before the last involves the expected-Hessian form of the information matrix. Both are nonsingular under the usual regularity conditions.

We now return to the question of when the IQMLE is as efficient as the MLE. Denote the asymptotic variance matrix of IQMLE by $\mathbb{V}_{\text{IQMLE}}$. Logically, this is the question of when the copula scores are partially redundant for α, given the marginal scores.

Definition 5.1. Suppose that α is identified by some moment conditions g_1. Now we consider estimation based on g_1 and some additional moment condition g_2. Then g_2 is *partially redundant* for α given g_1 if the estimate of α based on g_1 and g_2 is no more efficient than the estimate of α based on g_1 only.

Compared to (impartial) redundancy, the word "partially" refers to the possibility that g_2 may increase efficiency of estimation for some parameter other than α (in our case, nuisance parameters in the copula scores).

Breusch *et al.* (1999) developed a very useful toolbox for analyzing redundancy of a set of moment conditions given another set of moment conditions. However, their analysis assumes nonsingular $\mathbf{V}$. Therefore, it is impossible to employ their results here but instead it is possible to compare $\mathbb{V}_{\text{IQMLE}}$ with the relevant block of $\mathbb{V}_{\text{MLE}}$ directly.

Theorem 5.3. $\mathbb{V}_{\text{MLE}}$ *for α and* $\mathbb{V}_{\text{IQMLE}}$ *are equal if and only if*

$$\mathbf{J} - \mathbf{V}_{21}^{\alpha}\mathbf{V}_{11}^{-1}\mathbf{V}_{12}^{\alpha} - \mathbf{E}\mathbf{F}^{-1}\mathbf{E}' = 0, \qquad (5.25)$$

where $\mathbf{V}_{21}^{\alpha} = \mathbf{V}_{12}^{\alpha}{}' = [-\mathbf{G}' \quad -\mathbf{G}]$ *and* $\mathbf{V}_{11} = \begin{bmatrix} \mathbf{A} & \mathbf{G} \\ \mathbf{G}' & \mathbf{B} \end{bmatrix}$.

The cumbersome expression in (5.25) has a simple interpretation in terms of singularity of $\mathbf{V}$. It states that the error in the linear projection of moment condition (C) on moment conditions (A), (B) and (D) is uncorrelated with moment condition (C). More specifically, (5.25) can be rewritten as follows:

$$\mathbb{E}\left\{ \left(\nabla_{\alpha}\ln c - \mathbf{\Omega}_{21}\mathbf{\Omega}_{11}^{-1} \begin{bmatrix} \nabla_{\alpha}\ln f_1 \\ \nabla_{\alpha}\ln f_2 \\ \nabla_{\theta}\ln c \end{bmatrix} \right) \nabla_{\alpha}'\ln c \right\} = 0,$$

where

$$\mathbf{\Omega}_{21} = [-\mathbf{G}' \quad -\mathbf{G} \quad \mathbf{E}], \quad \mathbf{\Omega}_{11} = \begin{bmatrix} \mathbf{A} & \mathbf{G} & \mathbf{0} \\ \mathbf{G}' & \mathbf{B} & \mathbf{0} \\ \mathbf{0} & \mathbf{0} & \mathbf{F} \end{bmatrix}$$

and the arguments of the moment functions have been suppressed for brevity. In other words, (C) has to be a linear combination of (A), (B) and (D) for the copula information to be redundant in terms of asymptotic efficiency of estimation of α. Thus, $\mathbf{V}$ has to be singular.

To understand why this is an important result, we consider a distinction not made in the Breusch *et al.* (1999) paper. Consider estimation of the mean of y, $\mathbb{E}y = \mu_o$. Suppose that we have moment conditions

$$\mathbb{E}(y - \mu_o) = 0 \qquad [``g_1"], \qquad (5.26)$$

$$\mathbb{E}[3(y - \mu_o)] = 0 \qquad [``g_2"]. \qquad (5.27)$$

Clearly g_2 is redundant given g_1 (or vice versa). This is a case of "numerical redundancy" in that one of the moment conditions is a linear combination of the others. The variance matrix $\mathbf{V}$ is singular.

This can be contrasted with "statistical redundancy" (but not "numerical redundancy") in the next example. Now suppose we have

$$\mathbb{E}(y - \mu_o) = 0 \qquad [\text{``}g_1\text{''}], \qquad (5.28)$$
$$\mathbb{E}[(y - \mu_o)^2 - \sigma_o^2] = 0 \qquad [\text{``}g_2\text{''}], \qquad (5.29)$$

where σ_o^2 is known. The matrix $\mathbf{V}$ is not singular. Here g_2 is redundant given g_1 if and only if $\mathbb{E}(y - \mu_o)^3 = 0$.

The point of the distinction is that numerical redundancy is obvious, while statistical (but not numerical) redundancy is subtle. One can be seen by inspection whereas the other requires calculation. According to Theorem 5.3 and the subsequent discussion, the only way the copula score can be redundant is numerical redundancy. If there are any algebraic terms in the copula score with respect to α that are not a linear combination of the other scores, they cannot be redundant, and the MLE must be strictly more efficient than the IQMLE.

In some cases, the copula may not contain parameters. In this case, in (5.21) and (5.22) the rows and columns of $\mathbf{D}$ and $\mathbf{V}$ that contain the terms $\mathbf{E}$ and $\mathbf{F}$ do not exist, and (C) is redundant given (A) and (B), so that the IQMLE is efficient, if and only if (C) is a linear combination of (A) and (B). A third possibility is that the copula does contain parameters (θ) but they are "known" (specified). The following Corollary indicates that the same singularity result holds.

Corollary 5.1. *If* (C) *is a linear combination of* (A) *and* (B) *with* θ *known then*

(1) $\mathbf{E} = \mathbf{0}$;
(2) $\mathbf{J} - \mathbf{V}_{21}^\alpha \mathbf{V}_{11}^{-1} \mathbf{V}_{12}^\alpha = \mathbf{0}$;
(3) *IQMLE is efficient.*

We now present five examples that show how the redundancy results can be used in practice.

Example 5.1. Bivariate Normal with common mean. Assume Normal marginal densities with $\sigma_1^2 = \sigma_2^2 = 1$ and

$\mu_1 = \mu_2 = \mu$

$$f_1(x_1; \mu) = \frac{1}{\sqrt{2\pi}} e^{-\frac{(x_1-\mu)^2}{2}},$$

$$f_2(x_2; \mu) = \frac{1}{\sqrt{2\pi}} e^{-\frac{(x_2-\mu)^2}{2}}.$$

Let the true joint density be Bivariate Normal, i.e.,

$$h(x_1, x_2; \mu, \theta) = \frac{1}{2\pi\sqrt{1-\theta^2}} e^{-\frac{(x_1-\mu)^2 + (x_2-\mu)^2 - 2\theta(x_1-\mu)(x_2-\mu)}{2(1-\theta^2)}}.$$

Then, the implied copula is the Normal copula

$$c(F_1(x_1; \mu), F_2(x_2; \mu); \theta) = \frac{1}{\sqrt{1-\theta^2}} e^{-\frac{\theta\left(\theta(x_1-\mu)^2 + \theta(x_2-\mu)^2 - 2(x_1-\mu)(x_2-\mu)\right)}{2(1-\theta^2)}},$$

where θ is the copula dependence parameter (Pearson's correlation coefficient).

The relevant moment conditions are

$$\mathbb{E}\{X_1 - \mu\} = 0, \tag{A}$$

$$\mathbb{E}\{X_2 - \mu\} = 0, \tag{B}$$

$$\mathbb{E}\left\{-\frac{((X_1-\mu)+(X_2-\mu))\theta}{\theta+1}\right\} = 0, \tag{C}$$

$$\mathbb{E}\left\{-\frac{\theta(X_1^2+X_2^2)+\mu(1-\theta)^2(X_1+X_2)-(1+\theta^2)X_1X_2+\theta(\theta^2-1)-\mu^2(1-\theta)^2}{(\theta-1)^2(\theta+1)^2}\right\} = 0. \tag{D}$$

Clearly (C) is a linear combination of (A) and (B), so it is a linear combination of (A), (B) and (D). Therefore the IQMLE is efficient.

Example 5.2. d-variate Normal with common mean. This example is an extension of the previous example. It shows how our efficiency results generalize to $d > 2$. As above, assume Normal marginals with $\sigma_t^2 = 1$ and $\mu_t = \mu$, $t = 1, \ldots, d$. Let the true joint density be d-variate normal

$$h(\mathbf{x}; \mu, \Sigma) = \frac{1}{(2\pi)^{d/2}|\Sigma|^{1/2}} e^{-\frac{1}{2}(\mathbf{x}-\mu)'\Sigma^{-1}(\mathbf{x}-\mu)},$$

where $\mathbf{x} = (x_1, \ldots, x_d)'$ and Σ is the correlation matrix of $\mathbf{x}$. Then, the implied copula is Normal:

$$
\begin{aligned}
c(F_1(x_1; \mu), \ldots, F_d(x_d; \mu); \Sigma) &= \frac{h(\mathbf{x}; \mu, \Sigma)}{\prod_{t=1}^{d} f_t(x_t; \mu)} \\
&= \frac{1}{|\Sigma|^{1/2}} e^{-\frac{1}{2}(\mathbf{x}-\mu)'(\Sigma^{-1}-\mathbb{I})(\mathbf{x}-\mu)},
\end{aligned}
$$

where $\mathbb{I}$ is the identity matrix of dimension d.

The first $d + 1$ moment conditions are

$$
\begin{aligned}
\mathbb{E}\{\mathbf{x} - \mu\} &= 0 \\
\mathbb{E}\{1_d'(\Sigma^{-1} - \mathbb{I})(\mathbf{x} - \mu)\} &= 0,
\end{aligned}
$$

where 1_d denotes a $d \times 1$ vector of ones.

Here again, the copula score for μ is a linear combination of the marginal scores. Even without writing out the copula score for Σ, we can conclude that the IQMLE based on the first d moment conditions is efficient for μ.

Example 5.3. Bivariate Normal regression. Let $\mathbf{y} = \mathbf{x}\beta + \epsilon$, where $\mathbf{y} = (x_1, x_2)'$ is 2×1 and $\mathbf{x} = (x_1, x_2)'$ is $2 \times k$. Suppose $\mathbf{x}$ is non-random. Let $\epsilon = (\epsilon_1, \epsilon_2)' \sim \mathbb{N}(\mathbf{0}, \Sigma)$, where

$$\Sigma = \begin{bmatrix} \sigma_1^2 & \theta \\ \theta & \sigma_2^2 \end{bmatrix}.$$

For simplicity, we consider the case that σ_1^2 and σ_2^2 are known. Then,

$$f_1(x_1; x_1, \beta) = \frac{1}{\sqrt{2\pi\sigma_1^2}} e^{-\frac{(x_1 - x_1\beta)^2}{2\sigma_1^2}},$$

$$f_2(x_2; x_2, \beta) = \frac{1}{\sqrt{2\pi\sigma_2^2}} e^{-\frac{(x_2 - x_2\beta)^2}{2\sigma_2^2}},$$

$$h(\mathbf{y}; \mathbf{x}, \beta, \theta) = \frac{1}{2\pi\sqrt{|\Sigma|}} e^{-\frac{1}{2}(\mathbf{y}-\mathbf{x}\beta)'\Sigma^{-1}(\mathbf{y}-\mathbf{x}\beta)}.$$

Then, the implied copula is Normal,

$$c(F_1(x_1; x_1, \beta), F_2(x_2; x_2, \beta); \theta)$$

$$= \frac{\sqrt{\sigma_1^2 \sigma_2^2}}{\sigma_1^2 \sigma_2^2 - \theta^2} e^{-\frac{\epsilon_1}{2}\left(\frac{\epsilon_1 \sigma_2^2 - \epsilon_2 \theta}{\sigma_1^2 \sigma_2^2 - \theta^2}\right) - \frac{\epsilon_2}{2}\left(\frac{\epsilon_2 \sigma_1^2 - \epsilon_1 \theta}{\sigma_1^2 \sigma_2^2 - \theta^2}\right)}$$

$$\times e^{\frac{1}{2}\left(\frac{\epsilon_1^2}{2\sigma_1^2} + \frac{\epsilon_2^2}{2\sigma_2^2}\right)},$$

where $\epsilon_i = y_i - x_i\beta$, $i = 1, 2$.

The relevant moment conditions are

$$\mathbb{E}\left\{\frac{x_1 \epsilon_1}{\sigma_1^2}\right\} = 0, \tag{A}$$

$$\mathbb{E}\left\{\frac{x_2 \epsilon_2}{\sigma_2^2}\right\} = 0, \tag{B}$$

$$\mathbb{E}\left\{-\frac{\theta(\sigma_1^2 \sigma_2^2 x_1 \epsilon_2 + \sigma_1^2 \sigma_2^2 x_2 \epsilon_1 - \sigma_1^2 \theta x_2 \epsilon_2 - \sigma_2^2 \theta x_1 \epsilon_1)}{\sigma_1^2 \sigma_2^2 (\sigma_1^2 \sigma_2^2 - \theta^2)}\right\} = 0, \tag{C}$$

$$\mathbb{E}\left\{\frac{\sigma_1^2 \sigma_2^2 \epsilon_1 \epsilon_2 + \theta^2 \epsilon_1 \epsilon_2 - \sigma_2^2 \theta \epsilon_1^2 - \sigma_1^2 \theta \epsilon_2^2 + \theta \sigma_1^2 \sigma_2^2 - \theta^3}{(\sigma_1^2 \sigma_2^2 - \theta^2)^2}\right\} = 0. \tag{D}$$

Now (C) is not a linear combination of (A), (B) and (D), because it contains terms $x_1 \epsilon_2$ and $x_2 \epsilon_1$ that are not in (A), (B) or (D). Therefore the copula scores are non-redundant and the MLE is strictly more efficient than the IQMLE. There are two exceptions. The first is the case that $\theta = 0$, in which case these terms disappear from the copula. The second is the "common regressors" case that $x_1 = x_2$, in which case these terms are equal to terms that do appear in (A) and (B). In either of these two cases, the IQMLE is efficient.

Example 5.4. Bivariate Normal with common variance. Assume Normal marginal densities with $\sigma_1^2 = \sigma_2^2 = \sigma^2$ and $\mu_1 = \mu_2 = 0$. We want to estimate σ^2.

$$f_1(x_1; \sigma) = \frac{1}{\sqrt{2\pi\sigma^2}} e^{-\frac{x_1^2}{2\sigma^2}},$$

$$f_2(x_2; \sigma) = \frac{1}{\sqrt{2\pi\sigma^2}} e^{-\frac{x_2^2}{2\sigma^2}}.$$

Again, let the true joint distribution be Bivariate Normal, i.e.,

$$h(x_1, x_2; \sigma, \theta) = \frac{1}{2\pi\sqrt{\sigma^4 - \theta^2}} e^{-\frac{x_1^2\sigma^2 - 2x_1 x_2\theta + x_2^2\sigma^2}{2(\sigma^4 - \theta^2)}}.$$

Then, the implied copula is Normal,

$$c(F_1(x_1; \sigma), F_2(x_2; \sigma); \theta) = \frac{\sigma^2}{\sqrt{\sigma^4 - \theta^2}} e^{-\frac{\theta\left(x_1^2\theta + x_2^2\theta - 2\sigma^2 x_1 x_2\right)}{2(\sigma^4 - \theta^2)\sigma^2}}.$$

The relevant moment conditions are

$$\mathbb{E}\left\{\frac{x_1^2 - \sigma^2}{2\sigma^4}\right\} = 0, \tag{A}$$

$$\mathbb{E}\left\{\frac{x_2^2 - \sigma^2}{2\sigma^4}\right\} = 0, \tag{B}$$

$$\mathbb{E}\left\{-\frac{\left((3\theta\sigma^4 - \theta^3)(x_1^2 + x_2^2) - 4\sigma^6 x_1 x_2 - 2\sigma^2\theta(\sigma^4 - \theta^2)\right)\theta}{2(\sigma^2 - \theta)^2(\sigma^2 + \theta)^2\sigma^2}\right\} = 0, \tag{C}$$

$$\mathbb{E}\left\{-\frac{\theta\sigma^2(x_1^2 + x_2^2) - (\theta^2 + \sigma^4)x_1 x_2 - \theta(\sigma^4 - \theta^2)}{(\sigma^2 + \theta)^2(\sigma^2 - \theta)^2}\right\} = 0. \tag{D}$$

In this example, (C) is not a linear combination of (A) and (B), so we cannot claim that the IQMLE is efficient with θ known. However, if θ is unknown, (C) is a linear combination of (A), (B), and (D), so the variance matrix $\mathbf{V}$ is singular, and the IQMLE is efficient.

Example 5.5. Eyraud-Farlie-Gumbel-Morgenstern copula with general marginals. For $i = 1, 2$, denote the marginal pdf's and cdf's by

$$f_i \equiv f_i(x_i; \alpha)$$

and

$$F_i \equiv F_i(x_i; \alpha) = \int_{-\infty}^{x_i} f_i(z; \alpha)dz,$$

respectively.

Assume the FGM copula. Then

$$c(u, v; \theta) = 1 + \theta - 2\theta u - 2\theta v + 4\theta uv.$$

Our moment conditions are now

$$\mathbb{E}\left\{\frac{1}{f_1}\frac{\partial f_1}{\partial \alpha}\right\} = 0, \qquad \text{(A)}$$

$$\mathbb{E}\left\{\frac{1}{f_2}\frac{\partial f_2}{\partial \alpha}\right\} = 0, \qquad \text{(B)}$$

$$\mathbb{E}\left\{\frac{2\theta f_1 + 2\theta f_2 - 4\theta f_1 F_2 - 4\theta F_2 F_1}{1 + \theta - 2\theta F_1 - 2\theta F_2 + 4\theta F_1 F_2}\right\} = 0, \quad \text{(C)}$$

$$\mathbb{E}\left\{\frac{1 - F_1 - F_2 + 4F_1 F_2}{1 + \theta - 2\theta F_1 - 2\theta F_2 + 4\theta F_1 F_2}\right\} = 0. \quad \text{(D)}$$

In general, (C) is *not* a linear combination of (A), (B) or (A), (B) and (D). Therefore the copula scores are not redundant in general and IQMLE is generally inefficient.

5.4.4. *Validity and robustness of copulas*

In this section, as previously, we assume that the marginal likelihoods are correctly specified and that the parameters of interest are the parameters (α) that appear in the marginal likelihoods. However, we now consider the possibility that the copula is specified incorrectly, so that the joint likelihood is specified incorrectly.

We will say that a copula is *valid* if the expectation of the copula scores [(C) and (D) in equation (5.20)] is zero, when evaluated at α_o and some value of θ. The true copula is always valid. But an incorrect copula may also be valid. If an incorrect copula is valid, then we will say that it is *robust*.

We will use the terminology "pseudo MLE" (PMLE) to refer to the estimator obtained by maximizing the (incorrect) likelihood based on the incorrect copula.[1] In general, we would expect the PMLE to be inconsistent, because the model is misspecified. However, there are exceptions. A trivial example is the QMLE based on the independence copula, which is consistent given correct specification of the marginal likelihoods even when there is non-independence. Therefore the QMLE based on independence is robust against all

---

[1]Another possible terminology is quasi MLE (QMLE) but that is usually reserved for cases where the estimator is consistent.

possible forms of dependence. Cases that the PMLE is robust correspond to cases where the assumed copula is robust, as defined in the previous paragraph.

Whether an incorrect copula is robust depends on the nature of the marginal likelihoods and of the true copula. Specifically, the validity of a copula depends on the nature of the model, since that determines the marginal likelihoods, and a copula that is robust in the context of one model may not be robust in the context of a different model.

We can note a few simple results.

(i) The independence copula is always valid, and therefore it is robust against all true copulas. It is also always redundant. (The copula score is identically zero.)

(ii) If a copula is redundant when it is true, then it is also robust against all true copulas. The reason is that the copula score with respect to α must be a linear combination of the remaining scores; the marginal scores evaluated at α_o have mean zero by assumption, and the copula score with respect to θ has zero mean when evaluated at any value of α, including α_o, for some value of θ.

(iii) If the true copula is redundant, then any other valid copula is also redundant. This is true because if the true copula is redundant, the IQMLE is efficient.

(iv) If the copula is valid, the (P)MLE will be consistent.

A non-redundant robust copula is possible as shown in the following example.

Consider scalar random variables X_1 and X_2, with means μ_1 and μ_2, which we wish to estimate. Suppose that (X_1, X_2) have joint cdf H, marginal cdfs F_1 and F_2, and copula C. The corresponding joint density, marginal densities and copula densities will be h, f_1 and f_2, and c.

Definition 5.2. (X_1, X_2) is **radially symmetric (RS)** about (μ_1, μ_2) if

$$H(\mu_1 + x_1, \mu_2 + x_2) = 1 - F_1(\mu_1 - x_1) - F_2(\mu_2 - x_2)$$
$$+ H(\mu_1 - x_1, \mu_2 - x_2),$$

or, equivalently,

$$h(\mu_1 + x_1, \mu_2 + x_2) = h(\mu_1 - x_1, \mu_2 - x_2), \qquad (5.30)$$

for all (x_1, x_2).

Definition 5.3. X_1 is **marginally symmetric about** μ_1 if

$$F_1(\mu_1 + x_1) = 1 - F_1(\mu_1 - x_1),$$

or, equivalently,

$$f_1(\mu_1 + x_1) = f_1(\mu_1 - x_1). \qquad (5.31)$$

Definition 5.4. C is **radially symmetric** if

$$C(1 - u, 1 - v) = 1 - u - v + C(u, v),$$

or, equivalently,

$$c(1 - v, 1 - u) = c(v, u), \qquad (5.32)$$

for all (u, v) in $[0, 1] \times [0, 1]$.

It is well known (see, e.g., Nelsen (2006)) that the joint distribution is radially symmetric if and only if the marginal distributions are symmetric and the copula is radially symmetric. Many commonly used distributions are RS. For example, bivariate Normal, bivariate Student-t, bivariate Cauchy and other elliptically contoured distributions are RS. For a discussion of the elliptically contoured family of distributions, see Mardia *et al.* (1979, Section 2.7.2). With reference to the other commonly considered families of copulas, the independence, EFGM, Normal, Plackett, and Frank families are RS, while the Logistic, Ali-Mikhail-Haq (AMH), Joe, Clayton and Gumbel families are not. Interestingly, Frank (1979) shows that the only Archimedean copula family that satisfies Definition 5.4 is the Frank family. Joe, AMH, Clayton and Gumbel are all Archimedean copulas that are not RS.

We now have the following result.

Theorem 5.4. *Suppose that the distribution of* (X_1, X_2) *is RS about* $\alpha = (\mu_1, \mu_2)$. *Then any RS copula is robust for estimation of* α.

To state the result a bit more explicitly, let k be any RS copula density (we use the notation "k" to distinguish it from the true copula "c".) Then, with $\alpha = (\mu_1, \mu_2)$,

$$\mathbb{E}\nabla_\alpha \ln k(F_1(\mu_1 + X_1), F_2(\mu_2 + X_2), \theta) = 0.$$

This is true for any value of θ (the nuisance parameters in the assumed copula k).

Therefore, so long as the marginal distributions are correctly chosen and the true joint distribution is RS, the misspecified copula k can be used to consistently estimate α.

The robustness result given in Theorem 5.4 does not address the issue of whether a non-redundant robust copula can exist. We now give an example that shows that it can. The example consists of logistic marginals, each of which contains a common location parameter μ which is the parameter of interest, and the FGM copula. Because the EFGM copula is RS, it satisfies the conditions of the theorem above. We will show that the EFGM copula, if correct, is not redundant for this problem.

Example 5.6. Eyraud-Farlie-Gumbel-Morgenstern copula and logistic marginals with common mean. Consider logistic marginals with the common location parameter μ. Let the true value of μ be zero. For $i = 1, 2$ the marginal pdf's and cdf's are, respectively,

$$f_i(y_i; \mu) = \frac{e^{-y_i + \mu}}{(1 + e^{-y_i + \mu})^2}$$

and

$$F_i(y_i; \mu) = \frac{1}{(1 + e^{-y_i + \mu})}.$$

Suppose the true copula is the EFGM copula. The copula pdf is

$$c(u, v; \theta) = 1 + \theta - 2\theta u - 2\theta v + 4\theta uv.$$

Note that the logistic distribution is symmetric about zero and that the EFGM copula is RS.

Our moment conditions are now

$$\mathbb{E}\frac{1-e^{-x_1+\mu}}{1+e^{-x_1+\mu}} = 0, \tag{A}$$

$$\mathbb{E}\frac{1-e^{-x_2+\mu}}{1+e^{-x_2+\mu}} = 0, \tag{B}$$

$$\mathbb{E}\left\{\frac{2\theta[(-1+2F_1)f_2+(-1+2F_2)f_1]}{1+\theta-2\theta F_1-2\theta F_2+4\theta F_1 F_2}\right\} = 0, \tag{C}$$

$$\mathbb{E}\left\{\frac{1-F_1-F_2+4F_1 F_2}{1+\theta-2\theta F_1-2\theta F_2+4\theta F_1 F_2}\right\} = 0. \tag{D}$$

Let a, b, c and d denote the moment functions in (A), (B), (C) and (D), respectively. Then, simple algebra shows that

$$c = \frac{\theta d}{2ab}[b(1-a^2)+a(1-b^2)]$$

and

$$d = \frac{ab}{1+\theta ab}.$$

Clearly, (C) is not a linear combination of (A), (B) or of (A), (B) and (D).

We conclude that in this problem the EFGM copula is non-redundant.

In the above example, the EFGM copula is non-redundant when it is true. By Theorem 5.4, it is robust against any radially symmetric true copula. Therefore, while redundancy of a copula implies robustness, the converse is not true.

5.4.5. *Efficiency and redundancy under misspecified but robust copulas*

We now consider the question of efficient estimation when the assumed copula is misspecified but robust. The main thing we wish to point out is that the PMLE is dominated by the efficient GMM estimator. The reason is that summation is no longer the appropriate weighting of the scores with respect to α. This is the same logical point as was made in Section 5.4.1, and indeed Section 5.4.1 is therefore a special case of this section.

We go back to the set of moment conditions (A)–(D) as in equation (5.20). However, the copula scores (C) and (D) are based on an incorrect but robust copula.

Lemma 5.2. *Denote the covariance matrix of the moment functions in* (5.20) *by* **C**, *and their expected derivative matrix by* **D**. *Then,*

$$V = \left[\begin{array}{cc|cc} A & G & -K & -P \\ G' & B & -L' & -Q' \\ \hline -K' & -L & N & Z \\ -P' & -Q & Z' & W \end{array} \right]$$

and

$$D = \left[\begin{array}{c|c} -A & 0 \\ -B & 0 \\ \hline K' + L - M & -S \\ -S' & -T \end{array} \right],$$

where **A**, **B**, **G** *are as in Lemma* 5.1, *and* **K**, **L**, **M**, **N**, **P**, **Q**, **S**, **T**, **W**, **Z** *are matrix-functions of* (α, θ) *defined in the Appendix.*

These expressions are not nearly as simple as the corresponding expressions in (5.21) and (5.22). Less information equalities apply in the present case than in the case that the copula is correctly specified.

Lemma 5.2 can be used to make the following important observation. The optimal GMM estimator using the moment conditions (5.20), where the assumed copula is robust but not correct, is not the same as the PML estimator. This is in contrast to the case that the copula is correctly specified, in which case the optimal GMM estimator and the PMLE were asymptotically equivalent. The simple form of the optimal set of linear combinations, $D'V^{-1}$, as given in (5.23) does not hold in the present case. In other words, the optimal weighting now does not correspond to summation of (A), (B) and (C), which is what PMLE does.

We will call the optimal GMM estimator based on (5.20) the *Improved PML Estimator* (IPMLE). The following theorem formally states its dominance of the PMLE in terms of efficiency.

Theorem 5.5. *Let* $\mathbb{V}_{\text{IPMLE}}$ *and* $\mathbb{V}_{\text{PMLE}}$ *denote the asymptotic variance matrices of the IPMLE and PMLE of* (α_o, θ_o), *respectively. Then,* $\mathbb{V}_{\text{PMLE}} - \mathbb{V}_{\text{IPMLE}}$ *is positive semi-definite.*

We now revisit the redundancy question that was addressed in Section 5.4.3 for correctly specified copulas. The IPMLE estimator must dominate the IQMLE estimator in terms of efficiency, but the question is when the efficiency difference is zero. Since $\mathbb{V}_{\text{IPMLE}}$ is only defined when $\mathbf{V}$ is nonsingular, thus we can apply the redundancy toolbox of Breusch *et al.* (1999).

In stating the next result, we must distinguish the covariance matrix of the moment conditions based on the incorrect but robust copula (which we will call $\mathbf{V}^k$) from the covariance matrix of the moment conditions based on the true copula (which we will call $\mathbf{V}$). The true copula is not involved in estimation, but it is involved in evaluating expectations because it is part of the true joint distribution.

Theorem 5.6. $\mathbb{V}_{\text{IPMLE}}$ *for* α *and* $\mathbb{V}_{\text{IQMLE}}$ *are equal if and only if*

$$\mathbf{M} - \mathbf{V}_{21}^{\alpha k}\mathbf{V}_{11}^{-1}\mathbf{V}_{12}^{\alpha} - \mathbf{ST}^{-1}(\mathbf{R} - \mathbf{V}_{21}^{\theta k}\mathbf{V}_{11}^{-1}\mathbf{V}_{12}^{\alpha}) = 0, \qquad (5.33)$$

where $\mathbf{V}_{21}^{\alpha k} = [-\mathbf{K}' \quad -\mathbf{L}]$, $\mathbf{V}_{21}^{\theta k} = [-\mathbf{P}' \quad -\mathbf{Q}]$, *and* $\mathbf{V}_{12}^{\alpha} = \begin{bmatrix} -\mathbf{G} \\ -\mathbf{G}' \end{bmatrix}$.

In (5.33), $\mathbf{M} - \mathbf{V}_{21}^{\alpha k}\mathbf{V}_{11}^{-1}\mathbf{V}_{12}^{\alpha}$ and $\mathbf{R} - \mathbf{V}_{21}^{\theta k}\mathbf{V}_{11}^{-1}\mathbf{V}_{12}^{\alpha}$ can be viewed as covariance matrices between copula moments based on the incorrect but robust copula and the error in the linear projection of the true copula moment on the marginal moments (A–B). More explicitly,

$$\mathbf{M} - \mathbf{V}_{21}^{\alpha k}\mathbf{V}_{11}^{-1}\mathbf{V}_{12}^{\alpha}$$

$$= \mathbb{E}\left\{\nabla_\alpha \ln k \left(\nabla_\alpha \ln c - \mathbf{V}_{21}\mathbf{V}_{11}^{-1}\begin{bmatrix} \nabla_\alpha \ln f_1 \\ \nabla_\alpha \ln f_2 \end{bmatrix}\right)'\right\},$$

$$\mathbf{R} - \mathbf{V}_{21}^{\theta k}\mathbf{V}_{11}^{-1}\mathbf{V}_{12}^{\alpha}$$

$$= \mathbb{E}\left\{\nabla_\theta \ln k \left(\nabla_\alpha \ln c - \mathbf{V}_{21}\mathbf{V}_{11}^{-1}\begin{bmatrix} \nabla_\alpha \ln f_1 \\ \nabla_\alpha \ln f_2 \end{bmatrix}\right)'\right\}. \qquad (5.34)$$

where k is the incorrect copula pdf. Clearly, when both of these matrices are zero, (5.33) holds for any $\mathbf{S}$. Also, if only (5.34) is zero and $\mathbf{S} = 0$, (5.33) holds for any $\mathbf{R}$ and $\mathbf{V}_{21}^{\theta k}$.

Corollary 5.2. *If the true copula score with respect to α is a linear combination of* (A) *and* (B) *with θ known then*

(1) $\mathbf{M} - \mathbf{V}_{21}^{\alpha k} \mathbf{V}_{11}^{-1} \mathbf{V}_{12}^{\alpha} = 0$;

(2) $\mathbf{R} - \mathbf{V}_{21}^{\theta k} \mathbf{V}_{11}^{-1} \mathbf{V}_{12}^{\alpha} = 0$;

(3) *IQMLE and IPMLE for α are equally efficient.*

Basically, Corollary 5.2 repeats fact (iii) noted in Subsection 5.4.4. If the true copula is redundant, then any other valid copula must also be redundant. What is important to note is that this does *not* imply that the copula score based on the incorrect but robust copula must be a linear combination of the other scores. In other words, with a correctly specified copula, the only way redundancy can occur is when there is linear dependency among the scores. This is not the case when the copula is incorrect but robust.

5.5. Robustness and efficiency of nonparametric copulas

A nonparametric copula is robust to misspecification since it uses no parametric copula specification. This attractive property is balanced against other issues with nonparametric estimation such as computational complexity and the dependence of convergence rates on dimension.

This section presents selected results on the use on Bernstein copula defined in Section 5.2.2. Section 5.5.1 discusses how it can be used to obtain a semiparametrically efficient estimator of the parameters in marginals, including a sparse estimator. Section 5.5.2 discusses Bayesian efficiency, i.e., speed of mixing, of a multivariate density estimator based on the Bernstein copula.

5.5.1. Efficient semiparametric estimation of parameters in marginals

As in Section 5.4, consider an d-variate random variable X with joint pdf $h(x_1, \ldots, x_d)$. Let $f_1(x_1), \ldots, f_d(x_d)$ denote the corresponding marginal pdf's. Assume that the marginals are known up to a parameter vector α, where α collects all the distinct parameters in the marginals such as the tail exponents if the marginals are power law. The dependence structure between the marginals is not parameterized. We observe an i.i.d. sample $\{\mathbf{x}_i\}_{i=1}^N = \{x_{1i}, \ldots, x_{di}\}_{i=1}^N$ and we are interested in estimating α efficiently without assuming anything about the joint distribution except for the marginals.

As an example, consider the setting of a standard panel (small d, large N). We have a well specified marginal for each of d cross sections and we are interested in efficient estimation of the parameters in the marginal distributions with no apriori knowledge of the form or strength of dependence between them. This or similar setting is often encountered in microeconomic and actuarial applications (see, e.g., Winkelmann (2012); Amsler *et al.* (2014); Frees and Valdez (1998)). In finance, a similar setting arises in the so called SCOMDY models and in other multivarariate GARCH-type models, where interest is in estimation of univariate conditional distribution parameters while the error terms are allowed to have arbitrary dependence (Chen and Fan (2006a,b); Hafner and Reznikova (2010)).

However, recent literature on semiparametric copula models has focused on the case when the marginals are specified nonparametrically and the copula function is given a parametric form (see, e.g., Chen *et al.* (2006); Segers *et al.* (2008)), which is an appropriate setting for many financial applications where it is important to parameterize dependence. In our setting, dependence is used solely to provide more precision in estimation of marginal parameters so we study the converse problem.

As discussed in Section 5.4, the parameters of the marginals can be consistently estimated by maximizing the likelihood under the assumption of independence between the marginals — this is the so called quasi maximum likelihood estimator, or QMLE. The copula term in (5.2) and (5.19) is zero in this case because the independence copula density is equal to one. However, QMLE is not efficient if marginals are not independent and for highly dependent marginals, the efficiency loss relative to the correctly specified full likelihood MLE is quite large. Joe (2005), for instance, reported up to 93% improvements in relative efficiency over QMLE in simulations when the full likelihood is correctly specified.

The situation when using copula terms in the likelihood does not improve asymptotic efficiency over QMLE is known as copula redundancy. In Section 5.4.3, we provided a necessary and sufficient condition for copula redundancy and showed that such situations are very rare. Essentially, a parametric copula is redundant for estimation of parameters in the marginals if and only if the copula score with respect to these parameters can be written as a linear combination of the marginal scores — a condition generally violated for most commonly used parametric copula families and marginal distributions. As a result, significant efficiency gains remain unexploited.

An alternative that is more efficient asymptotically is a fully parametric estimation of the entire multivariate distribution by full MLE. This means assuming a parametric copula specification in addition to the marginal distributions. It is now well understood that, unlike QMLE, FMLE is generally not robust to copula misspecification. That is, the efficiency gains will come at the expense of an asymptotic bias if the joint density is misspecified. Section 5.4.4 pointed out that there are robust parametric copulas, for which the pseudo MLE (PMLE) using an incorrectly specified copula family leads to a consistent estimation. However, copula robustness is problem specific and some robust copulas are robust because they are redundant. Therefore, finding a general class of robust non-redundant copulas remains an unresolved problem.

Panchenko and Prokhorov (2016) addressed this problem using a semiparametric approach. That is, they investigated whether it is possible to obtain a consistent estimator of α, which is relatively more efficient than QMLE, by modelling the copula term nonparametrically. They use sieve MLE (SMLE) to do that. The questions they asked are whether a sieve-based copula approximator is the robust non-redundant alternative to QMLE and PMLE and what is the semiparametric efficiency bound for the SMLE of α. Therefore their results relate to the literature on sieve estimation (see, e.g., Ai and Chen (2003); Newey and Powell (2003); Bierens (2014)) and on semiparametric efficiency bounds (see, e.g., Severini and Tripathi (2001); Newey (1990)).

Denote the true copula density by $c_o(\mathbf{u})$, $\mathbf{u} = (u_1, \ldots, u_d)$, and denote the true parameter vector by α_o. Let α_o belong to finite dimensional space $\mathscr{B} \subset R^p$ and $c_o(\mathbf{u})$ belong to an infinite-dimensional space $\mathscr{C} = \{c(\mathbf{u}) : [0,1]^d \to [0,1], \int_{[0,1]^d} c(\mathbf{u})d\mathbf{u} = 1, \int_{[0,1]^{d-1}} c_{J_N}(\mathbf{u}_{-\ell})d\mathbf{u}_{-\ell} = 1, \forall \ell\}$, where $\mathbf{u}_{-\ell}$ excludes u_ℓ. Given a finite amount of data, optimization over the infinite-dimensional space $\mathscr{C}$ is not feasible. The method of sieves is useful for overcoming this problem. Compared to other nonparametric methods such as kernels, local linear estimators, etc., the method of linear sieves is also quite simple — the infinite dimensional optimization is reduced to a regular parametric MLE.

Define a sequence of approximating spaces $\mathscr{C}_N$, called sieves, such that $\bigcup_N \mathscr{C}_N$ is dense in $\mathscr{C}$. Optimization is then restricted to the sieve space. Grenander (1981) is credited for observing that the MLE optimization, which is infeasible over an infinite dimensional space, is remedied if we optimize over a subset of the parameter space, known as the sieve space, and then allow the subset to grow with the sample size (see, e.g., Chen (2007), for a survey of sieve methods).

Panchenko and Prokhorov (2016) propose using the Bernstein polynomial (5.6) for $\mathscr{C}_N$. Write the sieve for $\Gamma = \mathscr{B} \times \mathscr{C}$ as $\Gamma_N = \mathscr{B} \times \mathscr{C}_N$, where $\mathscr{C}_N$ contains a generic vector of the Bernstein copula parameters θ. In the notation of (5.6), θ contains all weights $\omega_{\mathbf{k}}$. Let $\gamma = (\alpha', \theta')$, then the sieve MLE (SMLE) can be written as follows:

$$\hat{\gamma} = \arg\max_{\gamma \in \Gamma_N} \sum_{i=1}^{N} \ln h(\mathbf{x}_i; \gamma). \qquad (5.35)$$

In essence, an infinite-dimensional problem over a space of functions is reduced to a finite-dimensional problem over a sieve of that space. As pointed out in sieve MLE literature (see, e.g., Chen (2007)), this estimator is very easy to implement in practice — it is a standard finite dimensional parametric MLE once we decide on the number of sieve copula coefficients, and, as we discuss later, a consistent estimator of the SMLE asymptotic covariance matrix can be obtained in some cases using standard MLE.

Clearly, this estimator is based on a finite-dimensional approximation of the unspecified part of the joint distribution. As such, the estimator inherits the costs and benefits of the multivariate sieve MLE. A major benefit is the increased precision compared to quasi-MLE, permitted by the use of dependence information. Panchenko and Prokhorov (2016) show that this estimator reaches the semiparametric efficiency bound for estimation of α. Simulations show that potential efficiency gains are huge. The efficiency bound is determined by the dependence strength and the estimator can get fairly close to FMLE in terms of precision.

The gains come at an increased computational expense. Panchenko and Prokhorov (2016) report that the convergence is slow for the traditional sieves and the Bernstein polynomial is preferred to other sieves. The running times are greater than the FMLE assuming an "off-the-shelf" parametric copula family but far from being prohibitive (at least for the two dimensional problem they consider). Moreover, simulations reveal a small downward bias in SMLE, which seems to be caused by the sieve approximation error — it decreases as the number of sieve elements increases.

Methods to improve computational efficiency of SMLE focus on reducing the effective number of sieve parameters. Such methods involve penalized and restricted estimation and are particularly appealing for the Bernstein polynomial where the sparse portions of the sieve parameter space correspond to histogram cells with little or no mass.

Liu and Prokhorov (2016) use a version of Dantzig selector to reduce the dimension of the parameter space. The Dantzig selector is traditionally used for point estimation by least squares when the number of parameters exceeds the number of observations. Liu and

Prokhorov (2016) use it to select non-zero elements of the copula parameter vector θ.

The non-sparse estimator of Panchenko and Prokhorov (2016) has smaller standard errors asymptotically than the conventional QMLE but, in finite samples, the number of parameters in the sieve is close to the sample size and may exceed it. At the same time, most of the sieve parameters are usually close to zero. Liu and Prokhorov (2016) propose a l_1-norm shrinkage estimator that selects and eliminates such parameters. The penalized likelihood has the form

$$\min_{\alpha,\theta} \left\{ -\frac{1}{N} \ln h(\mathbf{x}_i; \alpha, \theta) + r \sum_{j=1}^{\dim\{\theta\}} |\theta_j| \right\}. \qquad (5.36)$$

In essence, this estimator uses the Dantzig selector to find the sparsest vector of the sieve parameters satisfying the first-order conditions of the MLE up to a given tolerance level. This can be seen if we rewrite Equation (5.36) as follows:

$$\min_{\alpha,\theta_N} \|\theta_N\|_{l_1}$$

$$\text{subject to } \left\| \frac{1}{N} \sum_{i=1}^{N} \nabla_{(\alpha,\theta)} \ln c_{\theta_N} \left(F(x_{i1}; \alpha), F(x_{i2}; \alpha) \right) \right\|_{l_\infty} \leq r$$

$$\text{and } \frac{1}{N} \sum_{i=1}^{N} \nabla_\alpha \ln f(x_{it}; \alpha) = 0, \ t = 1, 2, \qquad (5.37)$$

where $c_\theta(\mathbf{u}) = b_\mathbf{k}(\mathbf{u})$ is the Bernstein copula density, and θ_N means that θ depends on the sample size.

Simulations show that this estimator has finite-sample properties very similar to the non-sparse alternative, and substantially better than QMLE. Thus, the sparsity imposed by the Dantzig selector is innocuous with respect to the non-asymptotic behavior of the sieve MLE; it also permits a substantial increase in computational efficiency compared to the unrestricted sieve MLE. That paper also studies the parameter path behavior for various tolerance levels and considers a version of a double Dantzig selector which resolves the arbitrariness in choosing the tolerance level.

5.5.2. *Bayesian efficiency and consistency*

Another robust alternative to parametric MLE is offered by Bayesian nonparametrics. In a setting of a generic density estimation, Burda and Prokhorov (2014) propose a Bayesian infinite mixture model and an estimation procedure based on a Dirichlet process prior. The point is that the computational complexity in such models is not directly related to the dimensionality of the problem. A rapid increase of mixture components required to represent arbitrary dependence structures accurately in high dimensions is a practical problem that can be solved by choosing a convenient mixing kernel.

A typical mixing kernel used in multivariate density estimation is the Gaussian kernel. Instead one can decompose the joint parametric mixing kernel into flexible building blocks, each with a parsimonious representation. The Bernstein copula can be used to represent dependence in such decomposition. Only a few latent classes, or mixing components, are required for each of the marginals, regardless of the overall number of dimensions, which provides a substantial improvement in precision over the conventional multivariate parametric kernels.

Because this is a generic multivariate density estimation as opposed to the estimation of parameters in marginals, we will use a slightly different notation in this subsection. Let $\mathcal{X}$ be the sample space with elements x, Θ the space of the mixing parameter θ, and Φ the space of the hyper parameter ϕ. Let $\mathcal{D}(\mathcal{X})$ denote the space of probability measures F on $\mathcal{X}$. Denote by $\mathcal{M}(\Theta)$ the space of probability measures on Θ and let P be the mixing distribution on Θ with density p and a prior Π on $\mathcal{M}(\Theta)$ with weak support $supp(\Pi)$. Denote the prior for ϕ by μ, and the support of μ by $supp(\mu)$, with μ independent of P. Let $K(x; \theta, \phi)$ be a kernel on $\mathcal{X} \times \Theta \times \Phi$, such that $K(x; \theta, \phi)$ is a jointly measurable function with the property that for all $\theta \in \Theta$ and $\phi \in \Phi$, $K(\cdot; \theta, \phi)$ is a probability density on $\mathcal{X}$.

Π, μ and $K(x; \theta, \phi)$ induce a prior on $\mathcal{D}(\mathcal{X})$ via the map

$$(\phi, P) \mapsto f_{P,\phi}(x) \equiv \int K(x; \theta, \phi) dP(\theta). \qquad (5.38)$$

Denote such composite prior by Π^*.

Burda and Prokhorov (2014) consider a kernel with the structure

$$K(x; \theta, \phi) = K_c(F(x; \theta_m, \phi_m); \theta_c, \phi_c) K_m(x; \theta_m, \phi_m) \qquad (5.39)$$

where

$$K_m(x; \theta_m, \phi_m) = \prod_{s=1}^{d} K_{ms}(x_s; \theta_{ms}, \phi_{ms}) \qquad (5.40)$$

is the product of univariate kernels of the marginals in d dimensions, $K_c(F(x; \theta_m, \phi_m); \theta_c, \phi_c)$ is a copula density kernel, $\theta = \{\theta_m, \theta_c\}$, and $\phi = \{\phi_m, \phi_c\}$. The arguments of $K_c(\cdot)$ consist of a d-vector of distribution functions of the marginals

$$F(x; \theta_m, \phi_m) = \int_{-\infty}^{x} K_m(t; \theta_m, \phi_m) dt$$

with copula parameter vectors θ_c and ϕ_c. $F(x; \theta_m, \phi_m)$ collects the univariate marginals of the d-variate kernel in (5.39). In essence, these are univariate kernels such as Gaussian kernels, with mixing parameters θ_m and hyper parameters ϕ_m.

The copula counterpart of the mixed joint density (5.38) takes the form

$$f_{P,\phi}(x) = \int K_c(F(x; \theta_m, \phi_m); \theta_c, \phi_c) K_m(x; \theta_m, \phi_m) dP(\theta), \qquad (5.41)$$

where $P(\theta) = P_c(\theta_c) \times P_m(\theta_m)$. The mixing parameter θ enters through both the marginals and the copula which complicates the analysis relative to cases when $K(x; \theta, \phi)$ in (5.38) is a single kernel, such as the multivariate Gaussian.

Let $\Theta_m \times \Phi_m$ denote the space of parameters in marginals and let $\Theta_c \times \Phi_c$ denote the space of copula parameters. Then, $(\theta, \phi) \in \Theta_m \times \Theta_c \times \Phi_m \times \Phi_c \subset \Theta \times \Phi$. This means that this setup effectively assumes that the space of copula parameters and space of marginal parameters form a Cartesian product.

Petrone (1999a,b) proposed a class of prior distributions on the set of densities defined on $[0, 1]$ based on the univariate Bernstein polynomials and Petrone and Wasserman (2002) showed weak and strong consistency of the Bernstein polynomial posterior for the space of univariate densities on $[0, 1]$. Zheng *et al.* (2010) and Zheng (2011) extend the settings to the multivariate case, where $\mathbf{k}$ is an $\mathbb{N}^d$-valued random variable and P_c is a random probability distribution function, yielding $B_{\mathbf{k}, P_c}$ as a random function. A random Bernstein density $b_{\mathbf{k}, P_c}(\mathbf{u})$ thus features random mixing weights $w_{\mathbf{k}, P_c}(\mathbf{j})$ with $\mathbf{w}_{\mathbf{k}, P_c} = \{w_{\mathbf{k}, P_c}(\mathbf{j}) : j_s = 1, \ldots, k_s, s = 1, \ldots, d\}$ belonging to the $\prod_{s=1}^{d} k_s - 1$ dimensional simplex

$$
\mathsf{s_k} = \left\{ \mathbf{w}_{\mathbf{k}, P_c} : w_{\mathbf{k}, P_c}(\mathbf{j}) \geq 0, \sum_{j_1=1}^{k_1} \cdots \sum_{j_d=1}^{k_d} w_{\mathbf{k}, P_c}(\mathbf{j}) = 1 \right\}.
$$

Burda and Prokhorov (2014) adopt the multivariate Bernstein density function as a particular case of the copula density kernel in (5.39):

$$
K_c(F(x; \theta_m, \phi_m); \theta_c, \phi_c) = b_{\mathbf{k}, P_c}(F(x; \theta_m, \phi_m)). \tag{5.42}
$$

Let $\{u_{s1}, u_{s2}, \ldots\}$ denote a sequence of exchangeable random variables with values in $[0, 1]$ for $s = 1, \ldots, d$. Conditional on the marginal parameters θ_m and ϕ_m, $u_{si} = F_s(x_{si}; \theta_{ms}, \phi_{ms})$. A multivariate version of the hierarchical mixture model based on the Bernstein-Dirichlet prior proposed in Petrone (1999b), p. 383, can be specified as follows:

$$
u_{si} | y_{si}, P_c, k_s \sim \beta(u_{si}; j_s, k_s - j_s + 1)
$$
$$
\text{if } y_{si} \in ((j_s - 1)/k_s, j_s/k_s], \text{ for each } s = 1, \ldots, d
$$
$$
\mathbf{y}_i | P_c, \mathbf{k} \sim P_c
$$
$$
P_c | \mathbf{k} \sim DP(\alpha_c, P_{c0})
$$
$$
\mathbf{k} \sim \mu(\mathbf{k}),
$$

where $\mathbf{y}_i = (y_{1i}, \ldots, y_{di})$ are latent random variables determining the hidden labels associated with $\mathbf{u}_i = (u_{1i}, \ldots, u_{di})$. So in this case, $\theta_c = \{\mathbf{y}_i\}_{i=1}^{n}$ and $\phi_c = \mathbf{k}$. P_{c0}, a probability measure on $[0, 1]^d$ that is absolutely continuous with respect to the Lebesgue measure, is the baseline of the Dirichlet process $DP(\alpha_c, P_{c0})$ with concentration parameter α_c. Burda and Prokhorov (2014) set P_{c0} to be uniform on $[0, 1]^d$ and, following Petrone (1999a), $\alpha_c = 1$. As a prior for the discrete distribution $\mu(\mathbf{k})$, they further use the Dirichlet distribution $Dir(\{j_s/k_s\}_{j_s=1}^{k_s}; 1/k_s)$ for $k_s \leq k_s^{\max}$ for each $s = 1, \ldots, d$. The posterior then follows directly from Petrone (1999a) who also proposes a sampling algorithm for the posterior that we follow in the implementation.

For the marginal univariate kernel in (5.40), Burda and Prokhorov (2014) take a product of univariate Gaussian kernels

$$K_{ms}(x_{si}; \theta_{ms}, \phi_{ms}) = (2\pi)^{-1/2}\sigma_s^{-1}\exp(-(x_s - \nu_s)^2/(2\sigma_s^2)), \quad (5.43)$$

with $\theta_{ms} = \{\nu_s, \sigma_s^2\}$ and ϕ_{ms} being a vacuous parameter. The prior structure for the marginal is then

$$x_{si} \sim N(x_{si}; \theta_{ms})$$
$$\theta_{ms}|P_m \sim P_m$$
$$P_m \sim DP(\alpha_m, P_{m0})$$
$$\alpha_m \sim Gamma(\alpha_{m01}, \alpha_{m02})$$

with P_{m0} for $\{\nu_s, \sigma_s^2\}$ composed of $N(\nu_{s0\nu}, \sigma_{s0\nu}^2)$ and $InvGamma$ $(\gamma_{s01}, \gamma_{s02})$, respectively.

Burda and Prokhorov (2014) show by simulations that this factorization permits a reduction of estimation error measured by MAD by a third, compared to the multivariate normal mixing kernel, and that key features of the data generating process, missed using the conventional kernel, are captured using the Bernstein-Dirichlet prior. Burda and Prokhorov (2014) also derive weak posterior consistency of the copula-based mixing scheme for general kernel types under high-level conditions, and strong posterior consistency for the specific Bernstein-Gaussian mixture model.

5.6. Robustness of estimators to heavy tails

5.6.1. *Trimming*

The conventional way to estimate the parameters in marginals, α, and in the copula, θ, is by a two-step maximization of the components of the joint parametric likelihood (the Inference Function for Margins method) or by maximizing the joint likelihood based on pseudo-observations (the Canonical MLE) — both methods were discussed in Sections 5.2.1–5.2.3. These estimators presume existence of copula-based moments or marginal-based moments or both.

Improved parametric estimators discussed in Section 5.4 such as GMM and PMLE estimators also traditionally employ the assumption of existence of moments of sufficient order so that standard versions for the central limit theorem can be applied. In the context of copula estimation, robustness of these estimators, and inference based on them, to heavy tails remain unexplored.

Recent developments in weak limit theory for tail-weighted arrays of dependent and non-stationary data Hill (2009, 2011, 2015b) provide tailed-trimmed versions of likelihood-based inference including robust specification tests, which can be used to robustify these estimators to violations of the thin tail assumption.

Hill and Prokhorov (2016) consider estimation of GARCH models after relaxing the assumption of finite moments. The estimator they proposed, which encompasses many estimators of the so-called Generalized Empirical Likelihood (GEL) family, imbeds tail-trimmed moment equations into the standard GEL optimization problem. This allows for over-identifying conditions, asymptotic normality, efficiency and empirical likelihood based confidence regions for very heavy-tailed random volatility data.

In essence this is done by dropping a small, "asymptotically negligible," proportion of extreme observations. Hill and Prokhorov (2016) discuss a way to decide on the optimal trimming proportion optimizing higher order behavior of the resulting estimator. It is well known that GEL is first-order asymptotic equivalent to GMM, i.e., converges to the same asymptotic distribution at the $\sqrt{N}$-rate but specializes into many different estimators, such as Empirical Likelihood, Exponential Tilting, Continuously Updated estimators, so the results on optimal trimming of GEL are fairly general.

Consider the following strong-GARCH(1,1) model:

$$y_t = \sigma_t \epsilon_t \text{ where } \epsilon_t \text{ is i.i.d., } \mathbb{E}[\epsilon_t] = 0 \text{ and } \mathbb{E}[\epsilon_t^2] = 1$$

$$\sigma_t^2 = \omega + \alpha y_{t-1}^2 + \beta \sigma_{t-1}^2, \text{ where } \omega > 0, \ \alpha, \beta \geq 0, \text{ and } \alpha + \beta > 0.$$

$$(5.44)$$

Let $\theta = (\alpha, \beta, \omega)$ and write the QML score equations, with added overidentifying moment conditions, as follows:

$$m_t(\theta) = \left(\epsilon_t^2(\theta) - 1\right) \times x_t(\theta) \in \mathbb{R}^q, \; q \geq 3,$$

$$\text{where } x_t(\theta) \equiv \left[s_t'(\theta), w_t'(\theta)\right]' \quad \text{and} \quad s_t(\theta) \equiv \frac{1}{\sigma_t^2(\theta)} \frac{\partial}{\partial \theta} \sigma_t^2(\theta).$$

The point is to relax the assumptions ensuring that m_t have finite second-order moments. The function m_t could in principle be anything, including the moment functions used in the IFM, GMM, PMLE in previous sections.

For the GARCH model with bounded s_t, the sources of large m_t are ϵ_t and w_t and so Hill and Prokhorov (2016) trim the moment functions by their components separately.

Let $z_t(\theta)$ denote $\epsilon_t(\theta)$ or $w_{i,t}(\theta)$, and denote its absolute value and its order statistics as follows:

$$z_t^{(a)}(\theta) \equiv |z_t(\theta)| \quad \text{and} \quad z_{(1)}^{(a)}(\theta) \geq \cdots \geq z_{(n)}^{(a)}(\theta) \geq 0.$$

Let $\{k_N^{(\epsilon)}, k_{i,N}^{(w)}\}$ for $i \in \{1, \ldots, q-3\}$ be intermediate order sequences, i.e., such that $k_N^{(\epsilon)}$ and $k_{i,N}^{(w)}$ go to infinity but slower than the sample size N. Then, trimming is based on the indicator functions defined as follows:

$$\hat{\mathbb{I}}_{N,t}^{(\epsilon)}(\theta) \equiv \mathbb{I}\left(|\epsilon_t(\theta)| \leq \epsilon_{(k_N^{(\epsilon)})}^{(a)}(\theta)\right)$$

$$\hat{\mathbb{I}}_{i,N,t}^{(w)}(\theta) \equiv \mathbb{I}\left(|w_{i,t}(\theta)| \leq w_{i,(k_{i,N}^{(w)})}^{(a)}(\theta)\right) \quad \text{and}$$

$$\hat{\mathbb{I}}_{N,t}^{(x)}(\theta) \equiv \begin{cases} \prod_{i=1}^{q-3} \hat{\mathbb{I}}_{i,N,t}^{(w)}(\theta) & \text{if } q > 3 \\ 1 & \text{if } q = 3, \end{cases}$$

and on tail-trimmed variables and equations

$$\hat{\epsilon}_{N,t}^*(\theta) \equiv \epsilon_t(\theta)\hat{\mathbb{I}}_{N,t}^{(\epsilon)}(\theta)\hat{\mathbb{I}}_{N,t}^{(x)}(\theta) \quad \text{and} \quad \hat{w}_{N,t}^*(\theta) \equiv w_t(\theta)\hat{\mathbb{I}}_{N,t}^{(w)}(\theta)$$

and $\hat{x}_{N,t}^*(\theta) \equiv [s_t(\theta)', \hat{w}_{N,t}^*(\theta)]'$, (5.45)

$$\hat{m}_{N,t}^*(\theta) \equiv \left(\hat{\epsilon}_{N,t}^{*2}(\theta) - \frac{1}{N} \sum_{t=1}^{N} \hat{\epsilon}_{N,t}^{*2}(\theta) \right) \times \hat{x}_{N,t}^*(\theta).$$ (5.46)

The last equation shows recentering used to eliminate small sample bias caused by trimming.

The estimator than proceeds as follows. Let $\rho : \mathcal{D} \to \mathbb{R}_+$ be a twice continuously differentiable concave function, with domain $\mathcal{D}$ containing zero. Write $\rho^{(i)}(u) = (\partial/\partial u)^i \rho(u)$, $i = 0, 1, 2$, and $\rho^{(i)} = \rho^{(i)}(0)$, and assume the normalizations $\rho^{(0)} = \rho(0) = 0$ and $\rho^{(1)} = \rho^{(2)} = -1$. If $\rho(u) = -u^2/2 - u$ we have the Continuously Updated Estimator or Euclidean Empirical Likelihood; $\rho(u) = \ln(1 - u)$ for $u < 1$ leads to Empirical Likelihood; $\rho(u) = 1 - \exp\{u\}$ represents Exponential Tilting.

The GEL estimator with Imbedded Tail-Trimming (GELITT) proposed by Hill and Prokhorov (2016) solves the conventional saddle-point optimization problem (Smith (1997); Newey and Smith (2004)) using the trimmed moment functions:

$$\hat{\theta}_N = \arg\min_{\theta \in \Theta} \sup_{\lambda \in \hat{\Lambda}_N(\theta)} \left\{ \frac{1}{N} \sum_{t=1}^{N} \rho\left(\lambda' \hat{m}_{N,t}^*(\theta) \right) \right\} \quad \text{and}$$

$$\hat{\lambda}_N = \arg\sup_{\lambda \in \hat{\Lambda}_N(\hat{\theta}_N)} \left\{ \frac{1}{N} \sum_{t=1}^{N} \rho(\lambda' \hat{m}_{N,t}^*(\hat{\theta}_N)) \right\}.$$

Hill and Prokhorov (2016) show that this estimator restores Gaussian asymptotics and the implied probabilities from the tail-trimmed Continuously Updated Estimator elevate weight for usable large values, assign large but not maximum weight to extreme observations, and give the lowest weight to non-leverage points. Among other things, they provide robust versions of Generalized Empirical Likelihood Ratio, Wald, and Lagrange Multiplier tests, and an efficient and heavy tail robust moment estimator with an application to *expected shortfall* estimation for the Russian Ruble — US Dollar exchange rate and for the Hang Seng Index. Their bottom line is that tail-trimmed CUE-GMM dominates other estimators in terms of bias, MSE and approximate normality.

5.7. Concluding remarks

This chapter has discussed aspects of statistical estimation from the perspective of robustness to copula misspecification and heavy tails. We started with definitions of various parametric, semiparametric and nonparametric estimators. We then assumed that we have a likelihood based model, like a logit model, that is correctly specified, but we want to account for non-independence over time. We showed how the standard parametric estimators can be improved and provided conditions under which the improved QMLE (IQMLE) does or does not have a positive efficiency gain. An interesting unanswered question is whether this is the best we can do using only the assumption that the marginal distributions are correct. In other words, what is the semiparametric efficiency bound for estimation of α when the true copula is unknown?

We addressed this question using a nonparametric copula framework. This framework is the converse of the setting considered by Chen *et al.* (2006), who look at efficiency bounds when the copula has a known parametric form but the marginals are unknown. Given that we have correctly specified parametric marginals, we can certainly estimate the copula non-parametrically, and then the question was whether using this estimated copula improves efficiency relative to IQMLE. We gave a positive answer to that question.

We also considered MLE based on a joint distribution. Often the joint distribution will be constructed by choosing a copula. The further assumption (copula) that converts the marginal distributions into a joint distribution raises standard questions of efficiency and robustness. We looked at how to address these questions in a GMM framework. The GMM approach is potentially useful in some practical ways. It leads directly to a way of testing the validity of the copula — a topic addressed in Chapter 6 — and it leads to an improvement over the pseudo-MLE (PMLE) in the case that the copula is misspecified but robust. We also demonstrated that the

GMM approach is useful because it allows us to study questions of efficiency and robustness in a structured way. The MLE offers a positive efficiency gain over the IQMLE if the copula scores are not redundant, and the PMLE or GMM-based IPMLE are consistent if the copula score moment conditions are valid.

We focused on a model in which the marginal distributions are the same, and the same parameters α appear in all of them. In fact, the mathematics of the chapter allows the marginals to be different. How many of these results can be generalized to the case that different marginals have different parameters remains to be seen.

An important remaining question is to provide further characterizations of a copula which is robust. We give a non-trivial example in which an incorrect copula can be robust, but this depends on the nature of the marginal models, the assumed copula and the true copula. The Holy Grail would be to find a *magic copula* that is always robust and sometimes non-redundant, or to prove that a magic copula cannot exist.

The use of nonparametrics may be one way to arrive at such magic copulas. We have discussed how the Bernstein copula is an alternative to parametric copulas, which is robust and efficiency improving. We also looked at the Bayesian approach to estimating multivariate densities where Bernstein mixing kernel provides significant improvements over competition.

Finally, we discussed asymptotically negligible trimming as a way of robustifying parametric estimators. We looked at GEL but this methodology extends to many other estimators and tests of this chapter and of Chapter 6. Specifically, robust versions of copula-based estimators and tests can be based on the tailed-trimmed CMLE score function

$$
\nabla_\theta \ln c(\hat{F}(x_{i1}), \ldots, \hat{F}(x_{id}); \theta) \prod_{j=1}^{d} \mathbb{I}\left\{ x_{j(k_j N_j)}^{(-)} < x_{ij} < x_{j(k_j N_j)}^{(+)} \right\},
$$

$$(5.47)$$

where $\hat{F}(x_{ij}), i = 1, \ldots, N_j, j = 1, \ldots, d$, are the pseudo-observations corresponding to x_{ij}, $\mathbb{I}\{\}$ is the indicator function, and $x_{j(1)}^{(-)} < \ldots <$

$x_{j(N_j)}^{(-)} \leq 0$ and $0 < x_{j(1)}^{(+)} < \ldots < x_{j(N_j)}^{(+)}$ are the order statistics of negative and positive values of X_j, respectively. Then, the robustness properties of the resulting tests and their asymptotic distribution are determined by the asymptotic behaviour of k_{jN_j} — the number of trimmed observations. When $k_{jN_j} \to \infty$ along with N_j, which makes trimming negligible asymptotically, it is possible to restore the Gaussian asymptotics. We leave these extensions for future work.

5.8. Appendix: Proofs

Proof of Theorem 5.1. Let $\mathbb{V}_{\text{QMLE}}$ and $\mathbb{V}_{\text{IQMLE}}$ denote the asymptotic variance matrix of QMLE and IQMLE, respectively. Then,

$$\mathbb{V}_{\text{QMLE}} = [(A\mathbb{H}_*)'(A\mathbb{V}_*A')^{-1}(A\mathbb{H}_*)]^{-1}, \qquad (5.48)$$

while

$$\mathbb{V}_{\text{IQMLE}} = [\mathbb{H}_*'\mathbb{V}_*^{-1}\mathbb{H}_*]^{-1}. \qquad (5.49)$$

But $\mathbb{V}_{\text{QMLE}} - \mathbb{V}_{\text{IQMLE}}$ is positive semi-definite (PSD) if and only if $\mathbb{V}_{\text{IQMLE}}^{-1} - \mathbb{V}_{\text{QMLE}}^{-1} = \mathbb{H}_*'\mathbb{V}_*^{-1}\mathbb{H}_* - \mathbb{H}_*'A'(A\mathbb{V}_*A')^{-1}A\mathbb{H}_*$ is PSD. The last expression can be rewritten as $\mathbb{H}_*'\mathbb{V}_*^{-1/2}[I - \mathbb{V}_*^{1/2}A'(A\mathbb{V}_*^{1/2} \mathbb{V}_*^{1/2}A')^{-1}A\mathbb{V}_*^{1/2}]\mathbb{V}_*^{-1/2}\mathbb{H}_*$. This is PSD because the matrix in brackets is the PSD projection matrix orthogonal to $\mathbb{V}_*^{1/2}A'$. The expression is zero (the two estimators are equally efficient) if and only if $\mathbb{V}_*^{-1/2}\mathbb{H}_*$ is in the space spanned by $\mathbb{V}_*^{1/2}A'$, or, equivalently, $\mathbb{H}_*$ is in the space spanned by $\mathbb{V}_*A'$.

Proof of Theorem 5.2. (a) If x_{ij} is independent over i and j, then $\mathbb{V}_* = I_d \otimes V_o$, where $V_o = \mathbb{E}s_{it}(\theta_o)s_{it}(\theta_o)'$, and $\mathbb{H}_* = 1_d \otimes H_o$, where $H_o = \mathbb{E}\nabla_\theta' s_{it}(\theta_o)$. We have $\mathbb{V}_*A' = (I_d \otimes V_o)(1_d \otimes I_p) = 1_d \otimes V_o$. Now, H_o is in the space spanned by V_o since V_o is nonsingular, and so $\mathbb{H}_*$ is in the space spanned by $\mathbb{V}_*A'$. Note that we did not need the "correct distribution" assumption that $H_o = -V_o$.

(b) If scores are equicorrelated, $\mathbb{H}_* = 1_d \otimes H_o$ and $\mathbb{V}_*A' = (1 + \rho(d-1))1_d \otimes V_o$. So $\mathbb{H}_*$ is in the space spanned by $\mathbb{V}_*A'$ so long as V_o is nonsingular. Of course, ρ should be greater than $-\frac{1}{d-1}$ for $\mathbb{V}_{\text{QMLE}}$

(asymptotic variance of the QMLE estimator) to be positive definite. Again, note that no "correct distribution" assumption is used.

Proof of Theorem 5.1. By the information matrix equality (IME),

$$\mathbf{A} \equiv \mathbb{E}\left\{\nabla_\theta \ln f_1(Y_1;\theta)\nabla_\theta' \ln f_1(Y_1;\theta)\right\} = -\mathbb{E}\nabla_\theta^2 \ln f_1(Y_1;\theta). \quad (5.50)$$

Similar for $\mathbf{B}, \mathbf{F}$.

By the generalized IME (GIME — see, e.g., Tauchen (1985)),

$$\mathbf{E} \equiv \mathbb{E}\left\{\nabla_\theta \ln c(F_1(Y_1;\theta), F_2(Y_2;\theta); \rho)\nabla_\rho' \ln c(F_1(Y_1;\theta), F_2(Y_2;\theta); \rho)\right\}$$
$$= -\mathbb{E}\nabla_{\theta\rho}^2 \ln c(F_1(Y_1;\theta), F_2(Y_2;\theta); \rho) \quad (5.51)$$

and, for $i = 1, 2$,

$$\mathbb{E}\left\{\nabla_\theta \ln f_i(Y_i;\theta)\nabla_\rho' \ln c(F_1(Y_1;\theta), F_1(Y_2;\theta); \rho)\right\}$$
$$= -\mathbb{E}\nabla_{\theta\rho}^2 \ln f_i(Y_i;\theta) = \mathbf{0}.$$

Also by GIME and (5.19),

$$\mathbb{E}\left\{\nabla_\theta \ln f_i(Y_i;\theta)\nabla_\theta' \left[\ln f_1(Y_1;\theta) + \ln f_2(Y_2;\theta) + \ln c(.,.;\rho)\right]\right\}$$
$$= -\mathbb{E}\nabla_\theta^2 \ln f_i(Y_i;\theta)$$

for $i = 1, 2$, which, along with (5.50), implies that

$$\mathbf{G} \equiv \mathbb{E}\left\{\nabla_\theta \ln f_1(Y_1;\theta)\nabla_\theta' \ln f_2(Y_2;\theta)\right\}$$
$$= -\mathbb{E}\left\{\nabla_\theta \ln f_1(Y_1;\theta)\nabla_\theta' \ln c(F_1(Y_1;\theta), F_1(Y_2;\theta); \rho)\right\}$$

and

$$\mathbb{E}\left\{\nabla_\theta \ln f_2(Y_2;\theta)\nabla_\theta' \ln f_1(Y_1;\theta)\right\}$$
$$= -\mathbb{E}\left\{\nabla_\theta \ln f_2(Y_2;\theta)\nabla_\theta' \ln c(F_1(Y_1;\theta), F_1(Y_2;\theta); \rho)\right\} = \mathbf{G}'. \quad (5.52)$$

Finally, by GIME and (5.19),

$$\mathbb{E}\{\nabla_\theta \ln c(F_1(Y_1;\theta), F_1(Y_2;\theta); \rho)$$
$$\times \nabla_\theta'[\ln f_1(Y_1;\theta) + \ln f_2(Y_2;\theta) + \ln c(F_1(Y_1;\theta), F_1(Y_2;\theta); \rho)]\}$$
$$= -\mathbb{E}\nabla_\theta^2 \ln c(F_1(Y_1;\theta), F_1(Y_2;\theta); \rho).$$

With **G** as defined above and

$$\mathbf{J} \equiv \mathbb{E}\{\nabla_\theta \ln c(F_1(Y_1;\theta), F_1(Y_2;\theta); \rho) \nabla_\theta' \ln c(F_1(Y_1;\theta), F_1(Y_2;\theta); \rho)\},$$

this implies that

$$\mathbb{E}\nabla_\theta^2 \ln c(F_1(Y_1;\theta), F_1(Y_2;\theta); \rho) = \mathbf{G} + \mathbf{G}' - \mathbf{J}.$$

Proof of Theorem 5.3. From the discussion in the main text,

$$\mathbb{V}_{\mathrm{MLE}} = \begin{bmatrix} \mathbf{A}+\mathbf{B}+\mathbf{J}-\mathbf{G}-\mathbf{G}' & \mathbf{E} \\ \mathbf{E}' & \mathbf{F} \end{bmatrix}^{-1},$$

$$\mathbb{V}_{\mathrm{IQMLE}} = \left([-\mathbf{A}\ -\mathbf{B}] \begin{bmatrix} \mathbf{A} & \mathbf{G} \\ \mathbf{G}' & \mathbf{B} \end{bmatrix}^{-1} \begin{bmatrix} -\mathbf{A} \\ -\mathbf{B} \end{bmatrix} \right)^{-1}. \tag{5.53}$$

Using partitioned inverse formulas, the upper left $p \times p$ block of $\mathbb{V}_{\mathrm{MLE}}$ can be written as $\mathbf{\Sigma}^{-1}$, where $\mathbf{\Sigma} = \mathbf{A} + \mathbf{B} + \mathbf{J} - \mathbf{G} - \mathbf{G}' - \mathbf{E}\mathbf{F}^{-1}\mathbf{E}'$. Also,

$$\mathbb{V}_{\mathrm{IQMLE}}^{-1} = \left([-\mathbf{G}'\ -\mathbf{G}] + [\mathbb{I}\ \ \mathbb{I}] \begin{bmatrix} \mathbf{A} & \mathbf{G} \\ \mathbf{G}' & \mathbf{B} \end{bmatrix} \right) \begin{bmatrix} \mathbf{A} & \mathbf{G} \\ \mathbf{G}' & \mathbf{B} \end{bmatrix}^{-1}$$

$$\times \left(\begin{bmatrix} \mathbf{A} & \mathbf{G} \\ \mathbf{G}' & \mathbf{B} \end{bmatrix} \begin{bmatrix} \mathbb{I} \\ \mathbb{I} \end{bmatrix} + \begin{bmatrix} -\mathbf{G} \\ -\mathbf{G}' \end{bmatrix} \right) \tag{5.54}$$

$$= [-\mathbf{G}'\ -\mathbf{G}] \begin{bmatrix} \mathbf{A} & \mathbf{G} \\ \mathbf{G}' & \mathbf{B} \end{bmatrix}^{-1} \begin{bmatrix} -\mathbf{G} \\ -\mathbf{G}' \end{bmatrix}$$

$$-\mathbf{G}' - \mathbf{G} + \mathbf{A} + \mathbf{B}. \tag{5.55}$$

Thus, $\mathbb{V}_{\mathrm{IQMLE}}^{-1} = \mathbf{\Sigma}$ if and only if

$$\mathbf{J} - \mathbf{E}\mathbf{F}^{-1}\mathbf{E}' = [-\mathbf{G}'\ -\mathbf{G}] \begin{bmatrix} \mathbf{A} & \mathbf{G} \\ \mathbf{G}' & \mathbf{B} \end{bmatrix}^{-1} \begin{bmatrix} -\mathbf{G} \\ -\mathbf{G}' \end{bmatrix}.$$

Proof of Corollary 5.1.

(1) If (C) is a linear combination of (A) and (B), then covariances between moment functions in (D) and (C) are linear combinations of covariances between (D) and (A–B), which are all zero by Lemma 5.1.

(2) Rewrite $\mathbf{J} - \mathbf{V}_{21}^{\theta}\mathbf{V}_{11}^{-1}\mathbf{V}_{12}^{\theta}$ as

$$\mathbb{E}\left\{\left(\nabla_\theta \ln c - \mathbf{V}_{21}^{\theta}\mathbf{V}_{11}^{-1}\begin{bmatrix}\nabla_\theta \ln f_1\\\nabla_\theta \ln f_2\end{bmatrix}\right)\nabla_\theta' \ln c\right\}.$$

This is identically zero because, due to linearity of (C) in (A–B),

$$\nabla_\theta \ln c - \mathbf{V}_{21}^{\theta}\mathbf{V}_{11}^{-1}\begin{bmatrix}\nabla_\theta \ln f_1\\\nabla_\theta \ln f_2\end{bmatrix} = \mathbf{0}.$$

(3) By Theorem 5.3.

Proof of Theorem 5.4. We show that $\mathbb{E}\nabla_\theta \ln k(F_1(\mu_1+Y_1), F_2(\mu_2+Y_2); \rho) = 0$, where $\theta = (\mu_1, \mu_2)'$, holds for any RS copula density k.

By the chain rule, $\nabla_\theta \ln k(F_1(\mu_1 + Y_1), F_2(\mu_2 + Y_2); \rho)$ contains terms of the form

$$\frac{1}{k(F_1(\mu_1 + Y_1), F_2(\mu_2 + Y_2); \rho)}$$
$$\times \frac{\partial k(F_1(\mu_1 + Y_1), F_2(\mu_2 + Y_2); \rho)}{\partial F_i(\mu_i + Y_i)} \times f_i(\mu_i + Y_i), \quad (5.56)$$

$i = 1, 2$.

Due to MS of (Y_1, Y_2) and RS of K, $f_i(\mu_i + Y_i) = f_i(\mu_i - Y_i)$ and $k(F_1(\mu_1 + Y_1), F_2(\mu_2 + Y_2)) = k(1 - F_1(\mu_1 + Y_1), 1 - F_2(\mu_2 + Y_2)) = k(F_1(\mu_1 - Y_1), F_2(\mu_2 - Y_2))$. So the first term in (5.56) is the same whether evaluated at (Y_1, Y_2) or $(-Y_1, -Y_2)$. Similarly, the last term is the same whether evaluated at Y_i or $-Y_i$.

Furthermore,

$$\frac{\partial k(F_1(\mu_1 + Y_1), F_2(\mu_2 + Y_2); \rho)}{\partial F_i(\mu_i + Y_i)}$$

$$= \frac{\partial k(1 - F_1(\mu_1 + Y_1), 1 - F_2(\mu_2 + Y_2); \rho)}{\partial(1 - F_i(\mu_i - Y_i))}$$

$$= -\frac{\partial k(F_1(\mu_1 - Y_1), F_2(\mu_2 - Y_2); \rho)}{\partial F_i(\mu_i - Y_i)}.$$

Thus, $\nabla_\theta \ln k(F_1(\mu_1+Y_1), F_2(\mu_2+Y_2); \rho) = -\nabla_\theta \ln k(F_1(\mu_1-Y_1), F_2(\mu_2 - Y_2); \rho)$.

Denote $g(Y_1, Y_2) \equiv \nabla_\theta \ln k(F_1(\mu_1 + Y_1), F_2(\mu_2 + Y_2); \rho) \cdot h(\mu_1 + Y_1, \mu_2 + Y_2)$. From the above, it follows with RS of (Y_1, Y_2) that $g(-Y_1, -Y_2) = -g(Y_1, Y_2)$.

We thus have

$$\mathbb{E}\nabla_\theta \ln k(F_1(\mu_1 + Y_1), F_2(\mu_2 + Y_2); \rho) = \int_{-\infty}^{\infty} \int_{-\infty}^{\infty} g(Y_1, Y_2) dY_1 dY_2$$

$$= 0.$$

Proof of Lemma 5.2. By construction, blocks $\mathbf{A}, \mathbf{B}, \mathbf{G}$ of matrices $\mathbf{V}$ and $\mathbf{D}$ are the same as in Lemma 5.1. However, GIME does not apply now. Denote the misspecified copula density by k.

$$\mathbf{Z} \equiv \mathbb{E}\{\nabla_\theta \ln k(F_1(Y_1; \theta), F_2(Y_2; \theta); \rho)$$

$$\times \nabla'_\rho \ln k(F_1(Y_1; \theta), F_2(Y_2; \theta); \rho)\}$$

$$\neq -\mathbb{E}\left\{\nabla^2_{\theta\rho} \ln k(F_1(Y_1; \theta), F_2(Y_2; \theta); \rho)\right\} \equiv \mathbf{S}.$$

$$-\mathbf{P} \equiv \mathbb{E}\left\{\nabla_\theta \ln f_1(Y_1; \theta)\nabla'_\rho \ln k(F_1(Y_1; \theta), F_1(Y_2; \theta); \rho)\right\}$$

$$\neq -\mathbb{E}\nabla^2_{\theta\rho} \ln f_1(Y_1; \theta) = \mathbf{0}$$

and

$$-\mathbf{Q}' \equiv \mathbb{E}\left\{\nabla_\theta \ln f_2(Y_2; \theta)\nabla'_\rho \ln k(F_1(Y_1; \theta), F_1(Y_2; \theta); \rho)\right\}$$

$$\neq -\mathbb{E}\nabla^2_{\theta\rho} \ln f_2(Y_2; \theta) = \mathbf{0}.$$

$$\mathbf{G} \equiv \mathbb{E}\left\{\nabla_\theta \ln f_1(Y_1; \theta)\nabla'_\theta \ln f_2(Y_2; \theta)\right\}$$

$$\neq -\mathbb{E}\left\{\nabla_\theta \ln f_1(Y_1; \theta)\nabla'_\theta \ln k(F_1(Y_1; \theta), F_1(Y_2; \theta); \rho)\right\} \equiv \mathbf{K}$$

and

$$\mathbb{E}\left\{\nabla_\theta \ln f_2(Y_2; \theta)\nabla'_\theta \ln f_1(Y_1; \theta)\right\}$$

$$\neq -\mathbb{E}\left\{\nabla_\theta \ln f_2(Y_2; \theta)\nabla'_\theta \ln k(F_1(Y_1; \theta), F_1(Y_2; \theta); \rho)\right\} \equiv \mathbf{L}'.$$

However, by GIME and (5.19),

$$\mathbb{E}\nabla^2_\theta \ln k(F_1(Y_1; \theta), F_1(Y_2; \theta); \rho)$$

$$= -\mathbb{E}\{\nabla_\theta \ln k(F_1(Y_1; \theta), F_1(Y_2; \theta); \rho)$$

$$\times [\nabla'_\theta \ln f_1(Y_1;\theta) + \nabla'_\theta \ln f_2(Y_2;\theta)$$
$$+ \nabla'_\theta \ln c(F_1(Y_1;\theta), F_1(Y_2;\theta);\rho)]\}$$
$$\equiv \mathbf{K}' + \mathbf{L} - \mathbf{M}, \tag{5.57}$$

and

$$\mathbb{E}\nabla^2_{\rho\theta} \ln k(F_1(Y_1;\theta), F_1(Y_2;\theta);\rho)$$
$$= -\mathbb{E}\{\nabla_\rho \ln k(F_1(Y_1;\theta), F_1(Y_2;\theta);\rho)$$
$$\times [\nabla'_\theta \ln f_1(Y_1;\theta) + \nabla'_\theta \ln f_2(Y_2;\theta)$$
$$+ \nabla'_\theta \ln c(F_1(Y_1;\theta), F_1(Y_2;\theta);\rho)]\}$$
$$\equiv \mathbf{P}' + \mathbf{Q} - \mathbf{R}, \tag{5.58}$$
$$-\mathbf{T} \equiv \mathbb{E}\nabla^2_\rho \ln k(F_1(Y_1;\theta), F_1(Y_2;\theta);\rho)$$
$$= -\mathbb{E}\{\nabla_\rho \ln k(F_1(Y_1;\theta), F_1(Y_2;\theta);\rho)$$
$$\nabla'_\rho \ln c(F_1(Y_1;\theta), F_1(Y_2;\theta);\rho)\},$$

and

$$-\mathbf{S} \equiv \mathbb{E}\nabla^2_{\theta\rho} \ln k(F_1(Y_1;\theta), F_1(Y_2;\theta);\rho)$$
$$= -\mathbb{E}\{\nabla_\theta \ln k(F_1(Y_1;\theta), F_1(Y_2;\theta);\rho)$$
$$\nabla'_\rho \ln c(F_1(Y_1;\theta), F_1(Y_2;\theta);\rho)\}.$$

Also,

$$\mathbf{N} \equiv \mathbb{E}\{\nabla_\theta \ln k(F_1(Y_1;\theta), F_1(Y_2;\theta);\rho)$$
$$\nabla'_\theta \ln k(F_1(Y_1;\theta), F_1(Y_2;\theta);\rho)\} \neq \mathbf{M}$$

and

$$\mathbf{W} \equiv \mathbb{E}\{\nabla_\rho \ln k(F_1(Y_1;\theta), F_1(Y_2;\theta);\rho)$$
$$\nabla'_\rho \ln k(F_1(Y_1;\theta), F_1(Y_2;\theta);\rho)\} \neq \mathbf{T}.$$

Finally, by the well-known algebraic property of cross-partial derivatives,

$$\mathbf{S} = -\mathbf{P} - \mathbf{Q}' + \mathbf{R}'.$$

Proof of Theorem 5.6. By Theorem 8(C) of Breusch *et al.* (1999), the copula scores are redundant for θ given the marginal scores if and only if

$$\begin{bmatrix} \mathbf{K}' + \mathbf{L} - \mathbf{M} \\ \mathbf{P}' + \mathbf{Q} - \mathbf{R} \end{bmatrix} - \begin{bmatrix} -\mathbf{K}' & -\mathbf{L} \\ -\mathbf{P}' & -\mathbf{Q} \end{bmatrix} \mathbf{V}_{11}^{-1} \begin{bmatrix} -\mathbf{A} \\ -\mathbf{B} \end{bmatrix} = \begin{bmatrix} -\mathbf{S} \\ -\mathbf{T} \end{bmatrix} \mathbb{B},$$

for some matrix $\mathbb{B} : q \times p$.

This is equivalent to

$$-\mathbf{M} - [-\mathbf{K}' \quad -\mathbf{L}]\mathbf{V}_{11}^{-1}\begin{bmatrix} \mathbf{G} \\ \mathbf{G}' \end{bmatrix} = -\mathbf{S}\mathbb{B},$$

$$-\mathbf{R} - [-\mathbf{P}' \quad -\mathbf{Q}]\mathbf{V}_{11}^{-1}\begin{bmatrix} \mathbf{G} \\ \mathbf{G}' \end{bmatrix} = -\mathbf{T}\mathbb{B}.$$

$\mathbf{T}$ is symmetric and invertible, so we can substitute $\mathbb{B}$ from the latter equation into the former to obtain

$$\mathbf{M} - [-\mathbf{K}' \quad -\mathbf{L}]\mathbf{V}_{11}^{-1}\mathbf{V}_{12}^{\theta} = \mathbf{S}\mathbf{T}^{-1}(\mathbf{R} - [-\mathbf{P}' \quad -\mathbf{Q}]\mathbf{V}_{11}^{-1}\mathbf{V}_{12}^{\theta}),$$

which completes the proof.

Proof of Corollary 5.2.

(1) By (5.34), $\mathbf{M} - \mathbf{V}_{21}^{\theta k}\mathbf{V}_{11}^{-1}\mathbf{V}_{12}^{\theta}$ is identically zero under linearity of (C) in (A–B).
(2) As in 1.
(3) By Theorem 5.6.

Chapter 6

Copula Tests
Using Information Matrix

This chapter presents an application of copula robustness results discussed in Chapter 5. Specifically, we discuss tests of copula validity based on the information matrix equality. We consider tests that assume a parametric form for the marginals and those that use empirical marginal distributions.

6.1. Introduction

Consider two continuous random variables X_1 and X_2 with cdf's $F_1(x_1)$ and $F_2(x_2)$ and pdf's $f_1(x_1)$ and $f_2(x_2)$, respectively. Suppose the joint cdf of (X_1, X_2) is $H(x_1, x_2)$ and the joint pdf is $h(x_1, x_2)$. As defined in Section 1.2.1, a copula is a function $C(u, v)$ such that $H(x_1, x_2) = C(F_1(x_1), F_2(x_2))$ or, in densities if they exist, $h(x_1, x_2) = c(F_1(x_1), F_2(x_2)) f_1(x_1) f_2(x_2)$. The marginal densities f_1 and f_2 are now "extracted" from the joint density and the copula density c captures the dependence between X_1 and X_2. Proposition 1.1 showed that given H, F_1 and F_2 of continuous variables, there exists a unique C. So, given F_1 and F_2, the choice when constructing a joint distribution is which copula C to use.

Therefore, we are often interested in testing the validity of a copula. As discussed in Section 5.4.4, a copula is valid if it is correctly specified, but also if it is misspecified but the MLE based on it is robust. A robust copula has a score which has mean zero. So we start this chapter with a conditional moment test of whether the

229

copula score moment conditions hold, assuming that the moment conditions based on the marginal scores do hold — this is the framework considered in Section 5.4.

Let C_θ denote the chosen copula family with dependence parameter(s) θ. Numerous papers have used different copula families in applications from finance (e.g., Patton (2006); Breymann et al. (2003)), from risk management (e.g., Embrechts et al. (2002, 2003)) and from health and labor economics (Smith (2003); Cameron et al. (2004)). Theoretical results on parametric and semiparametric estimation of copula-based models are discussed by Genest et al. (1995a); Joe (2005); Chen and Fan (2006b); Prokhorov and Schmidt (2009); among others. We discussed many such results in Chapter 5. But the issue of copula specification testing — clearly relevant in any copula-based application — has not received as much attention in the literature as the estimation problem.

This issue can be viewed as an application of the robustness results of Chapter 5 because a robust copula as defined in Section 5.4.4 is a copula whose score is zero mean even when it is not necessarily correctly specified. On the other hand, a correct copula is a copula for which the information matrix equality has to hold, which is a testable assumption.

We consider copula robustness tests in Section 6.2 and copula goodness-of-fit tests in Section 6.3. Throughout, we follow the definitions of Section 5.4.4 and use the definition that a copula family is correctly specified if, for some θ_o, $C_{\theta_o}(F_1, F_2) = H$ and a copula family is robust if its score is zero mean even if it is not correctly specified (see Section 5.4.4 for examples).

Generally, we prefer that copula tests do not involve parametric specification of the marginal distributions because if they do then they essentially test a joint hypothesis of correct copula *and* marginal specifications. It is also desirable that copula test be applicable to any copula family without requiring any strategic choices and arbitrary parameters, e.g., the choice of a kernel and a bandwidth. Genest et al. (2009) call tests that have these desirable properties "blanket" goodness-of-fit tests.

There exist a number of copula tests (see Genest et al. (2009); Berg (2009), for recent surveys). However, only a few are "blanket."

For example, Klugman and Parsa (1999) propose tests that involve ad hoc categorization of the data; Fermanian (2005) and Scaillet (2007) propose tests that are based on kernels, weight functions and use the associated smoothing parameters; Panchenko (2005) proposes a test based on a V-statistic, whose asymptotic distribution is unknown and depends on the choice of bandwidth.

Prokhorov and Schmidt (2009) propose a conditional moment test for whether the copula-based score function has zero mean, which depends on parametric marginals and does not distinguish between the correct copula and any other copula that has a zero mean score function, i.e., is robust. Assuming the marginals are correctly specified, this is a test of copula validity rather than of copula correctness. Because it involves the assumption of correctly specified marginals, this test along with all other tests mentioned in the previous paragraph do not qualify as "blanket." This is the copula validity test we cover in Section 6.2.

Genest *et al.* (2009) report five testing procedures that qualify as "blanket" tests. These tests are based on the empirical copula and on Kendall's and Rosenblat's probability integral transformation of the data as in, e.g., Dobric and Schmid (2007); Breymann *et al.* (2003); Genest *et al.* (2006); Genest and Remillard (2008). Recently, Mesfioui *et al.* (2009) propose one more "blanket" test based on a sample equivalent of Spearman's dependence function. All of these tests are quite difficult computationally. They are not asymptotically pivotal and require a procedure such as parametric bootstrap to obtain approximate p-values.

The "blanket" tests of copula correctness we consider in Section 6.3 were proposed by Huang and Prokhorov (2014) and generalized by Prokhorov *et al.* (2015). They are based on the information matrix equality which equates the copula Hessian and the outer-product of a copula score. In essence, this is the White (1982) specification test adapted to the first-step nonparametric estimation of the marginal distributions.

One difficulty with these tests is that the first stage affects the asymptotic variance of the estimated Hessian and estimated outer-product in a nontrivial way. In Section 6.3.2, we show that the test statistic asymptotically has a χ^2 distribution, and in the Appendix we provide the necessary adjustments for the first-stage rank-based estimation.

Section 6.3.1 sets the stage by discussing the connection between copulas and the information matrix equality. This discussion overlaps with Section 5.4.4 and therefore is brief. Section 6.3.3 gives a summary of the generalizations proposed by Prokhorov *et al.* (2015). In Section 6.3.4, we report selected results of a power study. Section 6.4 concludes.

6.2. Tests of copula robustness

Prokhorov and Schmidt (2009) consider the situation where we are willing to assert the correctness of the marginal distributions but we are doubtful about the correctness of the joint distribution. Recall the moment equations in (5.20). The assertion that the marginals are correct means we are willing to assert that the marginal moment conditions (A) and (B) in (5.20) hold. We are doubtful about the validity of the copula moment conditions (C) and (D). We wish to test the validity of the copula by testing the validity of the copula moment conditions. We will discuss two ways of doing this. It is worth noting that in either case we are testing the *validity* of the copula as opposed to the *correctness* of the copula. The tests we discuss would not distinguish the case of a true copula from the case of an incorrect but robust copula.

6.2.1. *Test of overidentifying restrictions*

One test that can be used is the usual test of overidentifying restrictions. The Generalized Method of Moments (GMM) problem (5.20) is overidentified since we have $3p + q$ moment conditions (where p is the dimension of α and q is the dimension of θ) and $p + q$ parameters. Therefore, we can test the validity of the full set of moment conditions by the test of overidentifying restrictions (see, e.g., Hansen (1982); Newey and West (1987a)). If we assert that the marginal moment conditions are correct, then this is a test of the validity of the copula moment conditions.

To be explicit, we will need more notation. For $m = 1, 2$ and $i = 1, \ldots, N$, denote $f_{mi}(\alpha) = f_m(y_{mi}; \alpha)$, $c_i(\alpha, \theta) = c(F_1(y_{1i}; \alpha), F_2$

$(y_{2i}; \alpha); \theta)$,

$$\psi_i(\alpha, \theta) = \begin{bmatrix} \nabla_\alpha \ln f_{1i}(\alpha) \\ \nabla_\alpha \ln f_{2i}(\alpha) \\ \nabla_\alpha \ln c_i(\alpha, \theta) \\ \nabla_\theta \ln c_i(\alpha, \theta) \end{bmatrix}, \quad g_i(\alpha) = \begin{bmatrix} \nabla_\alpha \ln f_{1i}(\alpha) \\ \nabla_\alpha \ln f_{2i}(\alpha) \end{bmatrix},$$

$$r_i(\alpha, \theta) = \begin{bmatrix} \nabla_\alpha \ln c_i(\alpha, \theta) \\ \nabla_\theta \ln c_i(\alpha, \theta) \end{bmatrix}.$$

Note that ψ_i is a (3p+q)-vector. Let

$$\bar{\psi}(\alpha, \theta) \equiv \frac{1}{N} \sum_{i=1}^{N} \psi_i(\alpha, \theta), \quad \bar{g}(\alpha) \equiv \frac{1}{N} \sum_{i=1}^{N} g_i(\alpha),$$

$$\bar{r}(\alpha, \theta) \equiv \frac{1}{N} \sum_{i=1}^{N} r_i(\alpha, \theta).$$

Following our previous notation, let

$$\mathbf{V_o} \equiv \mathbb{E}\psi(\alpha_o, \theta_o)\psi(\alpha_o, \theta_o)',$$

$$\mathbf{V_{11}^o} \equiv \mathbb{E}g(\alpha_o)g(\alpha_o)',$$

$$\mathbf{V_{22}^o} \equiv \mathbb{E}r(\alpha_o, \theta_o)r(\alpha_o, \theta_o)',$$

$$\mathbf{V_{12}^o} = \mathbf{V_{21}^o}{}' \equiv \mathbb{E}g(\alpha_o)r(\alpha_o, \theta_o)',$$

$$\mathbf{D_o} \equiv \mathbb{E}\nabla_{(\alpha', \theta')'}\psi(\alpha_o, \theta_o),$$

$$\mathbf{D_{11}^o} \equiv \mathbb{E}\nabla_\alpha g(\alpha_o),$$

$$\mathbf{D_{21}^o} \equiv \mathbb{E}\nabla_\alpha r(\alpha_o, \theta_o),$$

$$\mathbf{D_{22}^o} \equiv \mathbb{E}\nabla_\theta r(\alpha_o, \theta_o),$$

where "∇." denotes derivatives and expectations are with respect to the true joint density $h(x_1, x_2)$.

Theorem 6.1. *Let* $(\breve{\alpha}, \breve{\theta})$ *denote the optimal GMM estimate of* (α, θ) *based on (5.20). Then*

$$N\bar{\psi}(\breve{\alpha}, \breve{\theta})'\mathbf{V_o^{-1}}\bar{\psi}(\breve{\alpha}, \breve{\theta}) \overset{a}{\sim} \chi_{2p}^2. \tag{6.1}$$

This test is a specification test which, given that the marginal distributions are correct, should capture copula misspecification (or, more precisely, invalidity of the copula). A consistent estimator of $\mathbf{V_o}$ such as

$$\breve{\mathbf{V}}_\mathbf{o} = \frac{1}{N} \sum_{i=1}^{N} \psi_i(\breve{\alpha}, \breve{\theta}) \psi_i(\breve{\alpha}, \breve{\theta})'$$

is usually used in (6.1). It is however important to note that the statistic in (6.1) can be used only if $\mathbf{V}$ is non-singular, i.e., if copula terms are not redundant as defined in Section 5.4.3.

At an intuitive level, this test is unappealing because it does not focus strongly on the moment conditions we are doubtful about. Also one could object that the number of degrees of freedom does not seem right. We maintain the correctness of the marginal moment conditions. Then the copula scores add $p+q$ moment conditions, but also add q nuisance parameters. So the number of restrictions to test, and therefore the "right" number of degrees of freedom, should be p, not $2p$. We achieve this with the following two-step procedure.

6.2.2. *Two step test*

Theorem 6.2. *Let $\hat{\alpha}$ be the optimal GMM estimate based on the marginal moment conditions $\mathbb{E}g(\alpha) = 0$. Let $\hat{\theta}$ be obtained by minimizing $\bar{r}(\hat{\alpha}, \theta)' \mathbf{B}_\mathbf{o}^{-1} \bar{r}(\hat{\alpha}, \theta)$, where*

$$\mathbf{B_o} = \mathbf{V_{22}^o} - \mathbf{D_{21}^o}(\mathbf{D_{11}^o}\mathbf{V_{11}^o}^{-1}\mathbf{D_{11}^o})^{-1}\mathbf{D_{11}^o}'\mathbf{V_{11}^o}^{-1}\mathbf{V_{12}^o}$$
$$- \mathbf{V_{21}^o}\mathbf{V_{11}^o}^{-1}\mathbf{D_{11}^o}(\mathbf{D_{11}^o}\mathbf{V_{11}^o}^{-1}\mathbf{D_{11}^o})^{-1}\mathbf{D_{21}^o}'$$
$$+ \mathbf{D_{21}^o}(\mathbf{D_{11}^o}\mathbf{V_{11}^o}^{-1}\mathbf{D_{11}^o})^{-1}\mathbf{D_{21}^o}'.$$

Then,

$$N\bar{r}(\hat{\alpha}, \hat{\theta})' \mathbf{B_o}^{-1} \bar{r}(\hat{\alpha}, \hat{\theta}) \stackrel{a}{\sim} \chi_p^2. \tag{6.2}$$

Similarly to Theorem 6.1, consistent estimates of the elements of $\mathbf{V_o}$ and $\mathbf{D_o}$ will be used in practice for calculating the test statistic in (6.2).

Essentially the above test is a conditional moment test of the type discussed by Tauchen (1985) and Wooldridge (1991). Formally, it is an extension of those tests because it needs to accommodate the presence of nuisance parameters that appear in the moments that are being tested and not in the moments that are maintained. That extension may be useful in other contexts.

6.3. Tests of copula correctness

The tests we discussed above do not distinguish between a true copula and an incorrect but robust copula. As mentioned in Prokhorov and Schmidt (2009), this should be possible however because the truth of the copula imposes a number of "information equalities" that imply restrictions on the matrices $\mathbf{D}$ and $\mathbf{V}$ given in Equations (5.21)–(5.22), and those restrictions are testable.

In order to have a "blanket" test, Huang and Prokhorov (2014) do not assume a parametric form of the marginals but use empirical marginals instead.

6.3.1. *Copulas and information matrix equivalence*

Consider an d-dimensional copula $C(u_1, \ldots, u_d)$ and d univariate marginals $F_j(x_j)$, $j = 1, \ldots, d$. Then, by Sklar's theorem, the joint distribution of $(X_1, \ldots, X_d)$ is given by

$$H(x_1, \ldots, x_d) = C(F_1(x_1), \ldots, F_d(x_d)).$$

Assume F_j is continuous, $j = 1, \ldots, d$, so $C(u_1, \ldots, u_j)$ is unique. Assume further that the copula density exists, then the joint density of $(X_1, \ldots, X_d)$ is

$$h(x_1, \ldots, x_d) = \left. \frac{\partial^d C(u_1, \ldots, u_d)}{\partial u_1 \ldots \partial u_d} \right|_{u_j = F_j(x_j), j=1, \ldots, d} \prod_{j=1}^d f_j(x_j)$$

$$= c(F_1(x_1), \ldots, F_d(x_d)) \prod_{j=1}^d f_j(x_j),$$

where $c(u_1, \ldots, u_d)$ is the copula density.

We are interested in goodness-of-fit testing of parametric copula families, so our copulas are parametric. For example, the d-variate Gaussian copula with $\frac{d(d-1)}{2}$ parameters defined in Equation (1.4) can be written as follows

$$\Phi_d(\Phi^{-1}(u_1), \ldots, \Phi^{-1}(u_d); R),$$

where Φ_d is the joint distribution function of d standard normal covariates with a given correlation matrix R and Φ^{-1} is the inverse of the standard normal cdf. For Gaussian copulas, the copula parameters are the distinct elements of R.

Let subscript θ denote the dependence parameter vector of a copula function and let p denote its dimension. It is well known that if there exists a value θ_o such that $H(x_1, \ldots, x_d) = C_{\theta_o}(F_1(x_1), \ldots, F_d(x_d))$ then we have a correctly specified likelihood model and, under regularity conditions, the MLE is consistent for θ_o. Moreover, in this case White's (1982) information matrix equivalence theorem holds: the Fisher information matrix can be equivalently calculated as minus the expected Hessian or as the expected outer product of the score function.

· We wish to apply the information matrix equivalence theorem to copulas. Assume that the copula-based likelihood is three times continuously differentiable and the relevant expectations exist. Differentiability three times is required since, aside from the Hessian used in calculating the statistics, there is also an asymptotic variance expression involving the third derivative of the log-copula density. Let $\mathbb{H}(\theta)$ denote the expected Hessian matrix of $\ln c_\theta$ and let $\mathbb{C}(\theta)$ denote the expected outer product of the corresponding score function [not to confuse with copula C]. Then,

$$\mathbb{H}(\theta) = \mathbb{E}\nabla_\theta^2 \ln c_\theta(F_1(x_1), \ldots, F_d(x_d))$$
$$\mathbb{C}(\theta) = \mathbb{E}\nabla_\theta \ln c_\theta(F_1(x_1), \ldots, F_d(x_d))\nabla_\theta' \ln c_\theta(F_1(x_1), \ldots, F_d(x_d)),$$

where "∇_θ" denotes derivatives with respect to θ and expectations are with respect to the true distribution H.

White's (1982) information matrix equivalence theorem essentially says that, under correct copula specification,

$$-\mathbb{H}(\theta_o) = \mathbb{C}(\theta_o).$$

The copula misspecification tests of Huang and Prokhorov (2014) use this equality. Specifically, they test

$$\mathcal{H}_0 : \mathbb{H}(\theta_o) + \mathbb{C}(\theta_o) = 0 \text{ against } \mathcal{H}_1 : \mathbb{H}(\theta_o) + \mathbb{C}(\theta_o) \neq 0. \quad (6.3)$$

6.3.2. Information matrix test

In practice, θ_o is not observed. Moreover, the matrices $\mathbb{H}(\theta)$ and $\mathbb{C}(\theta)$ contain the marginals F_j which are usually unknown. However, these quantities are easily estimated. In particular, it is common to use the empirical distribution function $\hat{F}_j$ in place of F_j, a consistent estimate $\hat{\theta}$ in place of θ_o, the sample averages $\bar{\mathbb{H}}$ and $\bar{\mathbb{C}}$ in place of the expectations $\mathbb{H}$ and $\mathbb{C}$.

Given N observations, the empirical distribution function is given by

$$\hat{F}_j(s) = N^{-1} \sum_{i=1}^{N} \mathbb{I}\{x_{ji} \leq s\},$$

where $\mathbb{I}\{\cdot\}$ is the indicator function and s takes values in the observed set of x_j. Then, $\hat{\theta}$ — a consistent estimator of θ_o sometimes called the Canonical Maximum Likelihood estimator (CMLE) — is the solution to

$$\max_{\theta} \sum_{i=1}^{N} \ln c_\theta(\hat{F}_1(x_{1i}), \ldots, \hat{F}_d(x_{di})).$$

Section 5.2 contains more detail on these estimators.

The following new notation is used for the sample counterparts:

$$\hat{\mathbb{H}}_i(\theta) = \nabla_\theta^2 \ln c_\theta(\hat{F}_1(x_{1i}), \ldots, \hat{F}_d(x_{di})),$$
$$\hat{\mathbb{C}}_i(\theta) = \nabla_\theta \ln c_\theta(\hat{F}_1(x_{1i}), \ldots, \hat{F}_d(x_{di})) \nabla'_\theta \ln c_\theta(\hat{F}_1(x_{1i}), \ldots, \hat{F}_d(x_{di})).$$

Then, the sample equivalents of $\mathbb{H}(\theta)$ and $\mathbb{C}(\theta)$ for arbitrary θ are

$$\bar{\mathbb{H}}(\theta) = N^{-1} \sum_{i=1}^{N} \hat{\mathbb{H}}_i(\theta),$$

$$\bar{\mathbb{C}}(\theta) = N^{-1} \sum_{i=1}^{N} \hat{\mathbb{C}}_i(\theta).$$

The test of Huang and Prokhorov (2014) is based on distinct elements of the testing matrix $\bar{\mathbb{H}}(\hat{\theta}) + \bar{\mathbb{C}}(\hat{\theta})$. Given that the dimension of θ is p, there are $p(p+1)/2$ such elements. Under correct copula specification, these are all zero. So our test is in essence a variant of the likelihood misspecification test of White (1982). What distinguishes our test is that we deal with a semiparametric likelihood specification — a parametric copula and nonparametric marginals — while White (1982) deals with a full but possibly incorrect parametric log-density. Correspondingly, the elements of the White (1982) testing matrix (he calls them "indicators") do not contain empirical marginal distributions as arguments and this precludes direct application of his test statistic in our setting.

White (1982) points out that it is sometimes appropriate to drop some of the indicators because they are identically zero or represent a linear combination of the others. When $p = 1$ — the case of bivariate one-parameter copula — this problem does not arise. Whether it arises in higher dimensional models is a copula-specific question that we do not address in this chapter. Assume that no indicators need be dropped.

Following White (1982), define

$$d_i(\theta) = vech(\mathbb{H}_i(\theta) + \mathbb{C}_i(\theta))$$

and

$$\hat{d}_i(\theta) = vech(\hat{\mathbb{H}}_i(\theta) + \hat{\mathbb{C}}_i(\theta)),$$

where $vech$ denotes vertical vectorization of the lower triangle of a matrix. Note that, in our setting, $d_i(\theta)$ depends on the unknown marginals while $\hat{d}_i(\theta)$ uses their empirical counterparts $\hat{F}_j, j = 1, \ldots, d$. Define the indicators of interest

$$\bar{D}_\theta \equiv \bar{D}(\theta) \equiv N^{-1} \sum_{i=1}^{N} \hat{d}_i(\theta).$$

Let $\bar{D}_{\hat{\theta}} = \bar{D}(\hat{\theta})$ and $D_\theta = \mathbb{E}d_i(\theta)$. Also note that, under correct specification, $D_{\theta_o} \equiv \mathbb{E}d_i(\theta_o) = 0$.

What is different in the present setting from White (1982) is that nonparametric estimates of the marginals are used to construct the joint density. It is well known that the empirical distribution converges to the true distribution at the rate $\sqrt{N}$ so the CMLE estimate $\hat{\theta}$ that uses empirical distributions $\hat{F}_j$ is still $\sqrt{N}$-consistent. The rate of convergence of the CMLE follows from Proposition 2.1 of Genest *et al.* (1995a), which, along with everything that follows, is subject to regularity conditions.

The regularity conditions are listed in many papers on semi-parametric copula estimation (see, e.g., Genest *et al.* (1995a); Shih and Louis (1995); Hu (1998); Tsukahara (2005); Chen and Fan (2006a,b)). They include compactness of the parameter set, smoothness of the marginals, existence and continuity of the relevant log-density derivatives and some other conditions that guarantee consistency of CMLE. Verification of these conditions for commonly used copula families is beyond the scope of this chapter. For many copulas, including those we use, this has already been done elsewhere (see, e.g., Hu (1998), Chapter 5).

The asymptotic variance matrix of $\sqrt{N}\hat{\theta}$ will be affected by the nonparametric estimation of the marginals. Therefore, the asymptotic variance of $\sqrt{N}\bar{D}_{\hat{\theta}}$ will also be affected. To derive the proper adjustments to the variance matrix, we use the results on semiparametric estimation of Newey (1994) and Chen and Fan (2006b). Specifically, Chen and Fan (2006b) derive the distribution of $\hat{\theta}$ given the empirical estimates $\hat{F}_j, j = 1, \ldots, d$. Our setting is complicated by the fact that the test statistic is a function of both $\hat{\theta}$ and $\hat{F}_j, j = 1, \ldots, d$. The main result is given in the following proposition while the derivation of the asymptotic distribution is deferred to the Appendix.

Proposition 6.1. *Under correct copula specification and suitable regularity conditions, the information matrix test statistic*

$$\mathscr{I}_d = N \bar{D}'_{\hat{\theta}} V_{\theta_o}^{-1} \bar{D}_{\hat{\theta}},$$

where V_{θ_o} is given in (6.12) in Appendix, is distributed asymptotically as $\chi^2_{p(p+1)/2}$.

The test statistic has a similar structure and identical asymptotic distribution to that of the White (1982) test. Indeed it is a variant of that test adjusted for the first step estimation of the marginals. It is known that the White (1982) test statistic goes to infinity almost surely when the H_o does not hold (see, e.g., Golden *et al.* (2013)). So we may expect our test to be consistent, too, but we do not pursue this point further here.

In practice, a consistent estimate of V_{θ_o} will be used. Under correct copula specification, such an estimate can be obtained by replacing θ_o and F_{ji} in (6.12) by their consistent estimates $\hat{\theta}$ and $\hat{F}_{ji}$.

Unlike available alternatives, this test statistic is simple, easy to compute and has a standard asymptotically pivotal distribution. It involves no strategic choices such as the choice of a kernel and associated smoothing parameters or any arbitrary categorization of the data. Essentially, this is White's information equivalence test with the complication of a first-step empirical distribution estimation. However, as such, it also inherits a number of drawbacks. One complication is the need to evaluate the third derivative of the log-copula density function. Lancaster (1984) and Chesher (1983) show how to construct simplified versions of the test statistic, which are asymptotically equivalent to White's original statistic but do not use the third-order derivatives. Probably the simplest form of the test is NR^2, where R^2 comes from the regression of a vector of ones on

$$\nabla_{\theta_j} \ln c_\theta(\hat{F}_1(x_{1i}), \dots, \hat{F}_d(x_{di})), \quad j = 1, \dots, p$$

and

$$\nabla^2_{\theta_j \theta_k} \ln c_\theta(\hat{F}_1(x_{1i}), \dots, \hat{F}_d(x_{di}))$$

$$+\nabla_{\theta_j} \ln c_\theta(\hat{F}_1(x_{1i}), \ldots, \hat{F}_d(x_{di}))$$

$$\times \nabla_{\theta_k} \ln c_\theta(\hat{F}_1(x_{1i}), \ldots, \hat{F}_d(x_{di}))$$

$$j = 1, \ldots, p, \quad k = 1, \ldots, p,$$

evaluated at $\hat{\theta}$.

An important problem is the well-documented poor finite sample properties of the test, especially of the NR^2 form (see, e.g., Taylor (1987); Hall (1989); Chesher and Spady (1991); Davidson and MacKinnon (1992)). Horowitz (1994), for example, points out to large deviations of the finite-sample size of various forms of the White test from their nominal size based on asymptotic critical values and suggests using bootstrapped critical values instead. It can be expected that this test will inherit this problem.

6.3.3. *Generalized information matrix tests*

Prokhorov *et al.* (2015) consider generalizations of the information matrix test for copulas proposed by Huang and Prokhorov (2014). The idea is that eigenspectrum-based statements of the IM equality reduce the degrees of freedom of the test's asymptotic distribution and lead to better size-power properties, even in high dimensions. The gains are especially pronounced for vine copulas, where additional benefits come from simplifications of score functions and the Hessian.

They consider the following tests:

(1) White Test $\mathscr{I}_d$: $vech(\mathbb{H}) + vech(\mathbb{C}) = 0_{p(p+1)/2}$, where *vech* denotes vertical vectorization of the lower triangle of a square matrix.
(2) Determinant White Test $\mathcal{T}_d^{(D)}$: $det(\mathbb{H} + \mathbb{C}) = 0$
(3) Trace White Test $\mathcal{T}_d^{(d)}$: $tr(\mathbb{H} + \mathbb{C}) = 0$
(4) IR Test $\mathcal{Z}_d$: $tr(-\mathbb{H}^{-1}\mathbb{C}) - p = 0$
(5) Log Determinant IR Test $\mathcal{Z}_d^{(D)}$: $\log(det(-\mathbb{H}^{-1}\mathbb{C})) = 0$
(6) Log Trace IMT Tr_d: $\log(tr(-\mathbb{H})) - \log(tr(\mathbb{C})) = 0$

(7) Log GAIC IMT $\mathcal{G}_d$: $\log[\frac{1}{p}(1_p)'(\Lambda(-\mathbb{H}^{-1}) \odot \Lambda(\mathbb{C}))] = 0$, where $\odot$ denotes the Hadamard product, Λ denotes the eigenvalue function and 1_p denotes a vector of ones with length p.

(8) Log Eigenspectrum IMT $\mathcal{P}_N$: $\log(\Lambda(-\mathbb{H}^{-1})) - \log(\Lambda(\mathbb{C}^{-1})) = 0_p$

(9) Eigenvalue Test $\mathcal{Q}_d$: $\Lambda(-\mathbb{H}^{-1}\mathbb{C}) = 1_p$

For each, they derive the asymptotic distribution, accounting for the non-parametric estimation of the marginals and apply a parametric bootstrap procedure, valid when asymptotic critical values are inaccurate.

In Monte Carlo simulations, they study the behavior of the new tests, compare them with several Cramer-von Mises type tests and confirm that the new tests have superior properties in high dimensions.

6.3.4. *Power study*

In this section, we discuss results of a size and power study reviewing the properties of the test statistic we derived in Proposition 6.1. We remark on how this test compares with other copula goodness-of-fit tests discussed in Genest *et al.* (2009) but we do not compare here the various alternative forms of the test statistic such as the NR^2 form. Neither do we compare this test with the tests proposed by Prokhorov *et al.* (2015).

We start by plotting size-power curves under various copula families (see, e.g., Davidson and MacKinnon (1998), for a comparison of this and other graphical ways of studying test properties). We generate K realizations of the test statistic $\mathscr{I}$ using a data-generating process (DGP). Denote these simulated values by $\mathscr{I}_j$, $j = 1, \ldots, K$. Our size-power curves are based on the empirical distribution function (EDF) of the simulated p-value of $\mathscr{I}_j$, $p_j \equiv p(\mathscr{I}_j)$, i.e., the probability that $\mathscr{I}$ is greater than or equal to $\mathscr{I}_j$ according to its simulated distribution. At any point y in the $(0, 1)$ interval, the EDF of the p-values is defined by

$$\hat{F}(y) \equiv \frac{1}{K} \sum_{j=1}^{K} \mathbb{I}(p_j \leq y).$$

We choose the following values for y_i, $i = 1, \ldots, m$:

$$y_i = 0.001, \; 0.002, \ldots, \; 0.010, \; 0.015, \; \ldots,$$

$$0.990, \; 0.991, \ldots, \; 0.999 \quad (m = 215),$$

where we follow Davidson and MacKinnon (1998) and use a smaller grid near 0 and 1 in order to study the tail behavior more closely.

The point of drawing size-power curves is to plot power against true, rather than nominal, size. Given the well-documented poor finite sample size property of the information matrix test, this is useful because we can display the test power in situations when the nominal size is definitely incorrect. Two values of the test statistic are computed: one under the null DGP (H_0) and the other under the alternative DGP (H_1). Let $F(y)$ and $F^*(y)$ be the probabilities of getting a p-value less than y under the null and the alternative, respectively, and let $\hat{F}(y)$ and $\hat{F}^*(y)$ be their empirical counterparts. Given the sample size T, the number of simulation replications K and the grid of size m, a size-power curve is the set of points $(\hat{F}(y_i), \hat{F}^*(y_i))$, $i = 1, \ldots, m$, on the unit square where the horizontal axis measures size and the vertical axis measures power.

We keep the grid the same, set $K = 10,000$, and vary the sample size N and the strength of dependence in the various null and alternative DGPs we consider. The various null and alternative copula families are selected from the list used by Genest *et al.* (2009) in a large scale Monte Carlo study and, as usual, the dependence strength is measured by Kendall's τ, where $\tau = 4\mathbb{E}[C_\theta(U, V)] - 1$. We follow Genest *et al.* (2009) and use the copula parameter obtained by inversion of Kendall's τ. In all considered families, the solution is known to be unique so this produces one parameter value under H_0 and one under H_1. To preserve space, we report curves for $N = 200, 300$ and $\tau = 0.25, 0.33, 0.5, 0.75$ only.

Figure 6.1 shows what happens as we change the strength of dependence holding N fixed at 300. Panel (a) displays the size-power curves under H_0: Normal copula and H_1: Clayton copula, panel (b) displays the curves for H_0: Normal and H_1: Frank, panel (c) is for the test of H_0: Clayton against H_1: Normal, and panel (d) is for H_0: Clayton against H_1: Frank. We can clearly see from the figure that as

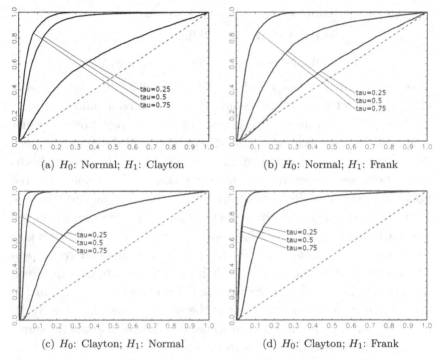

(a) H_0: Normal; H_1: Clayton   (b) H_0: Normal; H_1: Frank

(c) H_0: Clayton; H_1: Normal   (d) H_0: Clayton; H_1: Frank

Fig. 6.1.   Size-power curves for different levels of dependence: Kendall's $\tau = 0.25$, 0.5 and 0.75. Sample size is $N = 300$.

the strength of dependence increases, the power of the test becomes larger. This agrees with similar observations by Genest *et al.* (2009) made for other copula goodness-of-fit tests. Interestingly, there are areas on the plots where the test actually has power less than its size. This happens at small enough sizes to make this observation important but the same thing occasionally happens with other "blanket" tests under weak dependence (for $\tau = 0.25$, see, e.g., Genest *et al.* (2009), Table 1).

Figure 6.2 displays the size-power curves for different null and alternative DGPs holding both N and τ fixed. The set of nulls and alternatives we report includes H_0: Normal vs H_1: Clayton, H_0: Normal vs H_1: Frank, H_0: Clayton vs H_1: Normal, H_0: Clayton vs H_1: Frank. An interesting observation is that the size-adjusted power of the test varies greatly for the different nulls and

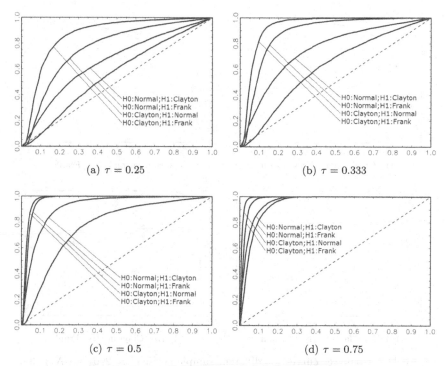

Fig. 6.2. Size-power curves for selected copulas. Sample size is $N = 300$.

alternatives — something that has been noted for other tests as well. If we further allow τ to increase holding sample size fixed, the variation in power becomes much smaller. It is interesting to observe that for the tests that involve the Clayton copula under H_0, the test has much more power than for the other models we consider. Again, this interesting observation coincides with results of Genest *et al.* (2009) obtained for other available "blanket" tests (see their Tables 1–3). Note that the ranking of power of the various tests changes as we change strength of dependence, but the two tests involving the Clayton null remain more powerful than the others.

Figure 6.3 shows how the size-power curves shift as the sample size changes from $N = 200$ to $N = 300$. The test in each panel is the same as in Figure 6.1. Not surprisingly, the power increases as the sample size grows. Plots for larger samples (not reported here) illustrate that as the sample size becomes larger, H_0 is rejected with

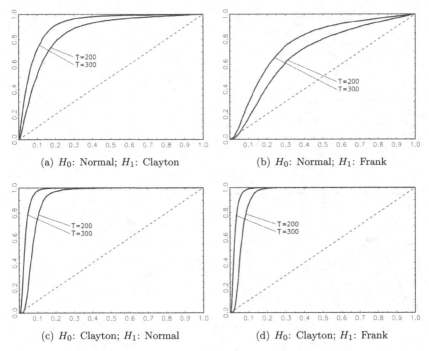

(a) H_0: Normal; H_1: Clayton (b) H_0: Normal; H_1: Frank

(c) H_0: Clayton; H_1: Normal (d) H_0: Clayton; H_1: Frank

Fig. 6.3. Size-power curves for different sample sizes: $N = 200$ and $N = 300$. Kendall's $\tau = 0.5$.

probability approaching one whenever H_1 is true, i.e., these tests are consistent.

To compare our test with other "blanket" tests in more detail and also to get an idea about the extent of size distortions, we construct a size and power table similar to those reported by Genest *et al.* (2009). Tables 6.1 and 6.2 report size and power of our test at the 5% significance level for $N = 200$ and $N = 1,000$. As before, we also vary Kendall's τ from 0.25 to 0.75. In each row, we report the percentage of rejections of H_0 associated with different tests for the bootstrap test (*Simul.*) and the asymptotic test (*Asy.*). For example, when testing for the Normal copula against Clayton at $N = 200$ and $\tau = 0.75$, the chance of the bootstrap test rejecting the incorrect null is approximately 34.6%.

Similar to analogous entries for other "blanket" tests, the frequencies reported in Tables 6.1 and 6.2 show that for these sample

Table 6.1. Power(Size) for N = 200 at nominal size 5%.

Copula under H_0	True Copula	$\tau = 0.25$		$\tau = 0.50$		$\tau = 0.75$	
		Simul.	Asy.	Simul.	Asy.	Simul.	Asy.
Normal	Clayton	4.9(5)	7.7(7)	21.7(5)	34.8(8)	34.6(5)	62.9(10)
	Frank	2.5(5)	4.0(7)	3.8(5)	7.8(7)	16.5(5)	42.0(9)
	Gumbel	6.8(5)	9.6(6)	9.2(5)	18.3(8)	9.1(5)	26.7(10)
Clayton	Normal	1.3(5)	12.2(10)	29.8(5)	85.1(11)	86.1(5)	99.2(11)
	Frank	4.2(5)	26.4(10)	41.6(5)	93.2(11)	64.2(5)	94.6(11)
	Gumbel	8.6(5)	36.5(10)	60.4(5)	96.5(12)	86.4(5)	98.4(10)
Frank	Normal	6.5(5)	8.2(6)	9.2(5)	14.6(9)	3.1(5)	8.0(10)
	Clayton	4.0(5)	5.3(6)	1.5(5)	5.7(9)	2.7(5)	22.4(10)
	Gumbel	4.8(5)	5.8(6)	1.8(5)	5.4(9)	1.0(5)	8.7(10)
Gumbel	Normal	2.9(5)	5.1(8)	1.3(5)	5.2(9)	1.0(5)	9.9(10)
	Clayton	16.9(5)	30.4(8)	37.5(5)	80.0(10)	79.1(5)	97.2(10)
	Frank	3.5(5)	8.0(8)	6.3(5)	31.2(9)	32.7(5)	80.2(10)

Table 6.2. Power(Size) for N = 1,000 at nominal size 5%.

Copula under H_0	True Copula	$\tau = 0.25$		$\tau = 0.50$		$\tau = 0.75$	
		Simul.	Asy.	Simul.	Asy.	Simul.	Asy.
Normal	Clayton	44.0(5)	5(6)	96.9(5)	98.8(7)	93.2(5)	99.0(12)
	Frank	10.7(5)	16.2(7)	65.2(5)	80.0(8)	90.3(5)	98.1(12)
	Gumbel	58.0(5)	63.4(6)	83.4(5)	92.3(8)	78.8(5)	94.7(11)
Clayton	Normal	83.5(5)	87.8(6)	100(5)	100(7)	100(5)	100(7)
	Frank	98.6(5)	99.3(7)	100(5)	100(7)	100(5)	100(7)
	Gumbel	99.6(5)	99.8(6)	100(5)	100(7)	100(5)	100(7)
Frank	Normal	10.1(5)	10.7(5)	21.2(5)	24.0(6)	4.3(5)	5.4(6)
	Clayton	8.5(5)	9.6(6)	17.2(5)	19.9(6)	93.5(5)	95.8(6)
	Gumbel	20.2(5)	21.4(5)	14.9(5)	17.3(6)	53.8(5)	64.4(7)
Gumbel	Normal	8.3(5)	9.4(6)	20.9(5)	25.7(6)	68.3(5)	72.8(6)
	Clayton	98.2(5)	98.6(6)	100(5)	100(6)	100(5)	100(6)
	Frank	50.8(5)	53.7(6)	99.2(5)	99.5(6)	100(5)	100(6)

sizes the test generally holds its nominal size. Indeed the frequencies listed in the *Simul.* columns are virtually equal to the nominal level of 5% no matter what sample size or copula family. This is hardly suprising since we are bootstrapping an asymptotically pivotal statistic using as many as 10,000 replications. In this setting, the bootstrap test is very close to the exact test for a sufficiently large number of replications, regardless of the specific null or the specific sample size (see, e.g., Hall and Hart (1990), Table 1). On the other hand, the frequencies shown in the *Asy.* columns are often substantially higher than 5%, suggesting oversize distortions. As expected, the distortions clearly reduce as the sample size increases.

Compared to equivalent entries in Tables 1 to 3 of Genest *et al.* (2009), the power of our test statistic is generally lower than that of the other "blanket" tests available in the literature. However, at the sample size equal to 1,000, our test power is usually reasonably high. Similar to other "blanket" tests, the performance of our test varies greatly with the DGPs. For some combinations of copulas under the null hypothesis and the alternative, the test's power is remarkably low. For example, if the null hypothesis is Frank and the true copula is Normal, the power of our test at $N = 1,000$ is as low as 4-6% even for $\tau = 0.75$. Interestingly, the power of other "blanket" tests is not very high for some combinations either, and for some combinations of copulas and some sample sizes, Genest *et al.* (2009) report even lower percentages of rejection. In such cases, it may make sense to use the generalized information matrix tests proposed by Prokhorov *et al.* (2015), which are more powerful.

6.4. Concluding remarks

We have discussed two kinds of tests. One uses the results in Chapter 5 to detect copulas whose score is zero mean. This means it does not distinguish between a correct copula and an incorrect but valid copula. Put differently, this is a test that detects robust copulas.

The other kind of tests considered detects correctness, or goodness-of-fit, of copulas. We have argued that these tests are "blanket" and have shown that they show reasonable properties. The main advantage of the test is its simplicity. Basically, it is the well-studied White specification test adapted to a two-step semiparametric estimation. As such, it inherits White test's benefits and costs. The most costly feature of the test is its poor behavior in samples smaller than $1,000$. Other potential criticisms include the test's inability to detect *all* deviations from the null in finite samples, its inability to differentiate between two well-performing alternatives, and its in-sample nature.

As in-sample procedures, this and other "blanket" tests can be argued to be susceptible to overfitting and data mining. However, recent studies in the setting of predictability tests tend to question the conventional wisdom that out-of-sample tests are more credible than in-sample (see, e.g., Inoue and Kilian (2005)).

The White test is simple compared to some of the other available "blanket" tests, which do not have such a simple asymptotic distribution. Obtaining up to three derivatives of the log-copula density is the main challenge in constructing the test statistic. However, for some families explicit formulas for the derivatives have been catalogued (see, e.g., Chen and Fan (2006b)) and, for others, symbolic algebra modules of modern software can be used to obtain them. Of course, there is always the brute force method of calculating the derivatives numerically. Moreover, the test has many asymptotically equivalent forms, some of which are derived specifically to reduce the order of derivatives and to make the finite sample behavior more appealing (see, e.g., Golden *et al.* (2013)). For example, the versions of Lancaster (1984) and Chesher (1983) do not require the third derivative while the first two derivatives of the likelihood often arise as byproducts of standard MLE optimization routines. Overall, the balance of costs and benefits speaks, we believe, in favor of this copula goodness-of-fit test, especially in large sample settings of a financial application.

We also discussed a battery of tests resulting from eigenspectrum-based versions of the information matrix equality applied to copulas. The benefit of this generalization is due to a reduction in degrees of freedom of the tests and to the focused hypothesis function used to construct them. For example, in testing goodness-of-fit of high-dimension, multi-parameter copulas we manage to reduce the information matrix based test statistic to an asymptotically χ^2 with one degree of freedom, and we can focus on the effect of larger or smaller eigenvalues by using specific functions of the eigenspectrum such as *det* or *trace*. However, only a few of these tests can be well approximated by their asymptotic distributions in realistic sample sizes so we need to resort to the bootstrap version of the tests.

Prokhorov *et al.* (2015) argue that the bootstrap versions of the generalized information matrix tests dominate other available tests of copula goodness-of-fit when copulas are high-dimensional and multi-parameter. They use this argument to motivate the use of Generalized Information Matrix Tests (GIMTs) on vine copulas, where additional simplifications result from the functional form of the Hessian and the score.

6.5. Appendix: Proofs

Proof of Theorem 6.1. See proof of Lemma 4.2 of Hansen (1982).

Proof of Theorem 6.2. First note that, by standard optimal GMM results, $\hat{\theta}$ satisfies

$$\sqrt{N}(\hat{\theta} - \theta_o) = -(\mathbf{D_{11}^o}'\mathbf{V_{11}^o}^{-1}\mathbf{D_{11}^o})^{-1}\mathbf{D_{11}^o}'\mathbf{V_{11}^o}^{-1}\sqrt{N}\bar{g}(\theta_o) + o_p(1). \tag{6.4}$$

The first-order condition for $\hat{\rho}$ can equivalently be written as

$$[\nabla'_\rho \bar{r}(\hat{\theta}, \hat{\rho})]'\mathbf{B_o}^{-1}\bar{r}(\hat{\theta}, \hat{\rho}) = 0,$$

$$\mathbf{D_{22}^o}'\mathbf{B_o}^{-1}\sqrt{N}\bar{r}(\hat{\theta}, \hat{\rho}) = o_p(1). \tag{6.5}$$

Now, by the mean-value theorem, we have

$$\sqrt{N}\bar{r}(\hat{\theta}, \hat{\rho}) = \sqrt{N}\bar{r}(\theta_o, \rho_o) + \mathbf{D_{21}^o}\sqrt{N}(\hat{\theta} - \theta_o) + \mathbf{D_{22}^o}\sqrt{N}(\hat{\rho} - \rho_o) + o_p(1). \tag{6.6}$$

Substituting (6.4) into (6.6), pre-multiplying by $\mathbf{D_{22}^o}'\mathbf{B_o}^{-1}$, and solving for $\sqrt{N}(\hat{\rho} - \rho_o)$ using (6.5) yields

$$\sqrt{N}(\hat{\rho} - \rho_o) = -(\mathbf{D_{22}^o}'\mathbf{B_o^{-1}}\mathbf{D_{22}^o})^{-1}\mathbf{D_{22}^o}'\mathbf{B_o^{-1}}\sqrt{N}\bar{r}(\theta_o, \rho_o)$$

$$+ (\mathbf{D_{22}^o}'\mathbf{B_o^{-1}}\mathbf{D_{22}^o})^{-1}\mathbf{D_{22}^o}'\mathbf{B_o^{-1}}\mathbf{D_{21}^o}$$

$$\times (\mathbf{D}_{11}^{o}{}'\mathbf{V}_{11}^{o}{}^{-1}\mathbf{D}_{11}^{o})^{-1}\mathbf{D}_{11}^{o}{}'\mathbf{V}_{11}^{o}{}^{-1}\sqrt{N}\bar{g}(\theta_o)$$
$$+ o_p(1). \tag{6.7}$$

Substituting (6.7) and (6.4) into (6.6) and simplifying results in

$$\sqrt{N}\bar{r}(\hat{\theta}, \hat{\rho}) = \mathbf{R_o}\sqrt{N}\bar{\phi}(\theta_o, \rho_o) + o_p(1), \tag{6.8}$$

where

$$\mathbf{R_o} = \mathbb{I} - \mathbf{D}_{22}^{o}(\mathbf{D}_{22}^{o}{}'\mathbf{B}_o^{-1}\mathbf{D}_{22}^{o})^{-1}\mathbf{D}_{22}^{o}{}'\mathbf{B}_o^{-1},$$
$$\bar{\phi}(\theta_o, \rho_o) = \bar{r}(\theta_o, \rho_o) - \mathbf{D}_{21}^{o}(\mathbf{D}_{11}^{o}{}'\mathbf{V}_{11}^{o}{}^{-1}\mathbf{D}_{11}^{o})^{-1}\mathbf{D}_{11}^{o}{}'\mathbf{V}_{11}^{o}{}^{-1}\bar{g}(\theta_o).$$

Note that $\sqrt{N}\bar{\phi}(\theta_o, \rho_o) \sim \mathbb{N}(\mathbf{0}, \mathbf{B_o})$, and thus $\mathbf{B_o}^{-1/2}\sqrt{N}\bar{\phi}(\theta_o, \rho_o) \sim \mathbb{N}(\mathbf{0}, \mathbb{I})$. Also, note that $\mathbf{R}_o'\mathbf{B}_o^{-1}\mathbf{R_o} = \mathbf{B}_o^{-\frac{1}{2}}[\mathbf{I} - \mathbf{B}_o^{-\frac{1}{2}}\mathbf{D}_{22}^{o}(\mathbf{D}_{22}^{o}{}'\mathbf{B}_o^{-1}\mathbf{D}_{22}^{o})^{-1}\mathbf{D}_{22}^{o}{}'\mathbf{B}_o^{-\frac{1}{2}}]\mathbf{B}_o^{-\frac{1}{2}}$.

Thus, the test statistic in (6.2) can be written as

$$N\bar{h}(\hat{\theta}, \hat{\rho})'\mathbf{B}_o^{-1}\bar{h}(\hat{\theta}, \hat{\rho}), \tag{6.9}$$

i.e., as a quadratic form in standard normals with the coefficient matrix

$$\mathbb{P} = \mathbf{I}_{p+q} - \mathbf{B}_o^{-1/2}\mathbf{D}_{22}^{o}(\mathbf{D}_{22}^{o}{}'\mathbf{B}_o^{-1}\mathbf{D}_{22}^{o})^{-1}\mathbf{D}_{22}^{o}{}'\mathbf{B}_o^{-1/2}. \tag{6.10}$$

This matrix is idempotent: it is the projection matrix orthogonal to $\mathbf{B}_o^{-\frac{1}{2}}\mathbf{D}_{22}^{o}$. The χ^2-test in (6.2) follows immediately because $tr(\mathbb{P}) = p + q - rank(D_{22}^{o}) = p$.

Proof of Proposition 6.1. We start with $N = 2$ for simplicity and later give the formulas for any N. Let $\hat{F}_{ji} = \hat{F}_j(x_{ji})$, $j = 1, 2$, $i = 1, \ldots, N$, be the empirical cdf's. Then,

$$\hat{d}_i(\theta) = vech[\nabla_\theta^2 \ln c(\hat{F}_{1i}, \hat{F}_{2i}; \theta)$$
$$+ \nabla_\theta \ln c(\hat{F}_{1i}, \hat{F}_{2i}; \theta)\nabla_\theta' \ln c(\hat{F}_{1i}, \hat{F}_{2i}; \theta)].$$

Provided that the derivatives and expectation exist, let

$$\nabla D_\theta = \mathbb{E}\nabla_\theta d_i(\theta)$$

and

$$\nabla\bar{D}_\theta = N^{-1}\sum_{i=1}^{N}\nabla_\theta\hat{d}_i(\theta).$$

First, expand $\sqrt{N}\bar{D}_{\hat{\theta}}$ with respect to θ:

$$\sqrt{N}\bar{D}_{\hat{\theta}} = \sqrt{N}\bar{D}_{\theta_o} + \nabla D_{\theta_o}\sqrt{N}(\hat{\theta} - \theta_o) + o_p(1).$$

Chen and Fan (2006a) show that

$$\sqrt{N}(\hat{\theta} - \theta_o) \to N(0, B^{-1}\Sigma B^{-1}),$$

where

$$B = -\mathbb{H}(\theta_0),$$

$$\Sigma = \lim_{N\to\infty} Var\left(\sqrt{N}A_N^*\right),$$

$$A_N^* = \frac{1}{N}\sum_{i=1}^{N}(\nabla_\theta \ln c(F_{1i}, F_{2i}; \theta_0) + W_1(F_{1i}) + W_2(F_{2i})).$$

Here terms $W_1(F_{1i})$ and $W_2(F_{2i})$ are the adjustments needed to account for the empirical distributions used in place of the true distributions. These terms are calculated as follows:

$$W_1(F_{1i}) = \int_0^1 \int_0^1 [\mathbb{I}\{F_{1i} \leq u\} - u]\nabla_{\theta,u}^2 \ln c(u, v; \theta_0)\, c(u, v; \theta_0)dudv,$$

$$W_2(F_{2i}) = \int_0^1 \int_0^1 [\mathbb{I}\{F_{2i} \leq v\} - v]\nabla_{\theta,v}^2 \ln c(u, v; \theta_0)\, c(u, v; \theta_0)dudv.$$

So,

$$\sqrt{N}(\hat{\theta} - \theta_o) = B^{-1}\sqrt{N}A_N^* + o_p(1).$$

Second, expand $\sqrt{N}\bar{D}_{\theta_0}$ with respect to F_{1i} and F_{2i}:

$$\sqrt{N}\bar{D}_{\theta_0} \simeq \frac{1}{\sqrt{N}}\sum_{i=1}^{N} d_i(\theta_0) + \frac{1}{N}\sum_{i=1}^{N}\nabla_{F_{1i}}d_i(\theta_0)\sqrt{N}(\hat{F}_{1i} - F_{1i})$$

$$+ \frac{1}{N}\sum_{i=1}^{N}\nabla_{F_{2i}}d_i(\theta_0)\sqrt{N}(\hat{F}_{2i} - F_{2i}). \qquad (6.11)$$

Under suitable regularity conditions,

$$\frac{1}{N} \sum_{i=1}^{N} \nabla_{F_{1i}} d_i(\theta_0) \sqrt{N} (\hat{F}_{1i} - F_{1i})$$

$$\simeq \int_0^1 \int_0^1 \nabla_u vech[\nabla_\theta^2 \ln c(u, v; \theta_0)$$

$$+ \nabla_\theta \ln c(u, v; \theta_0) \nabla_\theta' \ln c(u, v; \theta_0)]$$

$$\times \sqrt{N} (\hat{F}_1(F_1^{-1}(u)) - u) c(u, v; \theta_0) du dv$$

$$= \frac{1}{\sqrt{N}} \sum_{i=1}^{N} \int_0^1 \int_0^1 [\mathbb{I}\{F_{1i} \le u\} - u]$$

$$\times \nabla_u vech[\nabla_\theta^2 \ln c(u, v; \theta_0)$$

$$+ \nabla_\theta \ln c(u, v; \theta_0) \nabla_\theta' \ln c(u, v; \theta_0)] c(u, v; \theta_0) du dv.$$

Denote

$$M_1(F_{1i}) = \int_0^1 \int_0^1 [\mathbb{I}\{F_{1i} \le u\} - u] \nabla_u vech[\nabla_\theta^2 \ln c(u, v; \theta_0)$$

$$+ \nabla_\theta \ln c(u, v; \theta_0) \nabla_\theta' \ln c(u, v; \theta_0)] c(u, v; \theta_0) du dv,$$

then

$$\frac{1}{N} \sum_{i=1}^{N} \nabla_{F_{1i}} d_i(\theta_0) \sqrt{N} (\hat{F}_{1i} - F_{1i}) = \frac{1}{\sqrt{N}} \sum_{i=1}^{N} M_1(F_{1i}).$$

Similarly, denote

$$M_2(F_{2i}) = \int_0^1 \int_0^1 [\mathbb{I}\{F_{2i} \le v\} - v] \nabla_v vech[\nabla_\theta^2 \ln c(u, v; \theta_0)$$

$$+ \nabla_\theta \ln c(u, v; \theta_0) \nabla_\theta' \ln c(u, v; \theta_0)] c(u, v; \theta_0) du dv,$$

then

$$\frac{1}{N} \sum_{i=1}^{N} \nabla_{F_{2i}} d_i(\theta_0) \sqrt{N} (\hat{F}_{2i} - F_{2i}) = \frac{1}{\sqrt{N}} \sum_{i=1}^{N} M_2(F_{2i}).$$

Therefore, Equation (6.11) can be rewritten as

$$\sqrt{N}\bar{D}_{\theta_0} = \frac{1}{\sqrt{N}}\sum_{i=1}^{N} d(\theta_0) + \sqrt{N}B_N^* + o_p(1),$$

where

$$B_N^* = \frac{1}{N}\sum_{i=1}^{N}[M_1(F_{1i}) + M_2(F_{2i})].$$

Finally, combining the expansions gives

$$\sqrt{N}\bar{D}_{\hat{\theta}} = \frac{1}{\sqrt{N}}\sum_{i=1}^{N} d(\theta_0) + \sqrt{N}B_N^* + \nabla D_{\theta_0}B^{-1}\sqrt{N}A_N^* + o_p(1).$$

So

$$\sqrt{N}\bar{D}_{\hat{\theta}} \to N(0, V_{\theta_0}),$$

or, equivalently,

$$N\bar{D}_{\hat{\theta}}'V_{\theta_0}^{-1}\bar{D}_{\hat{\theta}} \to \chi^2_{p(p+1)/2},$$

where

$$\begin{aligned}
V_{\theta_0} = E\{&d_i(\theta_0) + M_1(F_{1i}) + M_2(F_{2i}) \\
&+ \nabla D_{\theta_0}B^{-1}[\nabla_\theta \ln c(F_{1i}, F_{2i}; \theta_0) + W_1(F_{1i}) + W_2(F_{2i})]\} \\
\times \{&d_i(\theta_0) + M_1(F_{1i}) + M_2(F_{2i}) \\
&+ \nabla D_{\theta_0}B^{-1}[\nabla_\theta \ln c(F_{1i}, F_{2i}; \theta_0) + W_1(F_{1i}) + W_2(F_{2i})]\}'.
\end{aligned}$$

Extension to $d \geq 2$ is straightforward. Now

$$\begin{aligned}
d_i(\theta) = vech[&\nabla_\theta^2 \ln c(F_{1i}, F_{2i}, \ldots, F_{di}; \theta) \\
&+ \nabla_\theta \ln c(F_{1i}, F_{2i}, \ldots, F_{di}; \theta)\nabla_\theta' \ln c(F_{1i}, F_{2i}, \ldots, F_{di}; \theta)],
\end{aligned}$$

and the asymptotic variance matrix becomes

$$\begin{aligned}
V_{\theta_0} = E\Bigg\{ &d_i(\theta_0) + \nabla D_{\theta_0}B^{-1} \\
&\times \left[\nabla_\theta \ln c(F_{1i}, F_{2i}, \ldots, F_{di}; \theta_0) + \sum_{j=1}^{d} W_j(F_{ji})\right] + \sum_{j=1}^{d} M_j(F_{ji})\Bigg\}
\end{aligned}$$

$$\times \left\{ d_i(\theta_0) + \nabla D_{\theta_0} B^{-1} \right.$$

$$\left. \times \left[\nabla_\theta \ln c(F_{1i}, F_{2i}, \ldots, F_{di}; \theta_0) + \sum_{j=1}^{d} W_j(F_{ji}) \right] + \sum_{j=1}^{d} M_j(F_{ji}) \right\}',$$

$$\text{(6.12)}$$

where, for $j = 1, 2, \ldots, d$,

$$W_j(F_{ji}) = \int_0^1 \int_0^1 \cdots \int_0^1 [\mathbb{I}\{F_{ji} \leq u_j\} - u_j]$$
$$\times \nabla_{\theta, u_j}^2 \ln c(u_1, u_2, \ldots, u_d; \theta_0) c(u_1, u_2, \ldots, u_d; \theta_0) du_1 du_2 \cdots du_d,$$

and

$$M_j(F_{ji}) = \int_0^1 \int_0^1 \cdots \int_0^1 [\mathbb{I}\{F_{ji} \leq u_j\} - u_j]$$
$$\nabla_{u_j} vech[\nabla_\theta^2 \ln c(u_1, u_2, \ldots, u_d; \theta_0)$$
$$+ \nabla_\theta \ln c(u_1, u_2, \ldots, u_d; \theta_0)$$
$$\times \nabla_\theta' \ln c(u_1, u_2, \ldots, u_d; \theta_0)] c(u_1, u_2, \ldots, u_d; \theta_0) du_1 du_2 \cdots du_d.$$

Chapter 7

Summary and Conclusion

7.1. Summary

In this chapter, we provide a summary of the book's results and outline promising research areas. We start with short summary statements for Chapters 2–6.

The book's premise is that economic and financial data are heavy tailed, dependent and heterogenous. In Chapter 2, we started building a framework for the analysis of such data. As a building block, we used the power law family of distributions and considered the problem of portfolio risk diversification with independent components. It turns out that linear combinations of extremely heavy tailed heterogenous components, i.e., power law risks with tail exponents smaller than one, do not reduce the overall risk of a portfolio as measured by Value-at-Risk. On the contrary, diversification increases portfolio riskiness.

This counterintuitive result has profound implications for economics and finance but it was not clear whether it extended to the more realistic settings where portfolio components are allowed to have arbitrary dependence. Before addressing this problem, we looked at the nature of dependence, that is at how to generate arbitrary dependence patterns going from independence as a starting point. The characterizations and dependence concepts provided in Chapter 3 permit a construction of dependent variables using U-statistics based on independent variables. This allowed us to consider a wide range of dependence structures of interest to time series econometrics as well as to say important things about various dependence measures.

Chapter 4 considered limits to diversification under various dependence assumptions. Though we stopped short of a general result for any dependence structure, the chapter stated that the same threshold value of the tail index characterizes the situation where diversification is no longer beneficial for dependent risks. It turns out that for many kinds of dependence produced by common shocks and power-type copula families, diversification increases portfolio riskiness if tail exponents of the portfolio components are smaller than one.

In the second part of the book, we looked at estimation of models with heavy tails and copulas. Chapter 5 discussed modern approaches to estimating copula-based models under possible copula misspecification, including topics in Bayesian and frequentist nonparametrics. It also touched upon robustness of estimators to heavy tails. It turns out that a generic robust parametric copula that is efficiency improving is hard to find but nonparametric alternatives are attractive substitutes.

Chapter 6 provide a few approaches to testing whether a copula is robust and valid. We considered tests based on the information matrix equality, which follow from Chapter 5, and we discussed extensions to nonparametric marginals and high-dimensional copulas. The goal was to construct "blanket" tests, applicable to general dependence structures and requiring no "heroic" assumptions.

We have focused in the book on the concept of heavy tails as applied to the univariate margins. An alternative is to define a similar concept for the copula cdf or density. For example, Hua *et al.* (2014) use the concept of upper and lower tail order of a copula. The lower tail order κ, $\kappa \geq 1$, is defined by the following equation:

$$C(u, \ldots, u) \asymp u^{-\kappa} L(u),$$

where $L(u)$ is a slowly varying function; and the upper tail order is defined similarly (see, Joe, 2014, for details). We believe this approach could be very useful in considering robustness to joint tail behavior, rather than to marginal tail behavior.

7.2. Future research

Active areas of future research will no doubt include the development of estimation and testing methods suitable for very high-dimensional copulas. The challenge here is to allow for flexibility while preserving parsimony and computational feasibility.

Current methods to analyze high-dimensional joint distributions via extreme dependence modelling fall into two broad categories. The first uses copula functions to model innovations of the individual data generating processes (DGP), focusing on how to construct flexible, yet manageable copulas that accommodate observed regularities in the data such as tail dependence and asymmetry. This book is mostly an example of such an approach. These models are often nonparametric and thus are limited to dimensions not much greater than two, due to the well-known curse of dimensionality preventing efficient estimation of high-dimensional models.

The other category focuses on specific univariate and multivariate "off-the-shelf" distributions. These models typically assume a multivariate family of fat-tailed distributions such as Student-t for the innovation terms in a DGP (e.g., Bauwens *et al.* (2006)). Typically such models are not robust to misspecification in the sense of Chapter 5. Such models are parsimonious, allowing for efficient estimation even in high dimensions. However, they are inflexible and require pretesting.

As an example, consider the case of a d-dimensional vector with Student-t distribution. The scale matrix, representing dependence for this case, has $\frac{d(d-1)}{2}$ parameters so the estimation problem is of order $O(d^2)$. Yet the dependence it models does not even include asymmetry. In the extreme case, we can think of a nonparametric d-copula such as Bernstein which is dense in the space of copulas but in practice has k^d parameters, where k is the number of nodes in each dimension. No feasible samples can populate such parameter spaces densely enough even with $d = 10$.

Promising current approaches to break the curse of dimensionality use various ad hoc factorization or decomposition schemes. Vine-copulas decompose d-copulas into a product of $O(d^2)$ bivariate conditional copulas under certain simplifying assumptions (e.g., Joe *et al.* (2010)). Factor copulas assume a specific latent factor structure for dependence (e.g., Krupskii and Joe (2013)). Sequential copula constructions use fixed functional forms such as Archimedian or strong approximation assumptions (e.g., Anatolyev *et al.* (2014); Neo *et al.* (2016)).

Of particular interest are adaptive sparsity methods of the kind mentioned at the end of Section 5.5.1. They permit parsimony and flexibility at the same time, achieving adaptive sparsity in two ways. One is by imposing a penalty function, say, within an MLE framework, that adjusts the penalty coefficient depending on the situation (Shen (1997)). This has not been done for copulas and can offer a fundamental improvement over the state-of-the-art as the effective dimensionality is inversely related the penalty coefficient. Efficient convex optimisation algorithms would be applicable here (see, e.g., Birge and Massart (1997)). An important practical benefit is that it allows to estimate models were the number of estimated parameters is no less than the sample size.

Alternatively, adaptive copula sparsity can be achieved by using differently scaled polynomials in different parts of $[0, 1]$.[J] In practice, this means modifying the Bernstein-Dirichlet representation of Section 5.5.2 to a multi-scale mixture of the Bernstein polynomials.

Interesting research often raises more questions than it answers. We hope the reader finds the book interesting.

Bibliography

Aaronson, J., Gilat, D., and Keane, M. (1992). On the structure of 1-dependent Markov chains, *Journal of Theoretical Probability* **5**, pp. 545–561.

Acerbi, C. (2002). Spectral measures of risk: A coherent representation of subjective risk aversion, *Journal of Banking and Finance* **26**, pp. 1505–1518.

Acerbi, C. and Tasche, D. (2002). On the coherence of expected shortfall, *Journal of Banking and Finance* **26**, pp. 1487–1503.

Aguilar, M. and Hill, J. (2015). Robust score and portmanteau tests of volatility spillover, *Journal of Econometrics* **184**, pp. 37–61.

Ai, C. and Chen, X. (2003). Efficient estimation of models with conditional moment restrictions containing unknown functions, *Econometrica* **71**, 6, pp. 1795–1843.

Akaike, H. (1973). Information theory and an extension of the maximum likelihood principle, in B. N. Petrov and F. Caski (eds.), *Second International Symposium on Information Theory (Tsahkadsor, 1971)* (Akadémiai Kiadó, Budapest), pp. 267–281, reprinted in *Selected Papers of Hirotugu Akaike*, E. Parzen, K. Tanabe and G. Kitagawa (eds.), *Springer Series in Statistics: Perspectives in Statistics.* (Springer-Verlag, New York, NY, 1998), pp. 199–213.

Albrecher, H., Asmussen, S., and Kortschak, D. (2006). Tail asymptotics for the sum of two heavy-tailed dependent risks, *Extremes* **9**, pp. 107–130.

Alexits, G. (1961). *Convergence Problems of Orthogonal Series, International Series of Monographs in Pure and Applied Mathematics*, Vol. 20 (Pergamon Press).

Ali, S. M. and Silvey, S. D. (1966). A general class of coefficients of divergence of one distribution from another, *Journal of the Royal Statistical Society Series B* **28**, pp. 131–142.

Alink, S., Löwe, M., and Wüthrich, M. V. (2005). Analysis of the expected shortfall of aggregate dependent risks, *Astin Bulletin* **35**, pp. 25–43.

Amsler, C., Prokhorov, A., and Schmidt, P. (2014). Using copulas to model time dependence in stochastic frontier models, *Econometric Reviews* **33**, 5–6, pp. 497–522.

An, M. Y. (1998). Logconcavity versus logconvexity: A complete characterization, *Journal of Economic Theory* **80**, pp. 350–369.

Anatolyev, S., Khabibullin, R., and Prokhorov, A. (2014). An algorithm for reconstructing high dimensional distributions from distributions of lower dimensions, *Economics Letters* **123**, 3, pp. 257–261.

Andrews, D. (1991). Heteroskedasticity and autocorrelation consistent covariance matrix estimation, *Econometrica* **59**, pp. 817–858.

Andrews, D. W. K. (2005). Cross-section regression with common shocks, *Econometrica* **73**, pp. 1551–1585.

Ang, A. and Chen, J. (2002). Asymmetric correlations of equity portfolios, *Journal of Financial Economics* **63**, 3, pp. 443–494.

Arnold, B. (1987). *Majorization and the Lorenz Order: A Brief Introduction*, *Lecture Notes in Statistics* (Springer).

Artzner, P., Delbaen, F., Eber, J.-M., and Heath, D. (1999). Coherent measures of risk, *Mathematical Finance* **9**, pp. 203–228.

Asmussen, S. and Rojas-Nandayapa, L. (2008). Asymptotics of sums of lognormal random variables with gaussian copula, *Statistics and Probability Letters* **78**, p. 2709.

Axtell, R. L. (2001). Zipf distribution of U.S. firm sizes, *Science* **293**, pp. 1818–1820.

Bagnoli, M. and Bergstrom, T. (2005). Log-concave probability and its applications, *Economic Theory* **26**, pp. 445–469.

Bai, J. (2009). Panel data models with interactive fixed effects, *Econometrica* **77**, pp. 1229–1279.

Bauwens, L., Laurent, S., and Rombouts, J. V. (2006). Multivariate garch models: A survey, *Journal of Applied Econometrics* **21**, 1, pp. 79–109.

Beare, B. K. (2010). Copulas and temporal dependence, *Econometrica* **78**, pp. 395–410.

Berg, D. (2009). Copula goodness-of-fit testing: An overview and power comparison, *The European Journal of Finance* **15**, 7, pp. 675–701.

Bierens, H. J. (2014). Consistency and asymptotic normality of sieve ml estimators under low-level conditions, *Econometric Theory* **30**, pp. 1021–1076.

Birge, L. and Massart, P. (1997). From model selection to adaptive estimation, in D. Pollard, E. Torgersen, and C. Yang (eds.), *Festschrift for Lucien LeCam: Research Papers in Probability and Statistics* (Springer), pp. 55–87.

Birkes, D., Seely, J., and Azzam, A. -M. (1981). An efficient estimator of the mean in a two-stage nested model, *Technometrics* **23**, pp. 143–148.

Birnbaum, Z. W. (1948). On random variables with comparable peakedness, *Annals of Mathematical Statistics* **19**, pp. 76–81.

Blattberg, R. C. and Gonedes, R. C. (1974). A comparison of the stable and student distributions as statistical models for stock prices, *Journal of Business* **47**, pp. 244–280.

Blyth, S. (1996). Out of line, *Risk* **9**, pp. 82–84.

Borovskikh, Y. V. and Korolyuk, V. S. (1997). *Martingale Approximation* (VSP, Utrecht).

Bouchaud, J. -P. and Potters, M. (2004). *Theory of Financial Risk and Derivative Pricing: From Statistical Physics to Risk Management*, 2nd edn. (Cambridge University Press).

Bouezmarni, T. and Rombouts, J. V. K. (2010). Nonparametric density estimation for multivariate bounded data, *Journal of Statistical Planning and Inference* **140**, 1, pp. 139–152.

Bouyé, E., Durrleman, V., Nikeghbali, A., Riboulet, G., and Roncalli, T. (2000). Copulas for finance: A reading guide and some applications, *Crédit Lyonnais Working Paper*.

Boyer, B., Gibson, M., and Loretan, M. (1997). Pitfalls in tests for changes in correlations, *International finance discussion papers* Vol. 597 (Board of Governors of the Federal Reserve System).

Bretagnolle, J., Dacunha-Castelle, D., and Krivine, J. L. (1966). Lois stables et espaces L^p, *Annales de l'Institute H. Poincaré. Section B. Calcul des Probabilités et Statistique* **64**, pp. 1278–1302.

Breusch, T., Qian, H., Schmidt, P., and Wyhowski, D. (1999). Redundancy of moment conditions, *Journal of Econometrics* **91**, pp. 89–111.

Breymann, W., Dias, A., and Embrechts, P. (2003). Dependence structures for multivariate high-frequency data in finance, *Quantitative Finance* **3**, 1, pp. 1–14.

Bücher, A. and Volgushev, S. (2013). Empirical and sequential empirical copula processes under serial dependence, *Journal of Multivariate Analysis* **119**, pp. 61–70.

Burda, M. and Prokhorov, A. (2014). Copula based factorization in Bayesian multivariate infinite mixture models, *Journal of Multivariate Analysis* **127**, pp. 257–261.

Burns, S. (2014). Diversification analysis in value at risk models under heavy-tailedness and dependence. MSc Risk Management and Financial Engineering Thesis, Imperial College Business School.

Burton, R. M., Goulet, M., and Meester, R. (1993). On 1-dependent processes and k-block factors, *Annals of Probability* **21**.

Cambanis, S. (1977). Some properties and generalizations of multivariate Eyraud-Gumbel-Morgenstern distributions, *Journal of Multivariate Analysis* **7**, pp. 551–559.

Cambanis, S. (1991). On Eyraud-Farlie-Gumbel-Morgenstern random processes, in *Advances in Probability Distributions with Given Marginals, Mathematics and its Applications*, Vol. 7 (Kluwer), pp. 207–222.

Cambanis, S., Keener, R., and Simons, G. (1983). On α-symmetric multivariate distributions, *J. Multivariate Anal.* **13**, 2, pp. 213–233.

Cameron, A. C., Li, T., Trivedi, P. K., and Zimmer, D. M. (2004). Modelling the differences in counted outcomes using bivariate copula models with application to mismeasured counts, *Econometrics Journal* **7**, 2, pp. 566–584.

Carrasco, M. and Chen, X. (2002). Mixing and moment properties of various GARCH and stochastic volatility models, *Econometric Theory* **18**, 1, pp. 17–39.

Chan, W., Park, D. H., and Proschan, F. (1989). Peakedness of weighted averages of jointly distributed random variables, in L. J. Gleser, M. D. Perlman, S. J. Press, and A. R. Sampson (eds.), *Contributions to Probability and Statistics*, pp. 58–62.

Chavez-Demoulin, V., Embrechts, P., and Neslehova, J. (2006). Quantitative models for operational risk: Extremes, dependence and aggregation, *Journal of Banking and Finance* **30**, 10, pp. 2635–2658.

Chen, D., Mao, T., Pan, X., and Hu, T. (2012). Extreme value behavior of aggregate dependent risks, *Insurance: Mathematics and Economics* **50**, pp. 99–108.

Chen, X. (2007). Large sample sieve estimation of semi-nonparametric models, in J. J. Heckman and E. E. Leamer (eds.), *Handbook of Econometrics* (Elsevier), Vol. 6, pp. 5549–5632.

Chen, X. and Fan, Y. (2004). Evaluating density forecasts via copula approach, *Finance Research Letters* **1**, pp. 74–84.

Chen, X. and Fan, Y. (2006a). Estimation and model selection of semiparametric copula-based multivariate dynamic models under copula misspecification, *Journal of Econometrics* **135**, 1-2, pp. 125–154.

Chen, X. and Fan, Y. (2006b). Estimation of copula-based semiparametric time series models, *Journal of Econometrics* **130**, pp. 307–335.

Chen, X., Fan, Y., and Tsyrennikov, V. (2006). Efficient estimation of semiparametric multivariate copula models, *Journal of the American Statistical Association* **101**, pp. 1228–1240.

Chen, X., Wu, W. B., and Yi, Y. (2009). Estimation of copula-based semiparametric time series models, *Annals of Statistics* **37**, pp. 4214–4253.

Cherubini, U., Gobbi, F., Mulinacci, S., and Romagnoli, S. (2012). *Dynamic Copula Methods in Finance*, Wiley Finance Series (John Wiley and Sons Ltd).

Cherubini, U., Luciano, E., and Vecchiato, W. (2004). *Copula Methods in Finance*, Wiley Finance Series (John Wiley and Sons Ltd).

Chesher, A. (1983). The information matrix test: Simplified calculation via a score test interpretation, *Economics Letters* **13**, 1, pp. 45–48.

Chesher, A. and Spady, R. (1991). Asymptotic expansions of the information matrix test statistic, *Econometrica* **59**, 3, pp. 787–815.

Choroś B., Ibragimov, R., and Permiakova, E. (2010). Copula estimation. In P. Jaworski, F. Durante, W. Härdle and T. Rychlik (eds.), *Copula Theory and Its Applications, Lecture Notes in Statistics — Proceedings* **198** (Springer, Heidelberg), pp. 77–91.

Cochran, W. G. (1954). The combination of estimates from different experiments, *Biometrics* **10**, pp. 101–129.

Cont, R. (2001). Empirical properties of asset returns: Stylized facts and statistical issues, *Quantitative Finance* **1**, pp. 223–236.

Cotter, D. and Dowd, K. (2006). Extreme spectral risk measures: An application to futures clearinghouse margin requirements, *Journal of Banking and Finance* **30**, pp. 3469–3485.

Cover, T. M. and Thomas, J. A. (1991). *Elements of Information Theory, Wiley Series in Telecommunications* (John Wiley & Sons, Inc., New York, NY), a Wiley-Interscience Publication.

Cover, T. M. and Thomas, J. A. (2012). *Elements of Information Theory* (John Wiley & Sons).

Cuadras, C. (2009). Constructing copula functions with weighted geometric means, *Journal of Statistical Planning and Inference* **139**, 11, pp. 3766–3772.

Cuadras, C. and Diaz, W. (2012). Another generalization of the bivariate fgm distribution with two-dimensional extensions, *Acta et Contationes Universitatis Tartuensis de Mathematica* **16**, 1, pp. 3–12.

Danielsson, J., Jorgensen, B. N., Samorodnitsky, G., Sarma, M., and de Vries, C. G. (2013). Fat tails, VaR and subadditivity, *Journal of Econometrics* **172**, 2, pp. 283–291.

Darsow, W. F., Nguyen, B., and Olsen, E. T. (1992). Copulas and Markov processes, *Illinois Journal of Mathematics* **36**, pp. 600–642.

Davidson, R. and MacKinnon, J. G. (1992). A new form of the information matrix test, *Econometrica* **60**, 1, pp. 145–157.

Davidson, R. and MacKinnon, J. G. (1998). Graphical methods for investigating the size and power of hypothesis tests, *The Manchester School* **66**, 1, pp. 1–26.

Davis, R. A. and Mikosch, T. (1998). The sample autocorrelations of heavy-tailed processes with applications to ARCH, *Annals of Statistics* **26**, pp. 2049–2080.

de la Peña, V. H. (1990). Bounds on the expectation of functions of martingales and sums of positive rvs in terms of norms of sums of independent random variables, *Proceedings of the American Mathematical Society* **108**, pp. 233–239.

de la Peña, V. H. and Giné, E. (1999). *Decoupling, Probability and Its Applications* (Springer-Verlag, New York, NY).

de la Peña, V. H., Ibragimov, R., and Jordan, S. (2004). Option bounds, *Journal of Applied Probability* **41**, pp. 145–156.

de la Peña, V. H., Ibragimov, R., and Sharakhmetov, S. (2002). On sharp Burkholder-Rosenthal-type inequalities for infinite-degree U-statistics, *Annales de l'Institut Henri Poincaré. Probabilités et Statistiques* **38**, pp. 973–990, en l'honneur de J. Bretagnolle, D. Dacunha-Castelle, I. Ibragimov.

de la Peña, V. H., Ibragimov, R., and Sharakhmetov, S. (2003). On extremal distributions and sharp L–bounds for sums of multilinear forms, *Annals of Probability* **31**, pp. 630–675.

de la Peña, V. H., Ibragimov, R., and Sharakhmetov, S. (2006). *Characterizations of Joint Distributions, Copulas, Information, Dependence and Decoupling, with Applications to Time Series, Lecture Notes–Monograph Series*, Vol. 49 (Institute of Mathematical Statistics, Beachwood, OH), pp. 183–209.

de la Peña, V. H. and Lai, T. L. (2001). Theory and applications of decoupling, in C. A. Charalambides, M. V. Koutras, and N. Balakrishnan (eds.), *Probability and Statistical Models with Applications* (Chapman and Hall/CRC), pp. 117–145.

Deheuvels, P. (1979). La fonction de dépendance empirique et ses propriétés: Un test non paramétrique d'indépendance, *Académie Royale de Belgique. Bulletin de la Classe des Sciences 5ième Série* **65**, pp. 274–292.

Deheuvels, P. (1981). A Kolmogorov-Smirnov type test for independence and multiple samples, *Revue Roumaine de Mathématiques Pures et Appliquées* **26**, pp. 213–226.

Dharmadhikari, S. W. and Joag-Dev, K. (1988). *Unimodality, Convexity and Applications* (Academic Press, Boston, MA).

Dobric, J. and Schmid, F. (2007). A goodness of fit test for copulas based on Rosenblatt's transformation, *Computational Statistics and Data Analysis* **51**, 9, pp. 4633–4642.

Dornbusch, R., Park, Y. C., and Claessens, S. (2000). Contagion: How it spreads and how it can be stopped, *World Bank Research Observer* **15**, 2, pp. 177–197.

Doukhan, P. (1994). *Mixing: Properties and Examples, Lecture Notes in Statistics*, Vol. 85 (Springer-Verlag, New York, NY).

Doukhan, P., Fermanian, J. -D., and Lang, G. (2004). Copulas of a vector-valued stationary weakly dependent process, *Working Paper, CREST*.

Doukhan, P., Fermanian, J. -D., and Lang, G. (2009). An empirical central limit theorem with applications to copulas under weak dependence, *Statistical Inference for Stochastic Processes* **12**, pp. 65–87.

Dragomir, S. S. (2000). An inequality for logarithmic mapping and applications for the relative entropy, *Nihonkai Mathematical Journal* **11**, pp. 151–158.

Drouet, M. D. and Kotz, S. (2009). *Correlation and Dependence* (World Scientific).

Dungey, M., Fry, R., Gonzalez-Hermosillo, B., and Martin, V. L. (2005). Empirical modelling of contagion: A review of methodologies, *Quantitative Finance* **5**, 1, pp. 9–24.

Durante, F. and Sempi, C. (2010). Copula theory: An introduction, in F. Durante, W. Härdle, P. Jaworski, and T. Rychlik (eds.), *Copula Theory and Its Applications: Proceedings of the Workshop Held in Warsaw, 25–26 September 2009, Lecture Notes in Statistics — Proceedings*, Vol. 198 (Springer).

Eaton, M. L. (1970). A note on symmetric Bernoulli random variables, *Annals of Mathematical Statistics* **41**, pp. 1223–1226.

Eaton, M. L. (1974). A probability inequality for linear combinations of bounded random variables, *Annals of Statistics* **2**, pp. 609–614.

Edelman, D. (1986). Bounds for a nonparametric t table, *Biometrika* **73**, pp. 242–243.

Edelman, D. (1990). An inequality of optimal order for the tail probabilities of the t-statistic under symmetry, *Journal of the American Statistical Association* **85**, pp. 120–122.

Efron, B. (1969). Student's t-test under symmetry conditions, *Journal of the American Statistical Association* **64**, pp. 1278–1302.

El-Bassiouni, M. Y. and Abdelhafez, M. E. M. (2000). Interval estimation of the mean in a two-stage nested model, *Journal of Statistical Computation and Simulation* **67**, pp. 333–350.

Embrechts, P., Höing, A., and Juri, A. (2003). Using copulae to bound the value-at-risk for functions of dependent risks, *Finance and Stochastics* **7**, pp. 145–167.

Embrechts, P., Klüppelberg, C., and Mikosch, T. (1997). *Modelling Extremal Events for Insurance and Finance* (Springer-Verlag, Berlin).

Embrechts, P., McNeil, A., and Straumann, D. (2002). Correlation and dependence in risk management: Properties and pitfalls, in M. Dempster (ed.), *Risk Management: Value at Risk and Beyond* (Cambridge University Press), pp. 176–223.

Embrechts, P., Neslehova, J., and Wuethrich, M. V. (2009). Additivity properties for value-at-risk under archimedean dependence and heavy-tailedness, *Insurance: Mathematics and Economics* **44**, 2, pp. 164–169.

Fama, E. F. (1965a). Portfolio analysis in a stable paretian market, *Management Science* **11**, pp. 404–419.

Fama, E. F. (1965b). The behavior of stock market prices, *Journal of Business* **38**, pp. 34–105.

Fan, Y. and Patton, A. J. (2014). Copulas in econometrics, *Annual Review of Economics* **6**, 1, pp. 179–200.

Fang, K. T., Kotz, S., and Ng, K. W. (1990). *Symmetric Multivariate and Related Distributions, Monographs on Statistics and Applied Probability*, Vol. 36 (Chapman and Hall Ltd., London).

Feller, W. (1959). Non-Markovian processes with the semi-group property, *Annals of Mathematical Statistics* **30**, pp. 1252–1253.

Fermanian, J. -D. (2005). Goodness-of-fit tests for copulas, *Journal of Multivariate Analysis* **95**, 1, pp. 119–152.

Fermanian, J. -D., Radulović, D., and Wegkamp, M. (2004). Weak convergence of empirical copula process, *Bernoulli* **10**, pp. 847–860.

Fermanian, J. D. and Scaillet, O. (2003). Nonparametric estimation of copulas for time series, *Journal of Risk* **5**, pp. 25–54.

Fernandes, M. and Flôres, M. F. (2002). Tests for conditional independence, Markovian dynamics and noncausality, *Working Paper*, European University Institute.

Fisher, N. I. and Sen, P. K. (1994). *The Collected Works of Wassily Hoeffding* (Springer, New York, NY).

Fölmer, H. and Schied, A. (2002). Convex measures of risk and trading constraints, *Finance and Stochastics* **6**, pp. 429–447.

Frank, M. (1979). On the simultaneous associativity of $f(x, y)$ and $x + y - f(x, y)$, *Aequationes Mathematicae* **19**, pp. 194–226.

Frees, E. and Valdez, E. (1998). Understanding relationships using copulas, *North American Actuarial Journal* **2**, pp. 1–25.

Frittelli, M. and Gianin, E. R. (2002). Putting order in risk measures, *Journal of Banking & Finance* **26**, pp. 1473–1486.

Gabaix, X. (1999a). Zipf's law and the growth of cities, *American Economic Review* **89**, pp. 129–132.

Gabaix, X. (1999b). Zipf's law for cities: An explanation, *Quarterly Journal of Economics* **114**, pp. 739–767.

Gabaix, X. (2009). Power laws in economics and finance, *Annual Review of Economics* **1**, 1, pp. 255–294.

Gabaix, X., Gopikrishnan, P., Plerou, V., and Stanley, H. E. (2003). A theory of power-law distributions in financial market fluctuations, *Nature* **423**, pp. 267–270.

Gabaix, X., Gopikrishnan, P., Plerou, V., and Stanley, H. E. (2006). Institutional investors and stock market volatility, *Quarterly Journal of Economics* **121**, pp. 461–504.

Gabaix, X. and Ibragimov, R. (2011). Rank−1/2: A simple way to improve the OLS estimation of tail exponents, *Journal of Business and Economic Statistics* **29**, pp. 24–39.

Gaenssler, P. and Stute, W. (1987). *Seminar on Empirical Processes, DMV Seminar*, Vol. 9 (Birkhäuser Verlag, Basel).

Gaißer, S., Ruppert, M., and Schmid, F. (2010). A multivariate version of Hoeffding's phi-square, *Journal of Multivariate Analysis* **101**, 10, pp. 2571–2586.

Garcia, R., Renault, É., and Tsafack, G. (2007). Proper conditioning for coherent VaR in portfolio management, *Management Science* **53**, pp. 483–494.

Genest, C., Ghoudi, K., and Rivest, L. -P. (1995a). A semiparametric estimation procedure of dependence parameters in multivariate families of distributions, *Biometrika* **82**, 3, pp. 543–552.

Genest, C., Quesada-Molina, J., and Rodriguez-Lallena, J. (1995b). *De l'impossibilite de construire des lois a marges multidimensionnelles donnees a partir de copules, Comptes rendus de l'Academie des sciences de Paris Series I*, Vol. 320 (Elsevier, Paris).

Genest, C., Quessy, J. -F., and Remillard, B. (2006). Goodness-of-fit procedures for copula models based on the probability integral transformation, *Scandinavian Journal of Statistics* **33**, pp. 337–366.

Genest, C. and Remillard, B. (2008). Validity of the parametric bootstrap for goodness-of-fit testing in semiparametric models, *Annales de lInstitut Henri Poincare - Probabilites et Statistiques* **44**, 6, pp. 1096–1127.

Genest, C., Rémillard, B., and Beaudoin, D. (2009). Goodness-of-fit tests for copulas: A review and a power study, *Insurance: Mathematics and Economics* **44**, 2, pp. 199–213.

Gijbels, I. and Mielniczuk, J. (1990). Estimating the density of a copula function, *Communications in Statistics, Theory and Methods* **19**, pp. 445–464.

Glasserman, P., Heidelberger, P., and Shahabuddin, P. (2002). Portfolio value-at-risk with heavy-tailed risk factors, *Mathematical Finance* **12**, pp. 239–269.

Gneiting, T. (1998). On α−symmetric multivariate characteristic functions, *Journal of Multivariate Analysis* **64**, pp. 131–147.

Godambe, V. P. (1960). An optimum property of regular maximum likelihood estimation, *The Annals of Mathematical Statistics* **31**, 4, pp. 1208–1211.

Godambe, V. P. (1976). Conditional likelihood and unconditional optimum estimating equations, *Biometrika* **63**, 2, pp. 277–284.

Golan, A. (2002). Information and entropy econometrics – editors view, *Journal of Econometrics* **107**, pp. 1–15.

Golan, A. and Perloff, J. M. (2002). Comparison of maximum entropy and higher-order entropy estimators, *Journal of Econometrics* **107**, pp. 195–211.

Golden, R., Henley, S., White, H., and Kashner, T. M. (2013). New directions in information matrix testing: Eigenspectrum tests, in X. Chen and N. R. Swanson (eds.), *Recent Advances and Future Directions in Causality, Prediction, and Specification Analysis* (Springer, New York, NY), pp. 145–177.

Gouriéroux, C. and Monfort, A. (1979). On the characterization of a joint probability distribution by conditional distributions, *Journal of Econometrics* **10**, pp. 115–118.

Granger, C. and Lin, J. L. (1994). Using the mutual information coefficient to identify lags in nonlinear models, *Journal of Time Series Analysis* **15**, pp. 371–384.

Granger, C. W., Teräsvirta, T., and Patton, A. J. (2006). Common factors in conditional distributions for bivariate time series, *Journal of Econometrics* **132**, 1, pp. 43–57.

Grenander, U. (1981). *Abstract Inference* (Wiley, New York, NY).

Grundy, B. D. (1991). Option prices and the underlying asset's return distribution, *Journal of Finance* **46**, pp. 1045–1069.

Guillaume, D., Dacorogna, M., Davé, R., Müller, U., and Olsen, R. (1997). From the bird's eye to the microscope: A survey of new stylized facts of the intra-daily foreign exchange markets, *Finance and Stochastics* **1**, pp. 95–129.

Hafner, C. M. and Reznikova, O. (2010). Efficient estimation of a semiparametric dynamic copula model, *Computational Statistics & Data Analysis* **54**, 11, pp. 2609–2627.

Hall, A. (1989). On the calculation of the information matrix test in the normal linear regression model, *Economics Letters* **29**, 1, pp. 31–35.

Hall, P. and Hart, J. D. (1990). Bootstrap test for difference between means in nonparametric regression, *Journal of the American Statistical Association* **85**, 412, pp. 1039–1049.

Hamao, Y., Masulis, R., and Ng, V. (1990). Correlations in price changes and volatility across international stock markets, *Review of Financial studies* **3**, 2, pp. 281–307.

Hamilton, J. D. (1994). *Time Series Analysis* (Princeton University Press, Princeton, NJ).

Hansen, L. (1982). Large sample properties of generalized method of moments estimators, *Econometrica* **50**, pp. 1029–1054.

Hayashi, F. (2000). *Econometrics* (Princeton University Press).

Hennessy, D. A. and Lapan, H. E. (2003). An algebraic theory of portfolio allocation, *Economic Theory* **22**, pp. 193–210.

Hill, J. (2009). On functional central limit theorems for dependent, heterogeneous arrays with applications to tail index and tail dependence estimation, *Journal of Statistical Planning and Inference* **139**, 6, pp. 2091–2110.

Hill, J. (2011). Tail and nontail memory with applications to extreme value and robust statistics, *Econometric Theory* **27**, 4, pp. 844–884.

Hill, J. (2015a). Expected shortfall estimation and gaussian inference for infinite variance time series, *Journal of Financial Econometrics* **13**, pp. 1–44.

Hill, J. (2015b). Robust estimation and inference for heavy tailed garch, *Bernoulli* **21**, pp. 1629–1669.

Hill, J. and Prokhorov, A. (2016). GEL estimation for heavy-tailed garch models with robust empirical likelihood inference, *Journal of Econometrics* **190**, 1, pp. 18–45.

Hoeffding, W. (1940). Masstabinvariante korrelationstheorie, in *Schrift. Math. Seminars*, Vol. 3 (Inst. Angew. Math. Univ. Berlin 5), pp. 181–233.

Hong, Y. and White, H. (2005). Asymptotic distribution theory for nonparametric entropy measures of serial dependence, *Econometrica* **73**, pp. 837–901.

Horowitz, J. L. (1994). Bootstrap-based critical values for the information matrix test, *Journal of Econometrics* **61**, 2, pp. 395–411.

Horta, P., Mendes, C., and Vieira, I. (2010). Contagion effects of the subprime crisis in the European NYSE Euronext markets, *Portuguese Economic Journal* **9**, 2, pp. 115–140.

Hu, H. -L. (1998). Large sample theory of pseudo-maximum likelihood estimates in semiparametric models, *Ph.D. Dissertation*, University of Washington.

Hu, L. (2006). Dependence patterns across financial markets: A mixed copula approach, *Applied Financial Economics* **16**, pp. 717–729.

Hua, L., Joe, H., and Li, H. (2014). Relations between hidden regular variation and the tail order of copulas, *Journal of Applied Probability* **51**, 37–57.

Huang, W. and Prokhorov, A. (2014). A goodness-of-fit test for copulas, *Econometric Reviews* **33**, 7, pp. 751–771.

Ibragimov, M. and Ibragimov, R. (2007). Market demand elasticity and income inequality, *Economic Theory* **32**, pp. 579–587.

Ibragimov, M., Ibragimov, R., and Kattuman, P. (2013). Emerging markets and heavy tails, *Journal of Banking and Finance* **37**, p. 2546–2559.

Ibragimov, M., Ibragimov, R., and Walden, J. (2015). *Heavy-Tailed Distributions and Robustness in Economics and Finance. Lecture Notes in Statistics 214* (Springer, New York, NY).

Ibragimov, R. (2007). Efficiency of linear estimators under heavy-tailedness: Convolutions of α−symmetric distributions, *Econometric Theory* **23**, pp. 501–517.

Ibragimov, R. (2009a). Copula-based characterizations for higher-order Markov processes, *Econometric Theory* **25**, pp. 819–846.

Ibragimov, R. (2009b). Portfolio diversification and value at risk under thick-tailedness, *Quantitative Finance* **9**, pp. 565–580.

Ibragimov, R., Jaffee, D., and Walden, J. (2009). Nondiversification traps in catastrophe insurance markets, *Review of Financial Studies* **22**, pp. 959–993.

Ibragimov, R., Jaffee, D., and Walden, J. (2011). Diversification disasters, *Review of Financial Economics* **99**, pp. 333–348.

Ibragimov, R., and Prokhorov, A. (2016). Heavy tails and copulas: Limits of diversification revisited, *Economics Letters* **149**, pp. 102–107.

Ibragimov, R. and Walden, J. (2007). The limits of diversification when losses may be large, *Journal of Banking and Finance* **31**, 8, pp. 2551–2569.

Ibragimov, R. and Walden, J. (2011). Value at risk and efficiency under dependence and heavy-tailedness: Models with common shocks, *Annals of Finance* **7**, 3, pp. 285–318.

Inoue, A. and Kilian, L. (2005). In-sample or out-of-sample tests of predictability: Which one should we use? *Econometric Reviews* **23**, 4, pp. 371–402.

Jagannathan, R. (1984). Call options and the risk of underlying securities, *Journal of Financial Economics* **13**, pp. 425–434.

Jansen, D. W. and Vries, C. G. D. (1991). On the frequency of large stock returns: Putting booms and busts into perspective, *The Review of Economics and Statistics* **73**, 1, pp. 18–24.

Jensen, D. R. (1997). Peakedness of linear forms in ensembles and mixtures, *Statistics and Probability Letters* **35**, pp. 277–282.

Joe, H. (1987). Majorization, randomness and dependence for multivariate distributions, *Annals of Probability* **15**, pp. 1217–1225.

Joe, H. (1989). Relative entropy measures of multivariate dependence, *Journal of the American Statistical Association* **84**, pp. 157–164.

Joe, H. (1997). *Multivariate Models and Dependence Concepts, Monographs on Statistics and Applied Probability*, Vol. 73 (Chapman & Hall).

Joe, H. (2005). Asymptotic efficiency of the two-stage estimation method for copula-based models, *Journal of Multivariate Analysis* **94**, 2, pp. 401–419.

Joe, H. (2014). *Dependence Modelling with Copulas, Monographs on Statistics and Applied Probability*, Vol. 134 (CRC Press).

Joe, H., Li, H., and Nikoloulopoulos, A. K. (2010). Tail dependence functions and vine copulas, *Journal of Multivariate Analysis* **101**, 1, pp. 252–270.

Johnson, N. L. and Kotz, S. (1975). On some generalized Farlie-Gumbel-Morgenstern distributions, *Communications in Statistics* **4**, pp. 415–424.

Kahneman, D. and Tversky, A. (1979). Prospect theory: An analysis of decision under risk, *Econometrica* **47**, pp. 263–292.

Karlin, S. (1968). *Total Positivity. Vol. I* (Stanford University Press, Stanford, CA).

Kendall, M. G. (1938). A new measure of rank correlation, *Biometrika* **30**, pp. 81–93.

Kimeldorf, G. and Sampson, A. (1975). Uniform representations of bivariate distributions, *Communications in Statistics* **4**, 7, pp. 617–627.

Klugman, S. A. and Parsa, R. (1999). Fitting bivariate loss distributions with copulas, *Insurance: Mathematics and Economics* **24**, 1-2, pp. 139–148.

Koch, G. G. (1967a). A general approach to the estimation of variance components, *Technometrics* **9**, pp. 93–118.

Koch, G. G. (1967b). A procedure to estimate the population mean in random effects models, *Technometrics. A Journal of Statistics for the Physical, Chemical and Engineering Sciences* **9**, pp. 577–585.

Kojadinovic, I. and Holmes, M. (2009). Test of independence among continuous random vectors based on Cramér-von Mises functionals of the empirical copula process, *Journal of Multivariate Analysis* **100**, 6, pp. 1137–1154.

Kortschak, D. and Albrecher, H. (2009). Asymptotic results for the sum of dependent non-identically distributed random variables, *Methodology and Computing in Applied Probability* **11**, pp. 443–494.

Kotz, S. and Seeger, J. P. (1991). A new approach to dependence in multivariate distributions, in G. Dall'Aglio, S. Kotz, and G. Salinetti, *Advances in Probability Distributions with Given Marginals* (Kluwer Academic Publishing, Dordrecht), pp. 113–127.

Krupskii, P. and Joe, H. (2013). Factor copula models for multivariate data, *Journal of Multivariate Analysis* **120**, pp. 85–101.

Kullback, S. and Leibler, R. A. (1948). The mathematical theory of communication, *Bell System Technical Journal* **27**, pp. 379–423.

Kullback, S. and Leibler, R. A. (1951). On information and sufficiency, *Annals of Mathematical Statistics* **22**, pp. 79–86.

Kuritsyn, Y. G. and Shestakov, A. V. (1984). On α-symmetric distributions, *Theory of Probability and Its Applications* **29**, 4, pp. 804–806.

Kwapien S. (1987). Decoupling inequalities for polynomial chaos, *Annals of Probability* **15**, pp. 1062–1071.

Kyle, A. S. and Xiong, W. (2001). Contagion as a wealth effect, *The Journal of Finance* **56**, 4, pp. 1401–1440.

Lancaster, H. O. (1958). The structure of bivariate distributions, *Annals of Mathematical Statistics* **29**, pp. 719–736.

Lancaster, T. (1984). The covariance matrix of the information matrix test, *Econometrica* **52**, 4, pp. 1051–1053.

Lapan, H. E. and Hennessy, D. A. (2002). Symmetry and order in the portfolio allocation problem, *Economic Theory* **19**, pp. 747–772.

Lee, L. -F. (1983). Generalized econometric models with selectivity, *Econometrica* **51**, pp. 507–512.

Lehmann, E. L. (1966). Some concepts of dependence, *Annals of Mathematical Statistics* **37**, p. 1137–1153.

Lentzas, G. and Ibragimov, R. (2008). Copulas and long memory, *Working Paper*, Harvard University.

Levy, P. (1949). Exemple de processus pseudo-markoviens, *Comptes Rendus de l'Académie des Sciences* **228**, pp. 2004–2006.

Lhabitant, F. S. (2000). Equally weighted index (hfrx), in G. N. Gregoriou (ed.), *Encyclopedia of Alternative Investments* (Chapman and Hall).

Li, D. (2000). On default correlation: A copula function approach, *Journal of Fixed Income* **9**, 4, pp. 43–54.

Lin, G. (1987). Relationships between two extensions of Farlie-Gumbel-Morgenstern distribution, *Annals of the Institute of Statistical Mathematics* **39**, 1, pp. 129–140.

Lin, W., Engle, R., and Ito, T. (1994). Do bulls and bears move across borders? International transmission of stock returns and volatility, *Review of Financial Studies* **7**, 3, pp. 507–538.

Liu, D. and Prokhorov, A. (2016). Sparse sieve mle, *University of Sydney Working Paper*.

Long, D. and Krzysztofowicz, R. (1995). A family of bivariate densities constructed from marginals, *Journal of American Statistical Association* **90**, pp. 739–746.

Longin, F. and Solnik, B. (2001). Extreme correlation of international equity markets, *The Journal of Finance* **56**, 2, pp. 649–676.

Lorentz, G. (1986). *Bernstein Polynomials* (University of Toronto Press).

Loretan, M. and Phillips, P. C. B. (1994). Testing the covariance stationarity of heavy-tailed time series, *Journal of Empirical Finance* **3**, pp. 211–248.

Low, L. (1970). An application of majorization to comparison of variances, *Technometrics* **12**, pp. 141–145.

Lowin, J. (2007). The Fourier Copula: Theory and Applications. Senior Honour Thesis, Harvard College. Available at http://dx.doi.org/10.2139/ssvn. 1804664

Lux, T. (1996). The stable Paretian hypothesis and the frequency of large returns: An examination of major German stocks, *Applied Financial Economics* **6**, pp. 463–475.

Ma, C. (1998). On peakedness of distributions of convex combinations, *Journal of Statistical Planning and Inference* **70**, pp. 51–56.

Mandelbrot, B. (1960). The pareto-levy law and the distribution of income, *International Economic Review* **1**, pp. 79–106.

Mandelbrot, B. (1963). The variation of certain speculative prices, *Journal of Business* **36**, pp. 394–419.

Mandelbrot, B. (1997). *Fractals and Scaling in Finance. Discontinuity, Concentration, Risk* (Springer-Verlag, New York, NY).

Mardia, K., Kent, J., and Bibby, J. (1979). *Multivariate Analysis*, Probability and Mathematical Statistics (Academic Press, London).

Mari, D. D. and Kotz, S. (2001). *Correlation and Dependence* (Imperial College Press, London).

Marshall, A. W. and Olkin, I. (1979). *Inequalities: Theory of Majorization and Its Applications* (Academic Press, New York, NY).

Marshall, A. W., Olkin, I., and Arnold, B. C. (2011). *Inequalities: Theory of Majorization and Its Applications*, 2nd edition. (Springer, New York, NY).

Massoumi, E. and Racine, J. (2002). Entropy and predictability of stock market returns, *Journal of Econometrics* **107**, pp. 291–312.

Matúš, F. (1996). On two-block-factor sequences and one-dependence, *Proceedings of the American Mathematical Society* **124**, pp. 1237–1242.

Matúš, F. (1998). Combining *m*-dependence with Markovness, *Annales de l'Institut Henri Poincaré. Probabilités et Statistiques* **34**, pp. 407–423.

McCulloch, J. H. (1996). Financial applications of stable distributions, in G. S. Maddala and C. R. Rao (eds.), *Handbook of Statistics, Vol. 14* (Elsevier, Amsterdam), pp. 393–425.

McCulloch, J. H. (1997). Measuring tail thickness to estimate the stable index alpha: A critique, *Journal of Business and Economic Statistics* **15**, 1, pp. 74–81.

McNeil, A., Frey, R., and Embrechts, P. (2005). *Quantitative Risk Management: Concepts, Techniques and Tools* (Princeton University Press, Princeton, NJ).

Medovikov, I. (2015). Non-parametric weighted tests for independence based on empirical copula process. *Journal of Computation and Simulation* **86**,1, pp. 105–121.

Medovikov, I. and Prokhorov, A. (2016). A new measure of vector dependence, with an application to financial contagion, *University of Sydney Working Paper BAWP-2016-04*.

Merton, R. C. (1973). Theory of rational option pricing, *Bell Journal of Economics* **4**, pp. 141–183.

Mesfioui, M., Quessy, J. -F., and Toupin, M. -H. (2009). On a new goodness-of-fit process for families of copulas, *Canadian Journal of Statistics* **37**, 1, pp. 80–101.

Mierau, J. and Mink, M. (2013). Are stock market crises contagious? The role of crisis definitions, *Journal of Banking and Finance* **37**, pp. 4765–4776.

Mikosch, T. and Stărică, C. (2000). Limit theory for the sample autocorrelations and extremes of a garch (1, 1) process, *The Annals of Statistics* **28**, 5, pp. 1427–1451.

Miller, D. J. and Liu, W. -H. (2002). On the recovery of joint distributions from limited information, *Journal of Econometrics* **107**, 1, pp. 259–274.

Mo, J. (2013). Diversification analysis in value at risk models under heavy-tailedness and dependence. MSc Risk Management and Financial Engineering Thesis, Imperial College Business School.

Molenberghs, G. and Lesaffre, E. (1994). Marginal modeling of correlated ordinal data using a multivariate plackett distribution, *Journal of the American Statistical Association* **89**, 426, pp. 633–644.

Mond, B. and Pečarić, J. (2001). On some applications of the ag inequality in information theory, *JIPAM. Journal of Inequalities in Pure and Applied Mathematics* **2**, 1.

Moscone, F. and Tosetti, E. (2009). A review and comparison of tests of cross-sectional independence in panels, *Journal of Economic Surveys* **23**, pp. 528–561.

Nelsen, R. B. (1996). Nonparametric measures of multivariate association, in *Distributions with Fixed Marginals and Related Topics (Seattle, WA, 1993)*, IMS Lecture Notes Monogr. Ser., Vol. 28 (Institute of Mathematical Statistics, Hayward, CA), pp. 223–232.

Nelsen, R. B. (1999). *An Introduction to Copulas, Lecture Notes in Statistics*, Vol. 139 (Springer).

Nelsen, R. B. (2006). *An Introduction to Copulas, Springer Series in Statistics*, Vol. 139, 2nd edn. (Springer).

Nelsen, R. B., Quesada-Molina, J. J., and Rodriguez-Lallena, J. A. (1997). Bivariate copulas with cubic sections, *Journal of Nonparametric Statistics* **7**, p. 205–220.

Neo, E., Matsypura, D., and Prokhorov, A. (2016). Estimation of hierarchical Archimedean copulas as a shortest path problem, *Economics Letters* **149**, pp. 131–134.

Nešlehova, J., Embrechts, P., and Chavez-Demoulin, V. (2006). Infinite mean models and the LDA for operational risk, *Journal of Operational Risk* **1**, pp. 3–25.

Newey, W. K. (1990). Efficient instrumental variables estimation of nonlinear models, *Econometrica* **58**, 4, pp. 809–837.

Newey, W. K. (1994). The asymptotic variance of semiparametric estimators, *Econometrica* **62**, pp. 1349–1382.

Newey, W. K. and Powell, J. L. (2003). Instrumental variable estimation of non-parametric models, *Econometrica* **71**, 5, pp. 1565–1578.

Newey, W. K. and Smith, R. J. (2004). Higher order properties of gmm and generalized empirical likelihood estimators, *Econometrica* **72**, 1, pp. 219–255.

Newey, W. K. and West, K. (1987a). Hypothesis testing with efficient method of moments estimation, *International Economic Review* **28**, pp. 777–787.

Newey, W. K. and West, K. D. (1987b). A simple, positive semidefinite, het-eroskedasticity and autocorrelation consistent covariance matrix, *Econometrica* **55**, pp. 703–708.

Panchenko, V. (2005). Goodness-of-fit test for copulas, *Physica A: Statistical Mechanics and Its Applications* **355**, 1, pp. 176–182.

Panchenko, V. and Prokhorov, A. (2016). Efficient estimation of parameters in marginals in semiparametric multivariate models, *University of Sydney Working Paper BAWP-2016-04*.

Patton, A. J. (2006). Modelling asymmetric exchange rate dependence, *International Economic Review* **47**, 2, pp. 527–556.

Patton, A. J. (2012). A review of copula models for economic time series, *Journal of Multivariate Analysis* **110**, pp. 4–18.

Pearson (1894). Contribution in the theory of evolution, xiii: On the theory of contingency and its relation to association and normal correlation, in *Drapers' Company Research Memoirs*, Biometric Series 1 (University College, London).

Peters, G. and Shevchenko, P. (2015). *Advances in Heavy Tailed Risk Modeling: A Handbook of Operational Risk* (John Wiley & Sons).

Petrone, S. (1999a). Bayesian density estimation using Bernstein polynomials, *Canadian Journal of Statistics* **27**, 1, pp. 105–126.

Petrone, S. (1999b). Random Bernstein polynomials, *Scandinavian Journal of Statistics* **26**, 3, pp. 373–393.

Petrone, S. and Wasserman, L. (2002). Consistency of Bernstein polynomial pos-teriors, *Journal of the Royal Statistical Society. Series B (Statistical Methodology)* **64**, 1, pp. 79–100.

Praetz, P. (1972). The distribution of share price changes, *Journal of Business* **45**, pp. 49–55.

Prokhorov, A., Schepsmeier, U., and Zhu, Y. (2015). Generalized information matrix test for copulas, *University of Sydney Working Paper BAWP-2015-05*.

Prokhorov, A. and Schmidt, P. (2009). Likelihood-based estimation in a panel setting: Robustness, redundancy and validity of copulas, *Journal of Econometrics* **153**, 1, pp. 93–104.

Proschan, F. (1965). Peakedness of distributions of convex combinations, *Annals of Mathematical Statistics* **36**, pp. 1703–1706.

Quesada-Molina, J. and Rodriguez-Lallena, J. (1994). Some advances in the study of the compatibility of three bivariate copulas, *Statistical Methods and Applications* **3**, 3, pp. 397–417.

Rachev, S. and Mittnik, S. (2000). *Stable Paretian Models in Finance* (Wiley, New York, NY).

Rachev, S. T., Menn, C., and Fabozzi, F. J. (2005a). *Fat-tailed and Skewed Asset Return Distributions: Implications for Risk Management, Portfolio Selection, and Option Pricing* (Wiley, Hoboken, NJ).

Resnick, S. (1987). *Extreme Values, Regular Variation and Point Processes* (Springer-Verlag, New York, NY).

Robinson, P. M. (1991). Consistent nonparametric entropy-based testing, *Review of Economic Studies* **58**, pp. 437–453.

Rodriguez, J. (2007). Measuring financial contagion: A copula approach, *Journal of Empirical Finance* **14**, pp. 401–423.

Rodriguez, R. J. (2003). Option pricing bounds: Synthesis and extension, *Journal of Financial Research* **26**, pp. 149–164.

Rosenblatt, M. (1952). Remarks on a multivariate transformation, *Annals of Mathematical Statistics* **23**, pp. 470–472.

Rosenblatt, M. (1960). An aggregation problem for Markov chains, in R. E. Machol (ed.), *Information and Decision Processes*, pp. 87–92.

Rosenblatt, M. (1971). *Markov Processes: Structure and Asymptotic Behavior* (Springer-Verlag).

Rosenblatt, M. and Slepian, D. (1962). nth order Markov chains with every n variables independent, *Journal of the Society for Industrial and Applied Mathematics* **10**, pp. 537–549.

Ross, S. A. (1976). A note on a paradox in portfolio theory, Mimeo, University of Pennsylvania.

Rothschild, M. and Stiglitz, J. E. (1970). Increasing risk: I. A definition, *Journal of Economic Theory* **2**, pp. 225–243.

Rüschendorf, L. (1976). Asymptotic distributions of multivariate rank order statistics. *Annals of Statistics* **4**, pp. 912–923.

Rüschendorf, L. (1985). Construction of multivariate distributions with given marginals, *Annals of Institute of Statistical Mathematics* **37**, pp. 225–233.

Samuelson, P. A. (1967). Efficient portfolio selection for pareto-levy investments, *Journal of Financial and Quantitative Analaysis* **2**, pp. 107–122.

Sancetta, A. and Satchell, S. (2004). The Bernstein copula and its applications to modeling and approximations of multivariate distributions, *Econometric Theory* **20**, 3, pp. 535–562.

Saposnik, R. (1993). A note on majorization theory and the evaluation of income distributions, *Economics Letters* **42**, 2, pp. 179–183.

Scaillet, O. (2007). Kernel-based goodness-of-fit tests for copulas with fixed smoothing parameters, *Journal of Multivariate Analysis* **98**, 3, pp. 533–543.

Scherer, F. M., Harhoff, D., and Kukies, J. (2000). Uncertainty and the size distribution of rewards from innovation, *Journal of Evolutionary Economics* **10**, pp. 175–200.

Segers, J. (2012). Asymptotics of empirical copula processes under non-restrictive smoothness assumptions, *Bernoulli* **18**, 3, pp. 764–782.

Segers, J., Akker, R. V. D., and Werker, B. (2008). Improving upon the marginal empirical distribution functions when the copula is known, *Discussion paper*, Tilburg University, Center for Economic Research.

Serfling, R. (1980). *Approximation Theorems of Mathematical Statistics* (Wiley).

Severini, T. A. and Tripathi, G. (2001). A simplified approach to computing efficiency bounds in semiparametric models, *Journal of Econometrics* **102**, 1, pp. 23–66.

Shaked, M. and Shanthikumar, J. G. (2007). *Stochastic Orders* (Springer, New York, NY).

Sharakhmetov, S. (1993). r-independent random variables and multiplicative systems, *Dopovidi Akademii Nauk Ukraini*, pp. 43–45.

Sharakhmetov, S. (2001). On a problem of n. n. leonenko and m. i. yadrenko. *Dopov. Nats. Akad. Nauk Ukr. Mat. Prirodozn. Tekh. Nauki*, pp. 23–27.

Sharakhmetov, S. and Ibragimov, R. (2002). A characterization of joint distribution of two-valued random variables and its applications, *Journal of Multivariate Analysis* **83**, pp. 389–408.

Shaw, J. (1997). Beyond VAR and stress testing, in *VAR: Understanding and Applying Value at Risk* (Risk Publications, London), pp. 211–224.

Shen, X. (1997). On methods of sieves and penalization, *The Annals of Statistics* **25**, pp. 2555–2591.

Shih, J. H. and Louis, T. A. (1995). Inferences on the association parameter in copula models for bivariate survival data, *Biometrics* **51**, 4, pp. 1384–1399.

Silverberg, G. and Verspagen, B. (2007). The size distribution of innovations revisited: An application of extreme value statistics to citation and value measures of patent significance, *Journal of Econometrics* **139**, 2, pp. 318–339.

Sklar, A. (1959). Fonctions de répartition à n dimensions et leurs marges, *Publications de l'Institut de Statistique de l'Université de Paris* **8**, pp. 229–231.

Smith, M. D. (2003). Modelling sample selection using archimedean copulas, *The Econometrics Journal* **6**, 1, pp. 99–123.

Smith, R. J. (1997). Alternative semi-parametric likelihood approaches to generalised method of moments estimation, *The Economic Journal* **107**, 441, pp. 503–519.

Soofi, E. S. and Retzer, J. J. (2002). Information indices: Unification and applications, *Journal of Econometrics* **107**.

Spearman, C. (1904). The proof and measurement of association between two things, *The American Journal of Psychiatry* **15**, pp. 72–101.

Stock, J. H. and Watson, M. W. (2006). *Introduction to Econometrics* (Addison Wesley).

Szegö, G. P. (ed.). (2004). *Risk Measures for the 21st Century* (Wiley, Chichester).

Tasche, D. (2002). Expected shortfall and beyond, *Journal of Banking and Finance* **26**, pp. 1519–1533.

Tauchen, G. (1985). Diagnostic testing and evaluation of maximum likelihood models, *Journal of Econometrics* **30**, 1-2, pp. 415–443.

Taylor, L. W. (1987). The size bias of White's information matrix test, *Economics Letters* **24**, 1, pp. 63–67.

Tenbusch, A. (1994). Two-dimensional Bernstein polynomial density estimators, *Metrika* **41**, 1, pp. 233–253, metrika.

Tong, Y. (1994). Some recent developments on majorization inequalities in probability and statistics, *Linear Algebra and Its Applications* **199**, pp. 69–90.

Trivedi, P. K. and Zimmer, D. M. (2007). *Copula Modeling: An Introduction for Practitioners, Foundations and Trends in Econometrics*, Vol. 1 (Now Publishers Inc).

Tsallis, C. (1988). Possible generalization of Boltzmann-Gibbs statistics, *Journal of Statistical Physics* **52**, pp. 479–487.

Tsukahara, H. (2005). Semiparametric estimation in copula models, *The Canadian Journal of Statistics/La Revue Canadienne de Statistique* **33**, 3, pp. 357–375.

Uchaikin, V. V. and Zolotarev, V. M. (1999). *Chance and Stability: Stable Distributions and Their Applications* (VSP, Utrecht).

Ullah, A. (2002). Uses of entropy and divergence measures for evaluating econometric approximations and inference, *Journal of Econometrics* **107**, pp. 313–326.

Vitale, R. (1975). A Bernstein polynomial approach to density function estimation, in M. Puri (ed.), *Statistical Inference and Related Topics* (Academic Press, New York).

Wang, Y. (1990). Dependent random variables with independent subsets, *Canadian Mathematical Bulletin* **33**, pp. 22–27.

Weiler, H. and Culpin, D. (1970). Variance of weighted means, *Technometrics* **12**, pp. 757–773.

White, H. (1982). Maximum likelihood estimation of misspecified models, *Econometrica* **50**, 1, pp. 1–26.

Winkelmann, R. (2012). Copula bivariate probit models: With an application to medical expenditures, *Health Economics* **21**, 12, pp. 1444–1455.

Wooldridge, J. (1991). Specification testing and quasi-maximum likelihood estimation, *Journal of Econometrics* **48**, pp. 29–55.

Zastavnyi, V. P. (1993). Positive definite functions depending on the norm, *Russian Journal of Mathematical Physics* **1**, pp. 511–522.

Zheng, Y. (2011). Shape restriction of the multi-dimensional Bernstein prior for density functions, *Statistics and Probability Letters* **81**, 6, pp. 647–651.

Zheng, Y., Zhu, J., and Roy, A. (2010). Nonparametric Bayesian inference for the spectral density function of a random field, *Biometrika* **97**, pp. 238–245.

Zimmer, D. (2012). The role of copulas in the housing crisis, *Review of Economics and Statistics* **94**, 2, pp. 607–620.

Zimmer, D. M. and Trivedi, P. K. (2006). Using trivariate copulas to model sample selection and treatment effects: Application to family health care demand, *Journal of Business and Economic Statistics* **24**, pp. 63–76.

Zolotarev, V. (1991). Reflection on the classical theory of limit theorems, *Theory of Probability and Its Applications* **36**, pp. 124–137.

Zolotarev, V. M. (1986). *One-Dimensional Stable Distributions* (American Mathematical Society, Providence, RI).

Index

Printed in the United States
By Bookmasters